Multifunctional Application of Electrospun Fiber

Multifunctional Application of Electrospun Fiber

Guest Editor

Rui Zhao

MDPI

Basel • Beijing • Wuhan • Barcelona • Belgrade • Novi Sad • Cluj • Manchester

Guest Editor
Rui Zhao
Faculty of Chemistry
Northeast Normal University
Changchun
China

Editorial Office
MDPI AG
Grosspeteranlage 5
4052 Basel, Switzerland

This is a reprint of the Special Issue, published open access by the journal *Polymers* (ISSN 2073-4360), freely accessible at: https://www.mdpi.com/journal/polymers/special_issues/KPG3G6QID6.

For citation purposes, cite each article independently as indicated on the article page online and as indicated below:

Lastname, A.A.; Lastname, B.B. Article Title. *Journal Name* **Year**, *Volume Number*, Page Range.

ISBN 978-3-7258-6157-6 (Hbk)
ISBN 978-3-7258-6158-3 (PDF)
https://doi.org/10.3390/books978-3-7258-6158-3

Contents

About the Editor

Rui Zhao

Rui Zhao received his PhD from the Department of Chemistry, Jilin University, in 2018. He is currently an Assistant Professor in the Faculty of Chemistry, Northeast Normal University. He has broad research interests in functional fiber materials, and is particularly passionate about the function-oriented development of advanced adsorbents and membranes for challenging chemical separations and environment-related applications.

polymers

MDPI

Editorial

Multifunctional Application of Electrospun Fiber

Rui Zhao

Faculty of Chemistry, Northeast Normal University, Changchun 130024, China; zhaor814@nenu.edu.cn

1. Introduction

Nanofibers have become an important development in materials science and engineering fields, owing to their exceptional material properties, such as their small size, high modulus, wide range of applications, etc. [1]. The developed fabrication techniques for nanofibers are diverse, including template synthesis, self-assembly, meltblowing, stretching, phase separation, and electrospinning [2–5]. Among these techniques, electrospinning stands out for its significant comprehensive advantages and has emerged as a preferred choice for both research and industrialization. Its exceptional material diversity is evident by its ability to process almost all spinnable polymers (synthetic/natural polymers, ceramic precursors, composite materials, etc.) while precisely controlling fiber diameter, morphology, orientation, porosity, and membrane thickness by adjusting solution parameters, process parameters, and environmental conditions [6–8]. Accordingly, this method ultimately enables the effective fabrication of organic, inorganic, and organic/inorganic composite nanofibers. Moreover, electrospinning exhibits notable cost-effectiveness, characterized by low equipment investment and operational costs, as well as high material utilization efficiency, making it highly competitive for industrial applications [9,10].

Compared to traditional fiber materials, electrospun nanofibers exhibit unique appeal because of their capacity for precise material customization according to broad and diverse application requirements. This has revealed significant application potential, attracting substantial attention and sustained investment in research from material scientists and engineers worldwide. The distinctive features of electrospun nanofibers are the key to their popularity: an extremely high specific surface area provides a number of active interfaces for adsorption, catalytic reactions, and sensing; a highly tunable fiber structure (from nano- to micro-scale) and surface morphology (e.g., smooth, rough, porous, core–shell, heterogeneous structures) enable the accurate regulation of substance transport, selective separation, and cellular behavior [11–13]; the unique morphological characteristics allow researchers to design and fabricate ultra-lightweight, highly porous materials with fine structural controllability and multifunctional integration [14,15]. Meanwhile, these materials also exhibit good mechanical properties, outstanding chemical stability, and specific thermal properties. As a result, electrospun nanofibers are widely used in biomedical applications, environmental and energy technologies, electronics and smart wearables, and safety and protection fields, continuously driving innovation and application development in nanofiber materials.

Despite significant progress, several challenges remain in comprehensively realizing the industrial application of electrospun fibers. In terms of material performance, most electrospun fibers rely on single-polymer substrates, making it difficult to balance functional characteristics, mechanical strength, biocompatibility, and conductivity. For example, in the biomedical field, high tensile strength and degradability often present a contradiction. Additionally, their dynamic responsiveness to complex environments is weak.

Under conditions of drastic humidity or temperature changes, fiber structures are prone to swelling or embrittlement, limiting their use in smart wearables and protection in extreme environments. In terms of production efficiency, conventional single-nozzle spinning methods suffer from low productivity, susceptibility to clogging, and difficulties in parameter control. On the other hand, there is also a disconnect between clinical adaptation and industrial transformation. In medical implant applications, the degradation rate of fiber scaffolds often fails to match the pace of tissue regeneration, leading to premature failure or inflammatory responses. In functional device integration, poor interfacial compatibility between nanofibers and electronic components often results in an insufficient stability of self-powered and sensing-based smart devices, generally generating short service lives that fail to meet practical application requirements. These existing issues indicate that there are still numerous research gaps in the field of electrospun nanofibers, necessitating continuous and in-depth exploration by researchers. The studies in this Special Issue address the limitations of electrospun nanofibers in various fields by proposing practical solutions, thereby advancing the development of electrospinning technology and nanofiber materials.

2. Overview of Published Articles

The collective contributions in this Special Issue reflect the applications and advancements of electrospun nanofibers across multiple fields, e.g., biomedicine and drug delivery, tissue engineering and regenerative medicine, environmental remediation and adsorption materials, air filtration and protection, antibacterial and antifungal applications, flexible electronics and energy devices, etc. The following section provides a detailed introduction to the relevant research:

Luca Éva Uhljar et al. [16] successfully developed an orally disintegrating drug delivery system based on PVP nanofibers containing diclofenac sodium, which were prepared from dual-nozzle electrospinning combined with 3D printing. The formulation exhibits ultra-fast disintegration, immediate release, high drug loading efficiency, good stability, and biocompatibility, presenting a highly promising oral dispersible drug delivery system. Additionally, a novel in vitro dissolution method ("AS-to-FaSSGF") was developed to provide a more comprehensive understanding of drug dissolution in the digestive tract. This study presents a new process to obtain nanofibers via combining electrospinning and 3D printing. Moreover, the prepared materials are a promising drug release system with application potential.

Daniela Anahí Sánchez-Téllez et al. [17] developed sol–gel siloxane-crosslinked polyvinyl alcohol (PVA)–hyaluronan (HA) electrospun nanofibrous mats. This crosslinking effectively inhibited the hydrolytic degradation and metabolic breakdown of HA while simultaneously providing structural support to cells and enhancing the intrinsic osteogenic potential of HA. The nanofiber scaffold with high PVA content supported three-dimensional spherical osteoblast morphology (indicative of tissue growth), as well as sustained cell adhesion, proliferation, differentiation, and calcium deposition over 21 days independent of exogenous growth factors. The nanofiber scaffold also exhibited a controlled degradation rate and enhanced osteogenic performance. Siloxane crosslinking overcomes its application limitations of electrospinning HA, making this a breakthrough material for bone tissue regeneration engineering.

Fatma Sude Cetin et al. [18] successfully fabricated chitosan/polyethylene oxide/bacterial cellulose (CS/PEO/BC) composite nanofibers via electrospinning. CS/PEO/BC nanofibers showed a significant improvement in tensile strength (approximately 3.4 MPa, about 59% higher than chitosan/polyethylene oxide nanofibers), ultra-high water absorption capacity exceeding 200%, excellent biocompatibility (99.26% cell viability on day 7), and good biodegradability. More importantly, CS/PEO/BC nanofibers exhibited good antimicrobial activity against Escherichia coli and Staphylococcus aureus. Therefore, these nanofibers represent highly promising materials for advanced biomedical applications, particularly in tissue engineering.

Leire Murillo et al. [19] successfully constructed an antifungal functional material through a two-step process, involving electrospinning polyethylene (PEO)/chitosan (CS) nanofibers (124 $\pm$ 36 nm) followed by dip-coating for an in situ synthesis of silver nanoparticles (AgNPs). AgNPs increased the fiber diameter to 330 $\pm$ 106 nm, reduced roughness to 38.9 nm, and increased the contact angle from 23.4° to 97.7° (enhanced hydrophobicity). PEO/CS/AgNPs achieved a 39.7% inhibition rate after 13 days of fungal (Pleurotus ostreatus) exposure and formed significant inhibition zones after 10 days. The external loading of AgNPs (content 10.23%) and the synergistic antimicrobial mechanism provide a long-lasting antifungal solution for medical devices and food packaging.

Chunling Zheng et al. [20] designed a β-cyclodextrin/polyacrylic acid (β-CD/PA) composite nanofibrous membrane from the electrospinning process. Through host–guest inclusion and electrostatic synergy, the adsorption capacities of β-CD/PA nanofiber adsorbents toward basic red 9, basic red 14, basic red 46, basic blue 9, basic yellow 19, and basic yellow 28 were 86.7 mg/g, 21.5 mg/g, 18.9 mg/g, 44.5 mg/g, 116.5 mg/g, and 155.2 mg/g, respectively. The effects of different initial concentrations and pH values were also used to study adsorption performance. It performed excellently across a wide concentration range and at neutral pH. In addition, adsorption was followed pseudo-second-order kinetics, and the nanofiber adsorbent was easily recyclable and renewable, making it a promising renewable adsorbent for dye wastewater treatment.

Xingyu Fu et al. [21] designed quaternary ammonium group-functionalized electrospun nanofibers guided by theoretical calculations. They achieved an ultra-high adsorption capacity for free bilirubin (818.9 mg/g) and an adsorption capacity for albumin-bound bilirubin of 163.7 mg/g (removal rate > 90%). Their dynamic adsorption performance and blood compatibility are significantly superior to commercial activated carbon and resins. This study demonstrates the effectiveness of the theory-driven design method to develop electrospun nanofiber adsorbents and provides an efficient and safe adsorbent paradigm for blood toxin hemoperfusion treatment.

Huanliang Liu et al. [22] prepared polysulfone (PSF) fibrous membranes (diameter: ~1.17 μm) via one-step fast electrospinning using a N, N-dimethylformamide/ tetrahydrofuran as the solvent in a volume ratio of 3:1. Leveraging their highly porous structure, excellent hydrophobicity (contact angle > 110°), and mechanical toughness (tensile strength of 1.14 MPa, elongation at break of 116.6%), they achieved ultra-high particulate matter removal performance. The filtration efficiencies for PM2.5 and PM1.0 99.6% and 99.2%, respectively, had a low pressure drop of ~65 Pa. The PSF fibrous membranes maintained these filtration efficiencies over 28 days without decay. The simple preparation process, low cost, and high PM removal efficiency make the PSF fibrous membrane a highly promising replacement for commercial masks and air conditioning filters in various industrial filtration applications.

Sujin Ryu et al. [23] constructed a dual-layer biodegradable membrane which was integrated from electrospun cellulose acetate nanofibers (first layer) deposited on the following layer via electrospinning and zeolitic imidazolate framework (ZIF-8)-coated cotton fabric (second layer). The first layer was used to filtrate particulate matter (PM), and the second layer was used to remove toxic gases. At the optimal thickness of the electrospun fiber layer, the filtration membrane achieved 93.1% PM filtration efficiency, an ultra-low-pressure drop of 14.2 Pa, and a quality factor of 18.8 $\times$ 10^{-2} Pa^{-1}; meanwhile, it showed toxic gas adsorption removal rates > 99% for H_2S, formaldehyde, and NH_3. The dual-layer filtration membrane was also environmentally friendly and exhibited a

62.5% biodegradation rate in 45 days. This study provides an innovative material with a multilayered structure for personal protective equipment that combines efficient filtration, gas purification, and full lifecycle sustainability.

Baodong Sun et al. [24] optimized inorganic NiMoO4 nanofibers via electrospinning and high-temperature calcination at 500 °C. Their unique nanofiber structure could effectively suppress powder agglomeration, which enabled an ultra-high specific capacitance of 1947 F/g in supercapacitors and a capacity retention rate of 82.4% after 3000 cycles. Their efficient charge transfer and ion diffusion characteristics make them the ideal electrode material for high-performance portable energy storage devices.

Popat Mohite et al. [25] discussed the characteristics of gelatin (Ge) and chitosan (CH), their roles in formulation strategies, and characterization techniques for electrospun fibers, aiming to provide a comprehensive overview of the potential of Ge/CH-based nanofibers in drug delivery and biomedical applications. The Ge/CH-based electrospun nanofiber drug delivery system, leveraging its biomimetic ECM structure, natural biocompatibility, and degradability, shows breakthrough potential in orthopedic repair and wound healing. Optimizing electrospinning processes and functionalization (e.g., with targeting ligands/stimuli-responsive elements) can overcome challenges in fiber structure control, drug-loading stability, and clinical translation, offering a new generation of biomedical solutions for precise delivery and multi-drug combination therapy. This review also prospects the full potential of these nanofibers for managing a range of health complications.

Emanuele Alberto Slejko et al. [26] systematically summarized the preparation, properties, and applications of poly (3,4-ethylenedioxythiophene) (PEDOT)-based nanofibers. By analyzing recent advances in their synthesis, functionalization, and post-processing, the article elaborates on how to fully exploit the potential of this material in next-generation electronics and functional devices. Electrospun PEDOT nanofibers, achieved via coaxial spinning/carrier polymer strategies, offered tunable optoelectronic properties and flexible structures, overcoming processing challenges posed by rigid molecular chains. They provided a core material platform for wearable electronics and biosensing. Future studies need to address industrial-scale stability and roll-to-roll continuous production to deepen bioelectronic integration applications. The review also identifies current challenges and future directions, proposing feasible paths to address the issues of scalability, reproducibility, stability, and device integration.

Yaoyao Yang et al. [27] discussed in detail the application and development prospects of electrospinning technology in drug delivery for cancer treatment. Leveraging exceptional material compatibility, this technology could flexibly combine biocompatible organic compounds, drug molecules, and gene fragments to construct functionalized nanofiber carriers. Compared to oral and implantable drug delivery, electrospun carriers enabled localized, sustained, and controllable drug release, significantly enhancing tumor targeting. Key advantages included synergistic therapeutic mechanisms, multi-modal integration capability, and customization for the pathological microenvironment. These characteristics make electrospinning a breakthrough strategy meeting the demands of personalized cancer therapy, providing effective technical support for next-generation intelligent anti-tumor systems.

3. Summary and Future Outlook

These twelve articles collectively elucidate the current trends in electrospun fiber research. Through precise control of fiber morphology, composition, and functionalization, the high specific surface area and porous structure of the nanofibers are utilized to enhance performance, e.g., improving adsorption/drug-loading capacity, achieving low-resistance filtration, and utilizing biomimetic topological structures to promote cell proliferation and tissue regeneration. Synergistic composite effects overcome the limitations of single functionality. These groundbreaking advances collectively demonstrate that electrospinning, by precisely regulating fiber structure and function, has become an important platform to drive innovation across multiple fields. Each study provides a unique perspective, deepening the scientific and engineering understanding of the design and fabrication processes for next-generation electrospun materials. However, related research still needs to further address challenges such as stability in scaled-up production, compliance for clinical translation, and design for environmental degradation.

The future development of electrospinning technology should not only focus on fundamental research, such as new theories, materials, technologies, devices, and applications, but also expand the application scope of electrospun fibers. According to statistics, the demand for electrospun nanofibers is highest in filtration materials among various applications. Expanding the market for electrospun nanofibers in other fields requires the collective efforts of practitioners in the electrospinning industry. It is hoped that through a collaborative industry–academia–research model of "scientific breakthroughs in universities–industrialization practices in enterprises", more electrospun nanofiber materials can move from the laboratory to the marketplace, promoting the true realization of industrial value for electrospun nanofibers. Additionally, future research should focus on integrating electrospinning with artificial intelligence (AI), using AI to optimize various parameters and conditions and taking advantage of material genome technology to ultimately enhance production efficiency, unlock material properties, and overcome industrialization bottlenecks. In this Special Issue, we aim to attract broader participation in electrospinning research.

Acknowledgments: The Guest Editor would like to acknowledge all contributing authors for their submissions to this Special Issue. Particular gratitude is expressed to the Editors-in-Chief for their invitation to organize this Special Issue of *Polymers* and to the Section Managing Editors for their efficient and proactive management of the editorial process.

Conflicts of Interest: The authors declare no conflicts of interest.

References

1. Kanjwal, M.A.; Ghaferi, A.A. Hybrid Nanofibers Opportunities and Frontiers—A Review. *J. Environ. Chem. Eng.* **2022**, *10*, 108850. [CrossRef]
2. Nataraj, S.K.; Yang, K.S.; Aminabhavi, T.M. Polyacrylonitrile-Based Nanofibers—A State-of-the-Art Review. *Prog. Polym. Sci.* **2012**, *37*, 487–513. [CrossRef]
3. Kramar, A.; González-Benito, F.J. Cellulose-Based Nanofibers Processing Techniques and Methods Based on Bottom-Up Approach—A Review. *Polymers* **2022**, *14*, 286. [CrossRef]
4. Chung, M.; Nirmale, V.S.; Reddy, V.S.; Koutsos, V.; Ramakrishna, S.; Radacsi, N. Enhancing the Performance of Wearable Flexible Sensors via Electrospinning. *ACS Appl. Mater. Interfaces* **2025**, *17*, 39747–39771. [CrossRef]
5. Li, Y.; Zhu, J.; Cheng, H.; Li, G.; Cho, H.; Jiang, M.; Gao, Q.; Zhang, X. Developments of Advanced Electrospinning Techniques: A Critical Review. *Adv. Mater. Technol.* **2021**, *6*, 2100410. [CrossRef]
6. Xue, J.; Wu, T.; Dai, Y.; Xia, Y. Electrospinning and Electrospun Nanofibers: Methods, Materials, and Applications. *Chem. Rev.* **2019**, *119*, 5298–5415. [CrossRef] [PubMed]
7. Zhang, H.; Wang, J.; Teng, Y.; Jia, S.; Huang, H.; Li, Y.; Wang, C. Ce-MOF Composite Electrospinning as Antibacterial Adsorbent for the Removal of 2,4-Dichlorophenoxyacetic Acid. *Chem. Eng. J.* **2023**, *462*, 142195. [CrossRef]
8. Lu, X.; Wang, C.; Wei, Y. One-Dimensional Composite Nanomaterials: Synthesis by Electrospinning and Their Applications. *Small* **2009**, *5*, 2349–2370. [CrossRef] [PubMed]
9. Persano, L.; Camposeo, A.; Tekmen, C.; Pisignano, D. Industrial Upscaling of Electrospinning and Applications of Polymer Nanofibers: A Review. *Macromol. Mater. Eng.* **2013**, *298*, 504–520. [CrossRef]
10. Ji, D.; Lin, Y.; Guo, X.; Ramasubramanian, B.; Wang, R.; Radacsi, N.; Jose, R.; Qin, X.; Ramakrishna, S. Electrospinning of Nanofibres. *Nat. Rev. Methods Primers* **2024**, *4*, 1. [CrossRef]
11. Ren, Z.; Chen, D.; Shi, M.; Zhao, R. Tailoring the Coordination Microenvironment of Electrospun Nanofibers for the Separation of Thorium Ions from Ore Wastewater. *J. Colloid Interface Sci.* **2025**, *697*, 137957. [CrossRef]

12. Xu, M.; Li, W.; Zhong, M.; Yang, J.; Gao, M.; Pinna, N.; Lu, X. Trimetallic FeNiMo Nanofibers as High-Efficiency Electrocatalyst for Robust Oxygen Evolution. *ACS Mater. Lett.* **2024**, *6*, 3548–3556. [CrossRef]
13. Xu, P.J.; Gu, Y.X.; Xue, Y.; Sun, J.; Liao, W.Q.; Yang, Q.Q.; Zhou, Y.L. Advanced Biomimetic Materials in the Prevention of Tendon Adhesions: Design, Preparation, and Application of Hydrogel and Electrospun fiber Membranes. *Small* **2025**, *21*, 2411913. [CrossRef]
14. Ding, M.; Wang, Y.; Gong, X.; Luo, M.; Yin, X.; Yu, J.; Zhang, S.; Ding, B. Fluorine-Free Nanofiber/Network Membranes with Interconnected Tortuous Channels for High-Performance Liquid-Repellency and Breathability. *ACS Nano* **2025**, *19*, 5539–5548. [CrossRef] [PubMed]
15. Tian, Y.; Chen, Y.; Wang, S.; Wang, X.; Yu, J.; Zhang, S.; Ding, B. Energy-Harvesting Carbon Aerogel Nanofiber Metafabric for High-Efficiency Thermoregulation. *Adv. Funct. Mater.* **2025**, *35*, 2414229. [CrossRef]
16. Uhljar, L.É.; Jáger, T.; Hajdu, C.; Motzwickler-Németh, A.; Jójárt-Laczkovich, O.; Cseh, M.; Burian, K.; Ambrus, R. Diclofenac-Loaded Orodispersible Nanofibers Prepared by Double-Needle Electrospinning. *Polymers* **2025**, *17*, 1262. [CrossRef] [PubMed]
17. Sánchez-Téllez, D.A.; Baltierra-Uribe, S.L.; Vidales-Hurtado, M.A.; Valdivia-Flores, A.; García-Pérez, B.E.; Téllez-Jurado, L. Novel PVA–Hyaluronan–Siloxane Hybrid Nanofiber Mats for Bone Tissue Engineering. *Polymers* **2024**, *16*, 497. [CrossRef]
18. Cetin, F.S.; Avci, T.; Uygur, E.; Ilhan, E.; Kaya, E.; Tinaz, G.B.; Duta, L.; Dogan, C.; Gunduz, O. Development and Characterization of Antimicrobial Chitosan/Polyethylene Oxide/Bacterial Cellulose Nanofibers. *Polymers* **2025**, *17*, 693. [CrossRef]
19. Murillo, L.; Rivero, P.J.; Sandúa, X.; Pérez, G.; Palacio, J.F.; Rodríguez, R.J. Antifungal Activity of Chitosan/Poly(Ethylene Oxide) Blend Electrospun Polymeric Fiber Mat Doped with Metallic Silver Nanoparticles. *Polymers* **2023**, *15*, 3700. [CrossRef]
20. Zheng, C.; Zhao, W.; Tu, X.; Zhou, S. Electrospinning Membrane with Polyacrylate Mixed Beta-Cyclodextrin: An Efficient Adsorbent for Cationic Dyes. *Polymers* **2025**, *17*, 243. [CrossRef]
21. Fu, X.; Shi, M.; Chen, D.; Zhao, X.; Jiang, T.; Zhao, R. Theory-Driven Tailoring of the Microenvironment of Quaternary Ammonium Binding Sites on Electrospun Nanofibers for Efficient Bilirubin Removal in Hemoperfusion. *Polymers* **2024**, *16*, 1599. [CrossRef]
22. Liu, H.; Lai, W.; Shi, Y.; Tian, L.; Li, K.; Bian, L.; Xi, Z.; Lin, B. One-Step Fast Fabrication of Electrospun Fiber Membranes for Efficient Particulate Matter Removal. *Polymers* **2024**, *16*, 209. [CrossRef]
23. Ryu, S.; Kim, D.; Lee, H.; Kim, Y.; Lee, Y.; Kim, M.; Lee, H.; Lee, H. Biodegradable Nanofiber/Metal–Organic Framework/Cotton Air Filtration Membranes Enabling Simultaneous Removal of Toxic Gases and Particulate Matter. *Polymers* **2023**, *15*, 3965. [CrossRef] [PubMed]
24. Sun, B.; Wang, S.; Zhang, M. One-Dimensional Nickel Molybdate Nanostructures with Enhanced Supercapacitor Performance. *Polymers* **2023**, *15*, 4538. [CrossRef] [PubMed]
25. Mohite, P.; Puri, A.; Munde, S.; Dave, R.; Khan, S.; Patil, R.; Singh, A.K.; Tipduangta, P.; Singh, S.; Chittasupho, C. Potential of Chitosan/Gelatin-Based Nanofibers in Delivering Drugs for the Management of Varied Complications: A Review. *Polymers* **2025**, *17*, 435. [CrossRef] [PubMed]
26. Slejko, E.A.; Carraro, G.; Huang, X.; Smerieri, M. Advances in the Fabrication, Properties, and Applications of Electrospun PEDOT-Based Conductive Nanofibers. *Polymers* **2024**, *16*, 2514. [CrossRef]
27. Yang, Y.; Zhang, R.; Liang, Z.; Guo, J.; Chen, B.; Zhou, S.; Yu, D. Application of Electrospun Drug-Loaded Nanofibers in Cancer Therapy. *Polymers* **2024**, *16*, 504. [CrossRef]

polymers

MDPI

Article

Diclofenac-Loaded Orodispersible Nanofibers Prepared by Double-Needle Electrospinning

Luca Éva Uhljar [1], Tekla Jáger [1], Csongor Hajdu [1], Anett Motzwickler-Németh [1], Orsolya Jójárt-Laczkovich [1], Martin Cseh [2], Katalin Burian [3] and Rita Ambrus [1,*]

[1] Institute of Pharmaceutical Technology and Regulatory Affairs, Faculty of Pharmacy, University of Szeged, Eötvös Street 6, H-6720 Szeged, Hungary; jagertekla@gmail.com (T.J.); csongi982010@gmail.com (C.H.); nemeth.anett@szte.hu (A.M.-N.); jojartne.laczkovich.orsolya@szte.hu (O.J.-L.)

[2] 3D Center, Center of Excellence for Interdisciplinary Research, Development and Innovation, University of Szeged, Tisza Lajos Blvd. 107, H-6725 Szeged, Hungary; cseh.martin@szte.hu

[3] Department of Medical Microbiology and Immunobiology, University of Szeged, Dóm Square 10, H-6720 Szeged, Hungary; burian.katalin@med.u-szeged.hu

* Correspondence: ambrus.rita@szte.hu

Abstract: The main aim of this study was to develop a diclofenac-loaded, orodispersible formulation prepared by double-needle electrospinning. For the use of two needles, one above the other, a new needle holder was designed and 3D printed. During the optimization of the drug-free PVP carrier, the effect of the polymer concentration on the morphology and average fiber diameter was investigated. Electrospinning was possible for solutions with a PVP concentration between 7.5 and 15 w/w%. Too low viscosity led to smooth-surfaced nanoparticles, since electrospraying occurred. The optimal material properties and process parameters were used to prepare drug-loaded nanofibers. The morphology, crystallinity, chemical interactions, encapsulation efficiency, drug distribution, in vitro disintegration, in vitro dissolution, cytocompatibility, and 6-month stability were tested. According to the results, the electrospun formulation was an amorphous solid dispersion with excellent encapsulation efficiency. The drug distribution was homogeneous within the nanofiber matrix. The disintegration was completed in about 5 s in artificial saliva and about 41 s on an artificial tongue. The dissolution in artificial saliva was complete within 10 min. Overall, a promising formulation was developed with rapid disintegration, immediate drug release, and good stability. Additionally, a new in vitro dissolution method ("AS-to-FaSSGF") was developed to obtain a bigger picture of drug dissolution throughout the gastrointestinal tract.

Keywords: diclofenac; double-needle electrospinning; drug delivery system; electrospinning; fast-dissolving; nanofiber; NSAID; orodispersible

1. Introduction

Over the past decades, the demand for orally dispersible formulations has increased significantly due to their convenience and improved patient adherence [1]. Orodisperse systems rapidly disintegrate upon contact with the tongue. They can be easily administered without the need for water, making them ideal for situations with a lack of liquid (e.g., traveling) [2]. Also, orally disintegrating formulations offer a significant advantage for patients with swallowing difficulties, e.g., pediatric, elderly, and hospitalized patients, as well as individuals experiencing nausea, vomiting, or motion sickness [3]. According to the European Pharmacopoeia, an orodispersible dosage form is defined as having a disintegration time of less than 3 min [4]. In general, the disintegration time of dispersible formulations ranges from a few seconds to approximately one minute [5].

Nanofibrous matrices and films can be good candidates as orodispersible formulations due to their small fiber diameter, large specific surface area, and connected porosity [6]. Also, a wide variety of polymers can be electrospun, and the use of a hydrophilic polymer as a fiber-forming matrix also facilitates the formulation of fast-dissolving drug delivery systems [7]. Commonly used polymers are the polyvinylpyrrolidone (PVP) [8], polyvinyl alcohol (PVA) [9], hydroxypropyl methylcelluloses (HPMC) [10], polyethylene oxide (PEO) [11], gelatin [12], hydroxypropyl-beta-cyclodextrin (HP-βCD) [13], or combinations of these [14].

Electrospinning (ES) is the most widely used method for nanofiber (NF) production. It is an electrohydrodynamic process in which NFs are drawn from an ES solution under the influence of high voltage. The principle behind the process is that a strong electric field induces the ejection of a thin liquid jet from a droplet. As the jet travels toward the collector, it undergoes elongation, stretching, and solidification, forming NFs. The ES process and the produced NF morphology are influenced by several factors which can be grouped as material properties, process parameters, and ambient parameters [15]. Among the material properties, every aspect that affects the viscosity and conductivity of the ES solution is important. The most important properties are the polymer concentration, and the type of the ingredients. Among the process parameters, the key factors are the type and number of the spinneret, and the applied high voltage. The traditional ES contains one needle as a spinneret. This single-needle ES is the most used method in academic research; however, industrial-scale production requires multiple needles connected side by side [16]. The main reason for the need for the multi-needle ES is the low productivity of the single-needle method (capacity of ~0.01–0.1 g/h) [17]. However, the multi-needle ES is challenging because the closely flying jets interact and modify the electric field, making NF collection more difficult since it can lead to the formation of multiple fiber deposition areas on the collector, rather than a single, uniform patch [18].

This study aimed to produce and investigate a diclofenac-containing orodispersible formulation prepared by double-needle ES. To the best knowledge of the authors', this is the first reported double-needle ES method for diclofenac-loaded NFs. After optimization of the preparation method, the formulation was extensively tested to determine its suitability for the development of an orodispersible dosage form. As part of the investigation, a novel in vitro dissolution method, the so-called "AS-to-FaSSGF" method, was developed.

2. Materials and Methods

2.1. Materials

Diclofenac sodium (DS; Mw = 318.13) was donated by PannonPharma Llc. (Pécsvárad, Hungary) for research work. Polyvinylpyrrolidone (PVP; Mw = 1,300,000) and ethanol (96% purity) were purchased from Thermo Fisher Scientific Inc. (Budapest, Hungary).

Phosphate-buffer solution (PBS; pH 7.4) was prepared in-house and used as a solvent for the drug loading (DL%) and entrapment efficiency (EE%) measurements. For the in vitro disintegration and dissolution studies, different artificial secretions, namely artificial saliva (AS) and Fasted State Simulating Gastric Fluid (FaSSGF), were prepared in-house. The recipe of the AS solution was as follows; measures of 8 g sodium chloride, 2.38 g disodium hydrogen phosphate, and 0.19 g potassium dihydrogen phosphate (all purchased from Molar Chemicals Ltd., Halásztelek, Hungary) were dissolved in 1 L distilled water, and the pH was adjusted to 6.8 by the addition of phosphoric acid. The FaSSGF medium was prepared according to the instructions of the producer, Biorelevant.com Ltd. (London, UK). HPLC-grade methanol (Promochem LGC Standards GmbH., Wesel, Germany) and PBS (pH 2.5) were used as eluents during the separation and detection of the dissolved drug.

For the in vitro cytotoxicity study, Minimal Essential Medium (MEM) with Earle's salts (Merck KGaA, Darmstadt, Germany) and MTT solution (thiazolyl blue tetrazolium bromide; Merck KGaA, Darmstadt, Germany) were used.

All the chemicals, except the HPLC-grade methanol, were analytical grade, and purified water was used.

2.2. Preparation of the ES Solutions

The first step was to optimize the fabrication of drug-free PVP NFs. Five ES solutions with different PVP concentrations (5, 7.5, 10, 12.5, 15%) were prepared by dissolving the required amounts of PVP (Mw = 1,300,000) in ethanol. After overnight magnetic stirring, the viscosity of the ES solutions was measured using an IKA Rotavisc me-vi rotating viscosimeter (IKA-Werke GmbH & Co., Staufen, Germany), equipped with a type SP-11 spindle.

After finding the best polymer concentration, the drug was loaded into the NFs. The required amounts of PVP and DS were dissolved in ethanol and stirred overnight to obtain the ES solution of 10 *w/w*%. Additionally, the viscosity was measured as written above.

2.3. Nanofiber Preparation via Double-Needle Electrospinning

The NF preparation was carried out with the Spincube electrospinning machine (Spinsplit Llc., Budapest, Hungary) (Figure 1). The ES solution was filled into the 5 mL syringes and fed into the 22-gauge needles at a constant rate. As a collector, a static metal plate coated with aluminum foil was used. The needle–collector distance was 17 cm, the chamber temperature was 24–25 °C, and the relative humidity was between 36 and 44%. The voltage was set between 10 and 16 kV during the optimization process.

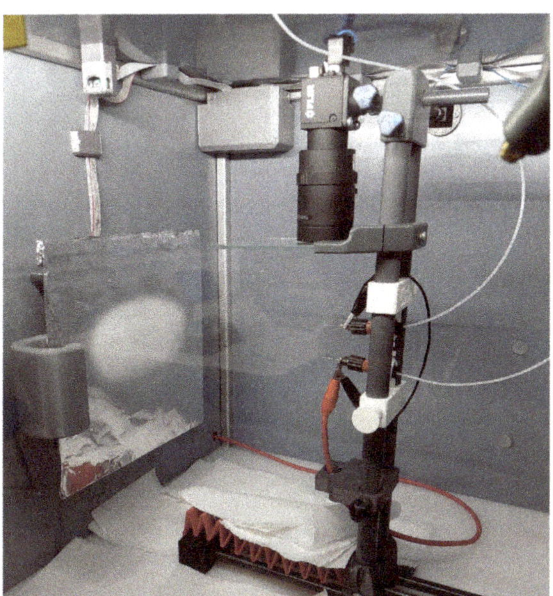

Figure 1. Electrospinning of PVP-based nanofibers using the double-needle method.

2.4. Design and Preparation of the 3D-Printed Needle Holder

One novelty of this research was the use of the double-needle ES setup. With the installation of the second nozzle into the instrument, the production rate could be doubled, or in other words, the preparation time would be halved. This required a new nozzle holder, which was designed using Shapr3D CAD software (version: 5.370; Shapr3D Zrt., Budapest, Hungary), as Figure 2A. shows. The design was exported in high-resolution .stl format and was sliced with Craftware software (version: 1.23; CraftUnique, Budapest, Hungary). The assembly was printed part-by-part with a CraftBot Plus Pro (CraftUnique, Budapest, Hungary) 3D printer with the following parameters: 0.2 mm layer height, along 15% parallel line infill with a 0.4 mm nozzle diameter and 60 mm/s draw speed. The PLA filament used was purchased from Herz Hungária Ltd. (Budapest, Hungary).

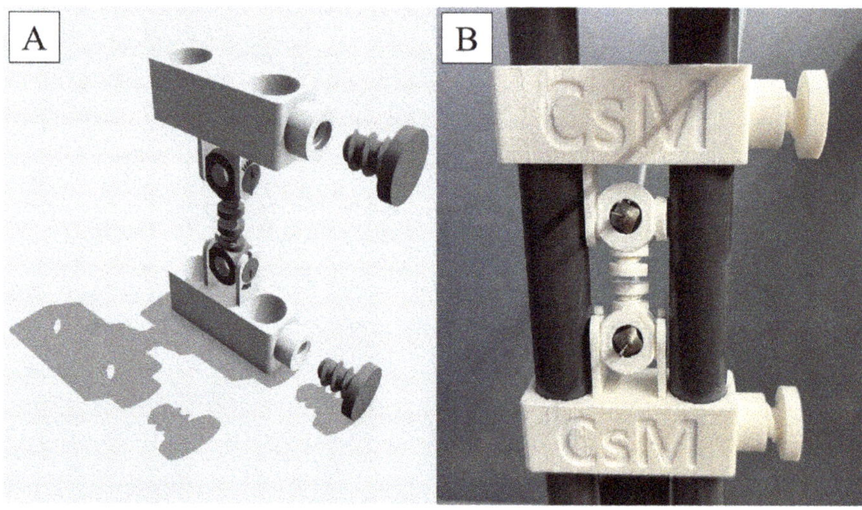

Figure 2. Design (**A**) and photo (**B**) of the 3D-printed needle holder for double-needle electrospinning.

2.5. Preparation of the Physical Mixture

As a reference for the XRPD, Raman Spectroscopy, and in vitro dissolution studies, untreated DS powder and physical mixtures were used. The composition of the mixture was the same as the NF, namely 10 *w*/*w*% DS and 90 *w*/*w*% PVP. The homogenization was carried out in a shaker mixer (Turbula System Schatz; Willy A. Bachofen AG Maschinenfabrik, Basel, Switzerland) under controlled conditions (50 rpm, 10 min).

2.6. Physicochemical Characterization of the Nanofibers

The morphology of the electrospun fibers was examined by scanning electron microscopy (SEM; Hitachi S4700, Hitachi Scientific Ltd., Tokyo, Japan). The samples were coated with a gold–palladium coating (Bio-Rad Sputter Coater 502, VG Microtech, Uckfield, UK) before capturing. The fiber diameter and fiber diameter distribution were determined from the SEM images using ImageJ 1.53a software (U. S. National Institutes of Health, Bethesda, MD, USA).

The crystallinity was investigated by X-ray powder diffraction (XRPD; BRUKER D8 Advance Diffractometer, BRUKER AXS GmbH, Karlsruhe, Germany). The fibrous samples were scanned at 40 kV and 40 mA over an angular range of 3–40°. Kα2 radiation was removed from the diffractogram, and background correction, smoothing, and evaluation were performed using DIFFRAC plus EVA software (version 5.2; Karlsruhe, Germany). The physical mixture, untreated polymer, and untreated drug were used as reference powders.

The presence of components was evaluated using Raman Spectroscopy (Thermo Fisher DXR Dispersive Raman microscope, Thermo Fisher Scientific, Waltham, MA, USA) equipped with a CCD camera and a diode laser operating at a wavelength of 780 nm.

Raman spectra were recorded in the spectral range of 800–1900 cm^{-1}, with an exposure time of 6 s. The same references were used as in the XRPD measurement.

The drug loading (DL%) and entrapment efficiency (EE%) of the NFs were calculated from absorbance. First, the NF samples were cut into 1 cm^2 pieces at random locations and weighted using a calibrated high-accuracy balance (Mettler-Toledo AX205 DeltaRange analytical balance; Mettler-Toledo International Inc., Columbus, OH, USA). Afterwards, the specimens were perfectly dissolved in 10 mL of pH = 7.4 PBS. Finally, the solutions were measured using a UV-Vis spectrophotometer (ABL&E-Jasco UV/VIS Spectrophotometer V-730, Budapest, Hungary) at 277 nm. The DS concentration was calculated using a calibration curve (y = 27.588x, R^2 = 0.9985). Measurements were performed six times, and the mean and standard deviations were reported.

The drug distribution within the NF matrix was studied using Raman Chemical Mapping (Thermo Fisher DXR Dispersive Raman microscope, Thermo Fisher Scientific, Waltham, MA, USA). Specimens of 1 cm^2 were randomly cut out from the NF matrix, and Raman spectra were collected from an 800 × 250 μm area with 30 measuring points in the spectral range of 3300–200 cm^{-1}. The exposure time was 6 s and an average of 12 scans was applied. Determining the distribution of DS in the specimens, the C-C double bond at 1600 cm^{-1} was selected for profiling.

2.7. In Vitro Disintegration Tests

Two methods were used to investigate the disintegration behavior of the NF samples. Firstly, free immersion into 25 mL of AS was executed in Petri dishes. Secondly, a more developed method was used, in which the physiological conditions of the surface of a moist tongue were simulated. This method was first reported by Bi et al. in 1996 [19], and is nowadays commonly used for the investigation of fast-disintegrating NFs [6,13,20]. In our test, the NF samples were examined with a slightly modified version of this technique. A filter paper with 7 cm diameter was soaked with AS, drained of excess moisture, and placed in a dry Petri dish. Then, a 2 cm^2 specimen of the NF mat was placed onto the wet paper. In both methods, the tests were performed in triplicate, the disintegration was recorded by a digital camera at 30 fps, and the disintegration time was calculated.

2.8. In Vitro Dissolution Tests

Two types of in vitro dissolution studies were executed to obtain a complete picture of the DS release from the NFs. Firstly, AS, as a single medium, was used in a thermostatic beaker with real-time UV probe detection (AvaLight DH-S-BAL spectrophotometer connected to an AvaSpec-2048L transmission immersion probe; AVANTES, Apeldoorn, The Netherlands). A measure of 2.5 mg untreated DS powder, 25 mg NF (containing 2.5 mg DS), and 25 mg physical mixture (containing also 2.5 mg DS) were placed into 200 mL AS and stirred with 200 rpm (DLAB MS-H280-Pro magnetic stirrer; DLAB Scientific Co., Ltd., Beijing, China) at 37 °C for 10 min. The UV absorbance was measured in real time at a 277 nm wavelength, but the averages calculated from the following points were plotted on the graph: 0.17, 0.33, 0.50, 0.67, 0.83, 1.00, 2.01, 3.01, 4.02, 5.02, 6.03, 7.03, 8.04, 9.04, 10.03 min.

In the case of orodispersible formulations, the disintegration and the dissolution mainly take place in the oral cavity. However, after disintegration, the saliva mixed with the medicine can be swallowed. This allows the drug to enter the stomach and meet with the gastric juice. Since testing of this idea is currently not found in the literature on orodispersible formulations, a new method for modeling this natural process had to be developed. So, a new "AS-to-FaSSGF" in vitro drug release method was developed as follows. The samples were placed into 5 mL of AS and gently shaken. After exactly one minute, the liquid was poured into the tank of a paddle-type dissolution apparatus (Hanson Research SR8-Plus Dissolution Test Station, Chatsworth, CA, USA) already containing 895 mL of FaSSGF medium. The paddles rotated at a speed of was 200 rpm, and the medium was 37 °C. The sampling points were at 1, 5, 10, 15, 30, 45, 60, and 90 min. At each sampling point, 1 mL aliquot was withdrawn, filtered, and measured by HPLC.

High-performance liquid chromatography (HPLC) was used to determine the dissolved DS concentration in the AS-to-FaSSGF method. Chromatographic separation was performed by an Agilent 1260 HPLC (Agilent Technologies, Santa Clara, CA, USA) using a Kromasil® C18 (4.6 × 150 mm, 5 µm; Nouryon, Bohus, Sweden) analytical column at 40 °C. The mobile phase was a 1:3 mixture of pH 2.5 PBS and methanol. The flow rate of the isocratic elution was 1 mL/min. The DS concentration was determined at 273 nm using a UV-Vis diode array detector. The retention time of the drug was 2.76 min. Data were evaluated using ChemStation B.04.03. Software (Agilent Technologies, Santa Clara, CA, USA).

To determine the kinetics of the drug dissolution, five different mathematical models (zero order, first order, Hixson–Crowell, Higuchi, and Korsmeyer–Peppas) were fitted to the in vitro dissolution curves. The model with the highest regression coefficient (R^2) was selected as the best at describing the kinetics.

2.9. Biocompatibility Studies

MTT assay was performed to evaluate the effect of the NF samples on cell viability. Caco-2 human colon carcinoma cell line (ATCC, Manassas, VA, USA) was cultured in 96-well plates at a seeding density of 4×10^4 cells per well. A measure of 100 µL MEM with Earle's salts medium was used, supplemented with 20% heat-inactivated fetal bovine serum (FBS), 2 mmol/L L-glutamine, non-essential amino acids, MEM vitamins, sodium pyruvate, and nystatin. Cells were incubated for 24 h at 37 °C in a humidified atmosphere containing 5% CO_2.

After the first incubation, the medium was aspirated, and the cells were treated with the test samples. All conditions were assessed in triplicate. The initial concentration of the DS was 1 mg/mL, and two-fold serial dilution was used. So, the final 1:512 dilution resulted in a 1.953 µg/mL DS concentration. Untreated cells served as the positive control, while MEM without samples was the negative control. The plates were incubated for a further 24 h at 37 °C.

After the second incubation, 20 µL of MTT was added to each well, and the plates were incubated for an additional 4 h. Finally, 100 µL of 10% SDS in 0.01 M HCl was added to each well, and the plates were incubated overnight.

Cell viability was assessed by measuring the optical density at 570 nm, with a reference wavelength of 650 nm, using an EZ READ 400 ELISA reader (Biochrom, Cambridge, UK). Viability was expressed as a percentage of the control cells, calculated according to previously described methods [21]. Statistical analyses were performed using GraphPad Prism 8.0.1 (GraphPad Software Inc., San Diego, CA, USA).

2.10. Stability Studies

The storage stability of the DS-loaded NFs was investigated after 6 months of storage in a desiccator. The fresh and the stored samples were investigated in three aspects. The morphology was analyzed using SEM, the crystallinity was checked by XRPD, and chemical stability was tested using Raman Spectroscopy. All methods are described above.

2.11. Statistical Analysis

Statistical analysis was conducted to evaluate significant differences in the measured data. A one-way analysis of variance (ANOVA), followed by a post hoc Tukey HSD test, was performed to compare the in vitro dissolution data. The dissolved DS amounts from the DS-PVP NF, the DS-PVP physical mixture, and the pure DS powder were compared in all time points. The results with $p < 0.05$ were assumed to be statistically significant.

3. Results

3.1. Morphology of the Drug-Free PVP Nanofibers

In the case of drug delivery systems, it is advisable first to develop the carrier system and subsequently load it with the active ingredient. This principle also applies to NFs. As a first step, the optimization of drug-free PVP NF production was carried out. The preparation of NFs is influenced by numerous parameters, among which the most critical are the viscosity of the precursor solution and the applied voltage. The viscosity is mainly determined by the molecular weight and concentration of the polymer. A sufficiently high viscosity enables the formation of uniform, homogenous, and smooth-surfaced NFs. In contrast, low viscosity results in bead-like structures or leads to dripping and/or electrospraying instead of electrospinning. On the other hand, the relationship between applied voltage and fiber diameter in electrospinning is complex and can vary depending on specific experimental conditions. While some studies suggest that increasing the applied voltage leads to thinner fibers due to greater stretching forces [22], others report that higher voltages can result in thicker fibers or bead formation due to jet instability [2]. The effect of increased voltage on the fiber diameter may depend on factors such as polymer concentration, solution viscosity, and ambient conditions.

In this study, the effect of the viscosity of the ES solution on the fiber diameter and morphology was studied. The voltage was set between 10 and 15 kV in an effort to yield the optimal preparation in each case.

Five different ES solutions (5, 7.5, 10, 12.5, 15 *w*/*w*%) were prepared by dissolving the PVP in ethanol. Then, the viscosity was measured, and the solutions were electrospun. As was expected, the viscosity of the ES solutions increased with the increase in the polymer concentration (Table 1). A similar positive correlation between viscosity and average fiber diameter was found. The diameter of the drug-free formulations increased as the ES solution became more concentrated. A steady increase was found between 5 and 12.5 *w*/*w*%. However, the difference between 12.5 and 15 *w*/*w*% was not significant (Figure 3).

Table 1. Viscosity and fiber diameter of the different drug-free formulations prepared by varying the PVP concentration.

Sample Name	PVP Concentration (*w*/*w*%)	Viscosity (mPa·s)	Fiber Diameter (nm)
PVP_5	5	21	N.A.
PVP_7.5	7.5	71	431 ± 143
PVP_10	10	164	886 ± 217
PVP_12.5	12.5	350	1427 ± 216
PVP_15	15	657	1564 ± 267

The SEM images showed that the 5 *w*/*w*% PVP concentration was below the limit for the fiber formation, resulting in electrospraying (Figure 3). Some fibers were present, but mostly spherical aggregates formed during the process. From the 7.5 *w*/*w*% PVP solution, red-blood-cell-shaped structures were observed, indicating insufficient viscosity. In the case of the 10 *w*/*w*% sample, beads and aggregates were observed occasionally but were still present. Additionally, the average fiber diameter exceeded 1 micrometer, so it is more appropriate to refer to the sample as microfibers rather than NFs. The morphology of the 12.5 *w/w*% fibers was good, with no beads, a uniform fiber diameter, and a smooth surface. The average fiber diameter was 1.427 ± 0.216 μm, classifying the fibers as microfibers. Finally, the 15 *w/w*% polymer solution was too concentrated, resulting in fibers with a ragged surface and an average diameter exceeding 1.564 ± 0.267 μm.

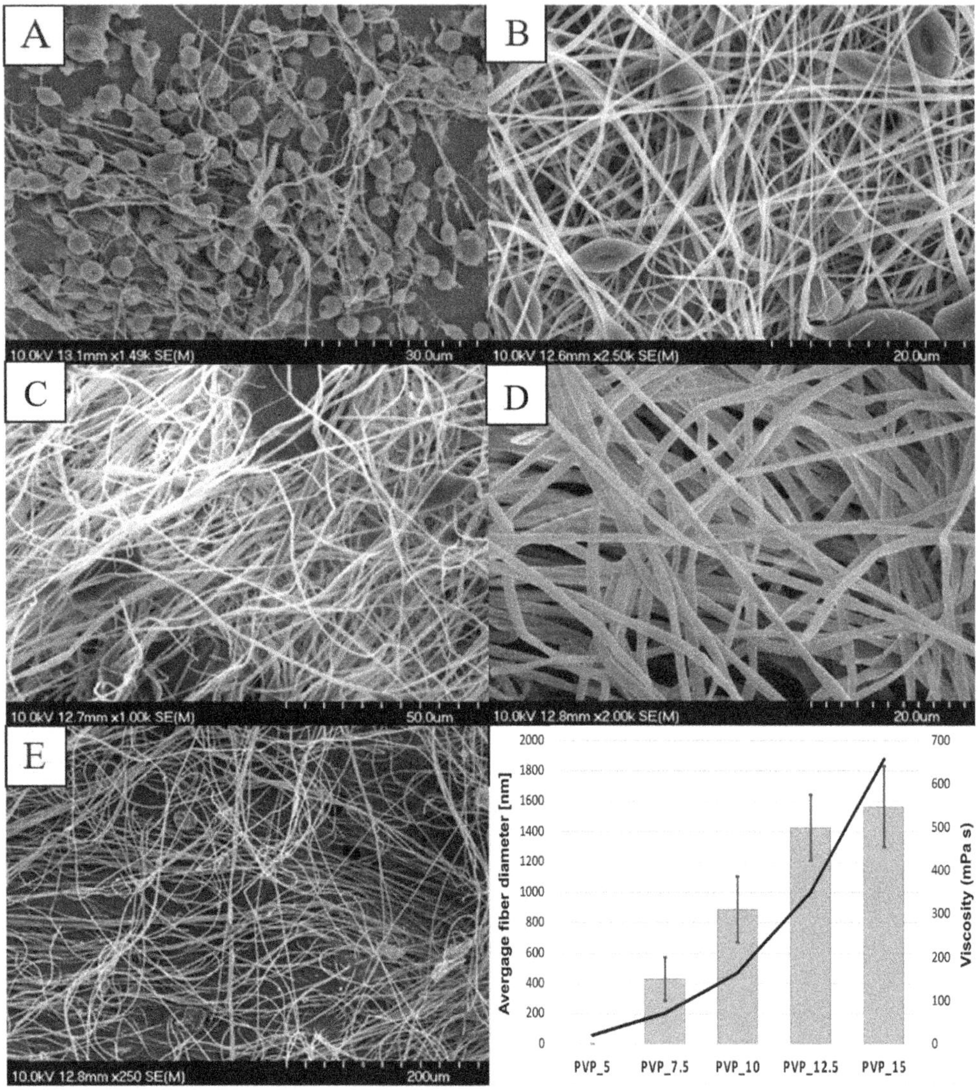

Figure 3. SEM images of the drug-free PVP nano- and microfibers with a graph showing the relationship between viscosity and fiber diameter. (**A**) PVP_5; (**B**) PVP_7.5; (**C**) PVP_10; (**D**) PVP_12.5; (**E**) PVP_15.

In conclusion, the viscosity of the ES solution significantly influenced the ES process. Fiber formation was inadequate at concentrations below 10% and above 12.5%. The optimal morphology of the drug carrier system was achieved using a 12.5 *w/w*% PVP solution filled in two syringes, connected to two 22 G needles. The voltage applied to the double-needle ES equipment was 16.5 kV with a 0.5 mL/h feeding rate.

3.2. Morphology of the DS-Loaded Nanofibers

The next step was the incorporation of DS into the microfibers. The aim was to achieve 10 w/w% drug content in the dry formulation. A content of 12.5 w/w% PVP and a sufficient amount of DS were perfectly dissolved in ethanol. The viscosity of the ES solution was 371 mPa·s, which was close to the drug-free solution (350 mPa·s). So, the addition of the DS did not result in a significant change in the viscosity. Therefore, the same process parameters were used during the double-needle ES.

The double-needle ES was successful as a relatively thick (0.53 ± 0.12 mm) and flexible fibrous membrane was collected. The handling of the membrane was convenient since it was easily removable from the aluminum foil and foldable to the desired size. The manufacturing of the NFs can be a challenge, so it is presumed that the very high molecular weight of the PVP was the reason for the flexible and adherent NF membrane.

The morphology of the DS-PVP fibers was appropriate, with long, bead-free, and smooth fibers (Figure 4). The fiber diameter was 896 ± 286 nm, which means a drastic decrease compared to the drug-free carrier (1427 ± 216 nm). Our hypothesis is that the addition of DS, being a salt, enhanced the conductivity of the ES solution. The increased charge in the jet led to greater elongation, resulting in the formation of thinner fibers.

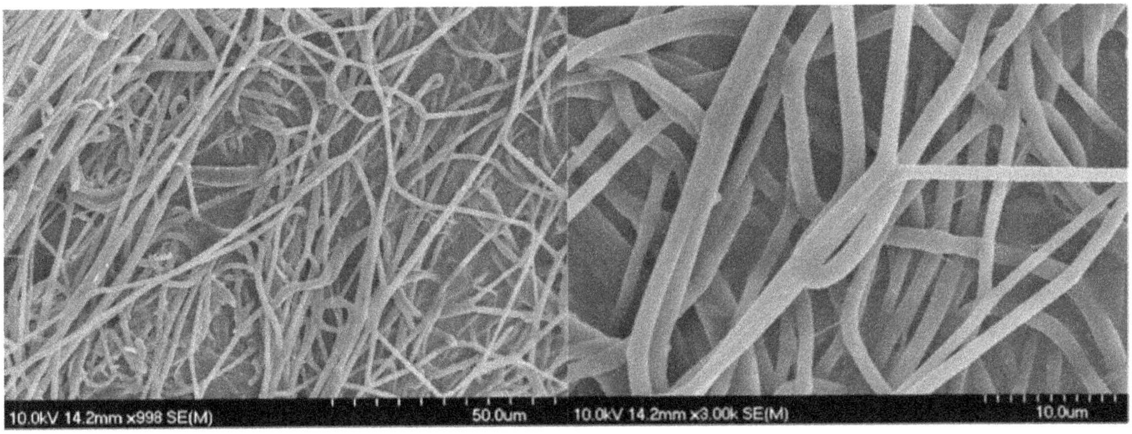

Figure 4. SEM images of the diclofenac sodium-loaded PVP nanofibers prepared by double-needle electrospinning.

3.3. Physicochemical Properties

Physicochemical properties of the electrospun formulation such as crystallinity, chemical bonding between components, drug content, and drug distribution were studied.

XRDP measurements confirmed the change in the crystallinity of the DS in the NFs. The characteristic peaks (6.5, 8.5, 11.2, 15.1, 16.2, 17.1, 20.0, 23.5, 25.1, 25.9, 27.1, 27.8, 37.8 2-theta) of the DS appeared in the diffractogram of the pure DS powder and the physical mixture but were completely absent from the fibrous sample (Figure 5). This proved that the NF sample was an amorphous solid dispersion where the DS was molecularly dispersed in the polymer.

The chemical structure and the presence of the DS were analyzed using Raman Spectroscopy. The characteristic fingerprint patterns of the pure DS and PVP powders were observed (Figure 6). In the case of the DS, the most intense, well-defined peak was a triplet around 1600 cm^{-1}, containing the ring stretching vibrations of the phenylacetate at 1606 cm^{-1} and the dichlorophenyl ring at 1586 cm^{-1}, and the asymmetric stretching vibration of the carboxylate at 1579 cm^{-1}. This triplet is characteristic of the DS [23].

In the case of pure PVP, the fingerprint pattern appeared between 1510 and 1410 cm^{-1}. Also, the characteristic pattern of both the drug and the polymer were distinctly visible in the spectrum of the physical mixture. On the spectrum of the NF, the characteristic peaks were also visible but appeared in lower intensity, indicating secondary interactions (e.g., hydrogen bond), and the formation of an amorphous solid dispersion.

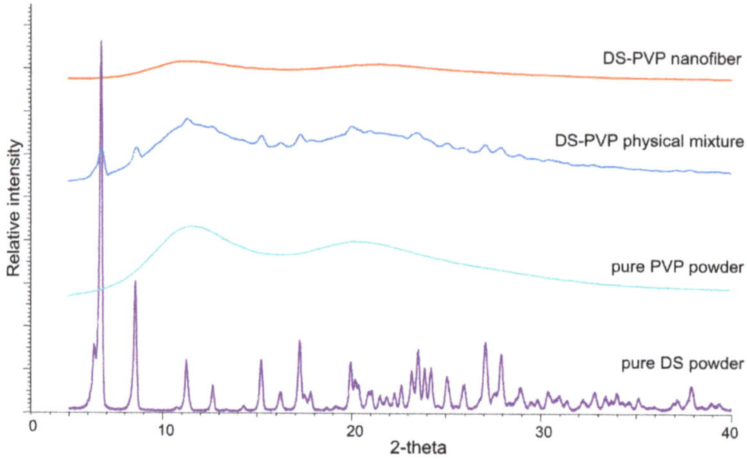

Figure 5. XRDP diffractograms of the crystalline diclofenac sodium (DS) powder and physical mixture and the amorphous PVP powder and DS-PVP nanofibers.

The DL% and EE% were monitored by UV absorbance. The measured DS content was 9.96 ± 0.26%, which means 99.57 ± 2.59% entrapment efficiency. Very high EE% is usual in the case of NFs, since this preparation method is basically a rapid solvent evaporation.

The drug distribution within the NF sample was analyzed using Raman mapping. Figure 7 shows the microscopic image displaying 30 measurement points alongside the corresponding chemical map. In the chemical map, warmer colors indicate a higher concentration of the DS. The balanced coloration of the map suggests that the drug is homogeneously distributed throughout the NF matrix.

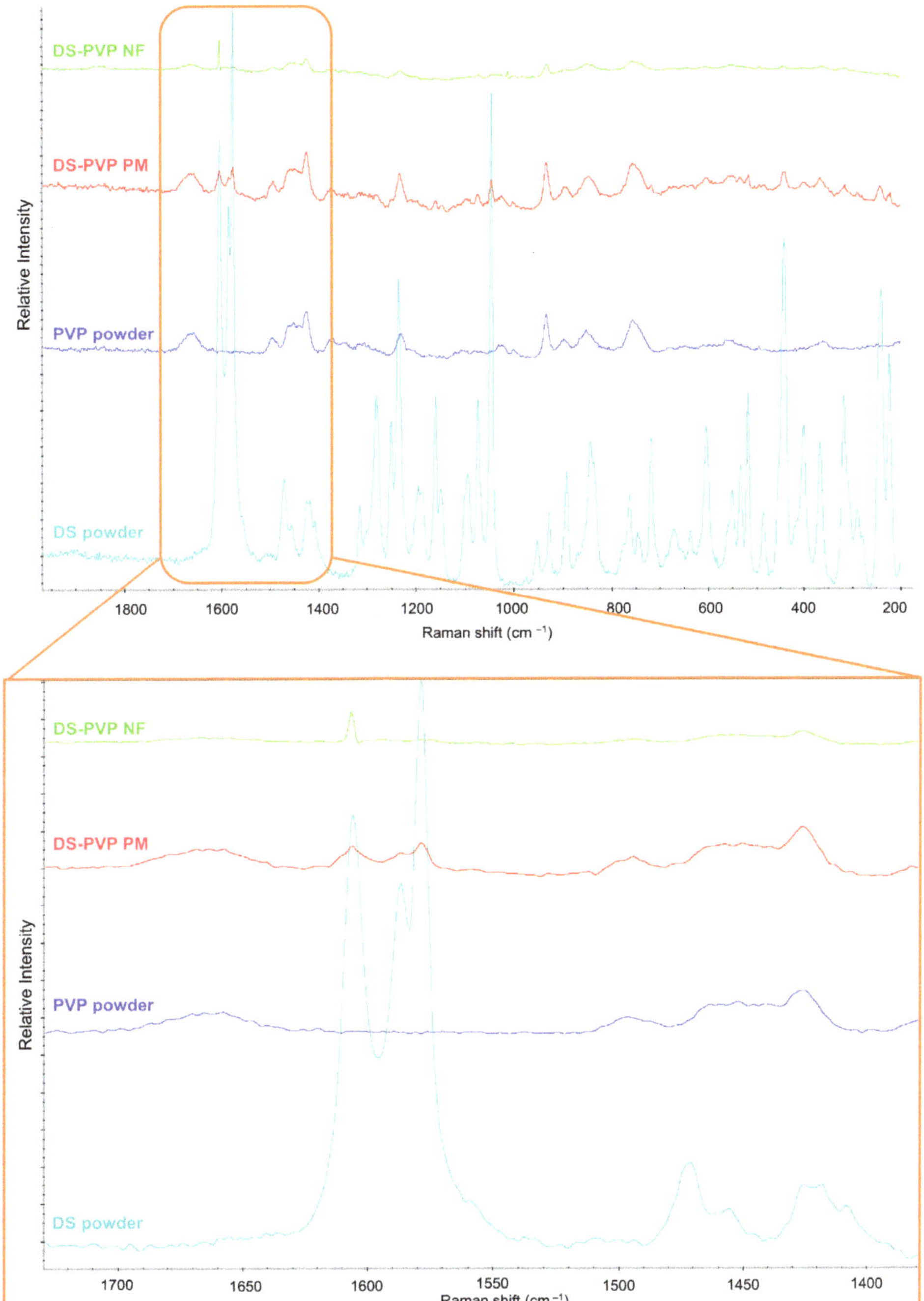

Figure 6. Raman spectra of the diclofenac sodium (DS), PVP powder, physical mixture, and the DS-PVP nanofiber.

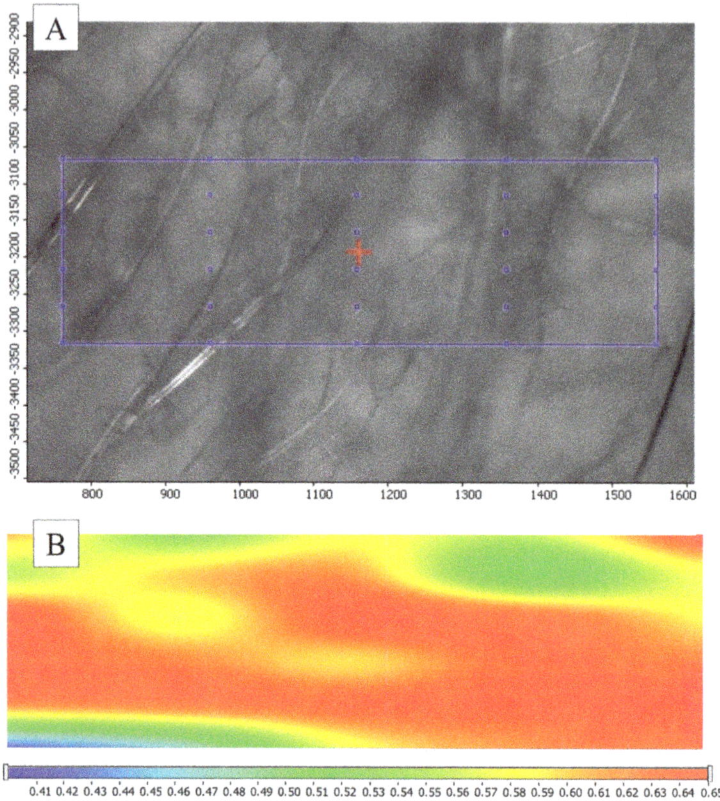

Figure 7. Microscopic image (**A**) and chemical mapping (**B**) of the diclofenac-loaded nanofibers.

3.4. Rapid Disintegration Proved by Two Different Methods

Rapid disintegration is a key characteristic of orodispersible formulations. In this study, two different in vitro methods were used to determine the disintegration time of the DS-PVP NF matrix.

First, the sample was immersed in 25 mL of AS (Figure 8A). Immediate disintegration was observed, with most of the formulation dissolving into the medium simultaneously upon wetting. The entire process was completed in 5.48 ± 1.53 s.

For orally disintegrating tablets, the U.S. Food and Drug Administration (FDA) requires a 30 s disintegration time using the United States Pharmacopeia (USP) method [24]. This method employs a constantly moving basket-rack assembly to test six tablets in one liter of distilled water. In contrast, our study used a smaller volume of AS and a stationary Petri dish for testing. Therefore, the disintegration time of the NF was notably fast.

The second in vitro disintegration test used wet filter paper to mimic the surface of the tongue. The same process was observed as the NF simultaneously wetted and disintegrated or dissolved on the surface of the filter paper (Figure 8B). The disintegration time was 41.38 ± 9.32 s. This result meets the requirements of a fast-disintegrating formulation.

Rapid disintegration is a critical parameter for an orodispersible NF matrix, as it facilitates the immediate release of the drug, making it readily available for dissolution and absorption.

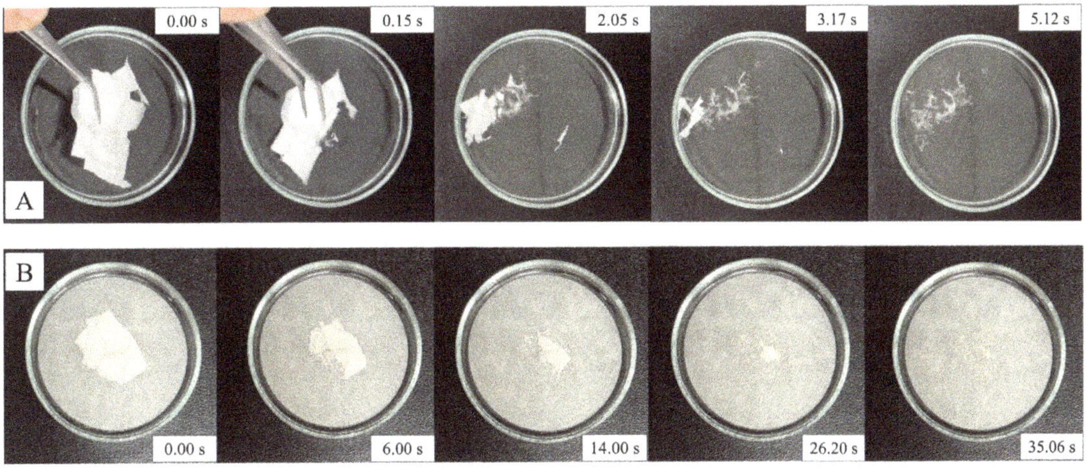

Figure 8. Two in vitro methods for the disintegration test of orodispersible nanofiber matrix loaded with diclofenac sodium. The processes of the free immersion method (**A**) and the artificial tongue method (**B**) are presented in five sequential frames.

3.5. In Vitro Dissolution in Artificial Saliva

The in vitro dissolution of the drug from the NF formulation was tested by two methods. First, AS, as a single medium, was used, and the drug release from the NFs was compared with the dissolution of the pure DS powder and DS-PVP physical mixture (Figure 9). The pure DS powder showed rapid and complete dissolution in the pH 6.8 medium; since the solubility of DS is pH-dependent, the drug is highly soluble in alkaline solutions [25].

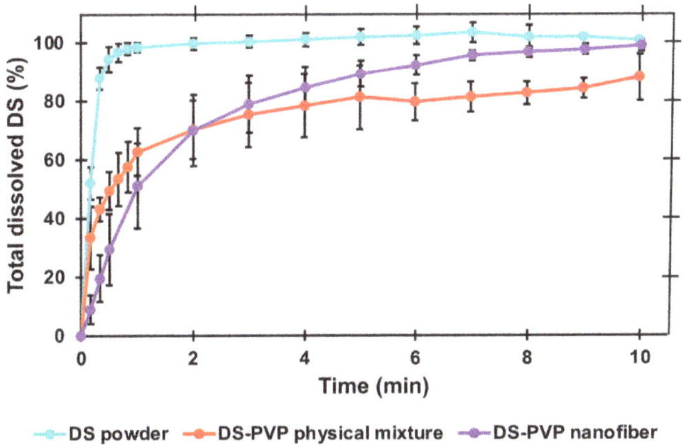

Figure 9. Dissolution in artificial saliva in the case of pure diclofenac sodium (DS), DS-PVP physical mixture, and 10 *w*/*w*% DS-PVP nanofiber matrix.

The presence of the PVP slightly prolonged the drug release by forming a viscous matrix, which slowed the diffusion of the DS. It was also confirmed by the kinetic models since the best-fitting mathematical model was the first order kinetics (R^2 = 0.9520) in the case of the DS powder. This met our expectations as the drug dissolution in this case is not influenced by any external factors, and the process is driven by the concentration gradient. On the other hand, the dissolution kinetics of the DS-PVP physical mixture followed the Korsmeyer–Peppas model ($R^2 = 0.9464$), in which both diffusion and polymer–drug interactions contribute to the overall release mechanism.

The data between the physical mixture and the NFs did not differ significantly until the fourth minute. From the fourth minute, the dissolution from the NFs was significantly higher ($p < 0.05$), and it even reached the curve of the pure DS at the last time point. Also, the shape of the curve was exponential, and the best-fitting dissolution kinetics was the first-order release ($R^2 = 0.9917$). The released drug amount reached 80% within 4 min and the complete dissolution was achieved within 10 min. The FDA dissolution criterion for immediate release is Q = 80% in 30 min [26]. Thus, the tested DS-PVP NF can be considered an immediate-release formulation.

3.6. In Vitro Dissolution Tested by "AS-to-FaSSGF" Method

The idea behind the second tested in vitro dissolution method was the swallowing of the drug mixed with saliva. This way, the DS could reach the stomach and interact with the gastric fluid. Since this aspect has not been widely studied, a novel in vitro drug release method, called "AS-to-FaSSGF", was developed to simulate the natural conditions of drug transit and release (Figure 10).

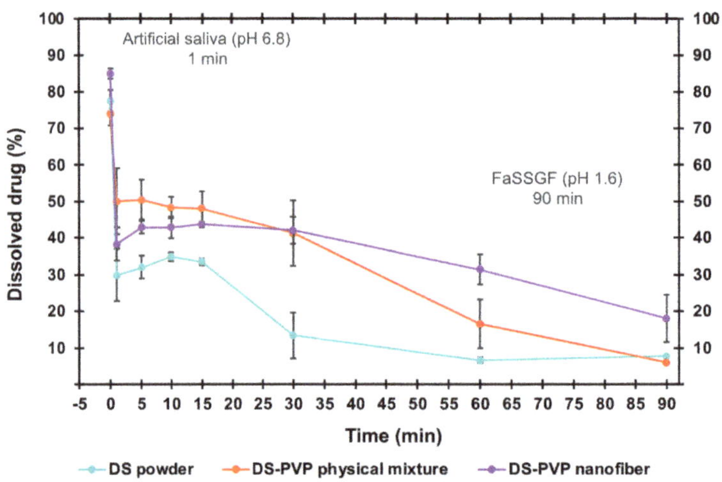

Figure 10. Result of the AS-to-FaSSGF dissolution method in the case of pure diclofenac sodium (DS), DS-PVP physical mixture, and 10 *w*/*w*% DS-PVP nanofiber matrix.

In the first minute of the "AS-to-FaSSGF" method, the dose was placed into 5 mL of pH 6.8 AS. The DS, due to its salt form, dissolved well in this medium. The dissolved drug amount was 85 ± 1%, 74 ± 3%, and 78 ± 3% in the case of the NF, physical mixture, and pure DS powder, respectively. The NF showed significantly better dissolution in AS in one minute.

After exactly one minute, the liquid was poured into 895 mL pH 1.2 FaSSGF. Since the solubility of the DS is reduced by lowering the pH, the drug precipitated in the second phase. Fast precipitation was observed after the pouring, then the concentration constantly decreased until the end of the measurement. Here, 18 ± 6.5%, 8 ± 0.1%, and 6 ± 0.3% of the initial DS remained dissolved in the case of the NF, physical mixture, and pure DS powder, respectively. The precipitation was visible to the bare eye.

Although the trend was the same for all three samples, there were significant differences between the curves. The DS released from the NF matrix was significantly higher ($p < 0.05$) than the pure drug between 5 and 90 min and the physical mixture at 60 and 90 min. The presence of the PVP improved the solubility, since the physical mixture showed significantly higher ($p < 0.05$) dissolution than the DS powder between 1 and 30 min.

Overall, the "AS-to-FaSSGF" in vitro dissolution method underlined the relevance of an orodispersible dosage form when formulating the DS, as the solubility decreased drastically in simulated gastric fluid.

3.7. Biocompatibility of the Nanofibers

The biocompatibility of the double-needle electrospun sample and the two pure powders was tested using an MTT assay on Caco-2 cells (Figure 11). According to the results, all three samples were cytocompatible up to 0.25 mg/mL (1:4 dilution). However, in higher concentrations, the DS was cytotoxic both in pure form and in the NF. The pure PVP was cytocompatible even at a 1 mg/mL concentration.

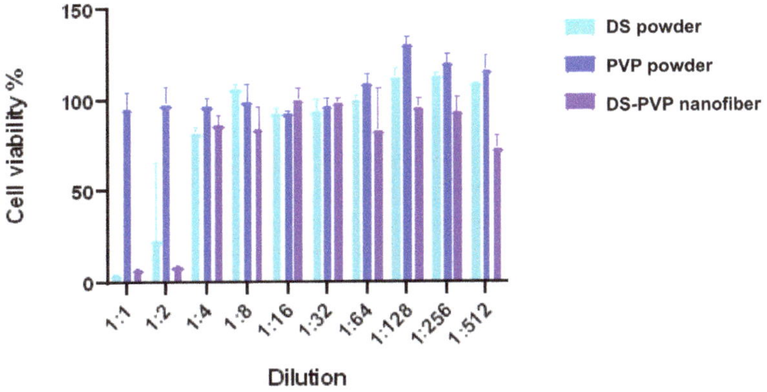

Figure 11. Result of the MTT test comparing the biocompatibility of the pure diclofenac sodium (DS), pure PVP, and the DS-PVP nanofiber matrix.

3.8. Storage Stability of the Diclofenac-Loaded Nanofibers

The storage stability of the NFs prepared by double-needle ES was investigated in three aspects: morphology, crystallinity, and chemical stability (Figure 12). All three methods confirmed that the NF matrix was stable for 6 months. The SEM analyses showed that the morphology was still satisfactory. The autonomy of the fibers was maintained, no fusion occurred, and their surface remained smooth (Figure 12A). The measured average fiber diameter did not differ significantly. The XRPD measurement confirmed that no recrystallization occurred during storage since the diffractogram remained flat without any distinctive peaks (Figure 12B). The Raman spectra of the fresh and the stored fibrous samples were similar, and the fingerprint pattern was maintained (Figure 12C). So, neither chemical degradation, nor any other chemical transformation, occurred. The DS-loaded NF matrix was chemically stable over the 6 months.

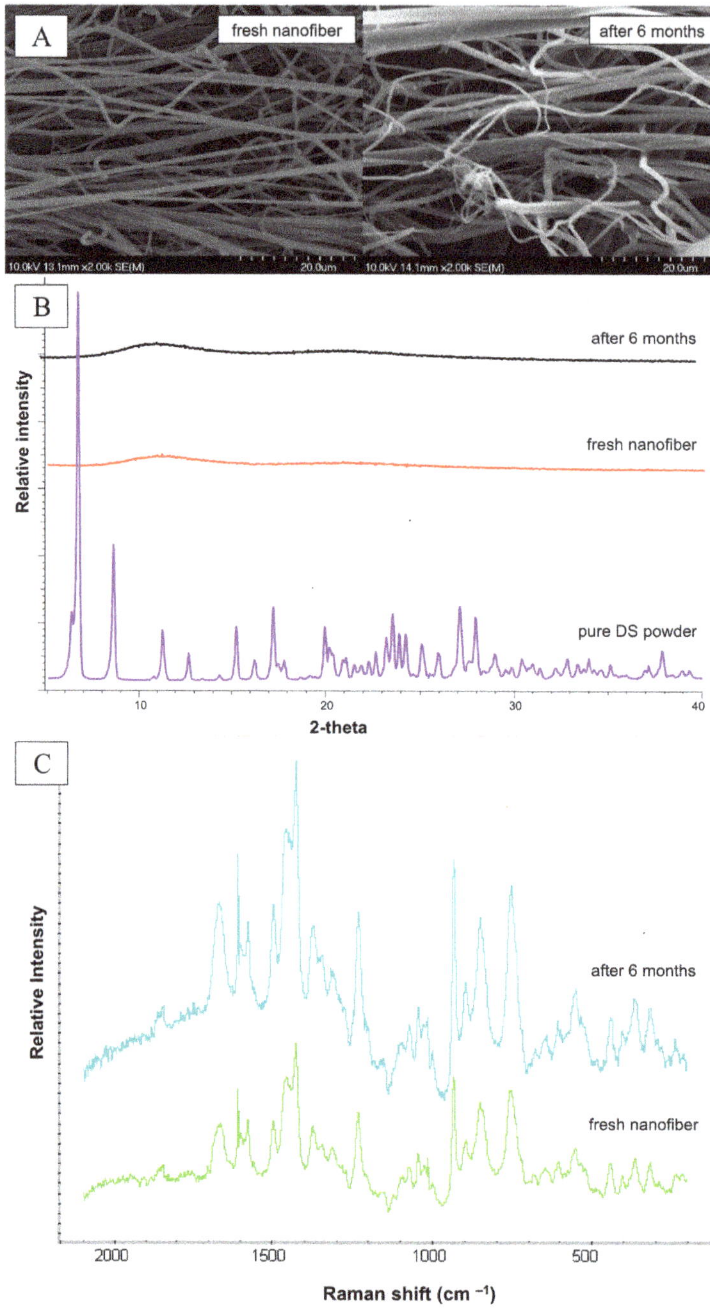

Figure 12. SEM images (**A**), XRPD diffractogram (**B**), and Raman spectra (**C**) of the fresh and the stored double-needle electrospun nanofiber at the beginning and end of the 6-month stability test.

4. Discussion

The most commonly used equipment for NF production is the needle-based ES. Besides its many advantages, one of its biggest disadvantages is its low productivity.

Improving this requires more needles connected side by side. In this way, industrial-scale production is also possible.

In this study, DS-loaded PVP NFs were prepared using a double-needle ES device. The needle holder was designed and 3D-printed. The angle formed by the needle tips was adjustable, ensuring that the NFs on the collector formed a single unified patch instead of two separate ones. The process was suitable for the preparation of PVP NF matrices from a 1,300,000 Mw PVP solution with a concentration between 7.5 and 15 $w/w\%$.

First, drug-free PVP solutions were electrospun to investigate the relationship between the polymer concentration and the fiber morphology. As the polymer concentration increased, the viscosity of the ES solution, as well as the average fiber diameter, increased. A concentration that was too low caused inappropriate viscosity, which led to electrospraying instead of electrospinning. From the 5 $w/w\%$ PVP solution, almost-spherical polymer nanoparticles were formed by the spraying. As the concentration was increased, the globes elongated first into red-blood-cell-shaped structures, then to proper fibers. Overall, the fiber formation was optimal in the case of 12.5 $w/w\%$ PVP solution with an applied voltage of 15 kV and a pumping rate of 0.5 mL/h.

Secondly, DS-loaded NFs were prepared by double-needle ES using the optimal process parameters. The addition of the drug drastically decreased the average fiber diameter. This was probably due to the changed material properties, since the addition of the DS salt increased the conductivity of the ES solution. Previously, our research group observed a similar result with an ES solution of the same polymer and ciprofloxacin [27].

Afterwards, the physicochemical properties of the DS-PVP NFs were investigated. The results proved that the electrospun formulation was an amorphous solid dispersion, the EE% was more than 99%, and the drug distribution was homogeneous within the NF matrix. These properties, and the general characteristics of NFs, such as high specific surface area, predicted rapid disintegration and immediate dissolution. Moreover, fast disintegration of the water-soluble PVP matrix was expected, further promoting the efficient release and subsequent absorption of the drug. The rapid disintegration was proved by two different methods, namely free immersion and simulated tongue. Both methods confirmed the fast disintegration of the nanofibrous formulation.

In vitro dissolution studies are cornerstones of pharmaceutical research. Orodispersible formulations require immediate release in the oral cavity. The DS-PVP NF matrix met the requirements of the USP, as in AS, the dissolution was complete within 10 min. Additionally, a new "AS-to-FaSSGF" dissolution method was developed to test the natural conditions of swallowing the drug. The method used two dissolution media, a small volume of AS and the generally used 900 mL of FaSSGF. Since the solubility of the DS is pH-dependent, the shift to acidic conditions caused a remarkable decrease in the solubility which led to visible precipitation. The DS amount that remained dissolved was significantly higher in the case of the NFs. Therefore, it can be beneficial to use it as an orodispersible formulation.

Finally, the storage stability of the DS-PVP NF matrices confirmed that the formulation was stable for 6 months. Additionally, the cytocompatibility of the ingredients and the electrospun fibers was proved.

Overall, the developed nanofibrous formulation can be a promising candidate as a novel orally disintegrating dosage form. The most common orally disintegrating dosage forms, which are on the market, are the orally disintegrating tablet (ODT) and orodispersible film (ODF). Generally, tablets are a widely favored dosage form due to their ease of manufacturing with uniform mass and consistent dosing. However, ODTs typically exhibit poor mechanical strength, as achieving rapid disintegration requires a porous structure [28]. Another drawback of ODTs is the need for a high amount of excipients to mask the bitterness of the active pharmaceutical ingredient [29]. A proposed solution to these issues is using flexible ODFs because the polymer matrix can effectively mask unpleasant tastes. However, a significant disadvantage of these films is their considerably slower drug release [29]. Orodispersible nanofiber membranes could be an alternative, as they combine the advantages of both dosage forms; they are more flexible than fragile, and have a porous structure, allowing for extremely rapid disintegration. Their potential drawback lies in their production costs, as nanofiber fabrication is a novel technology in the pharmaceutical industry compared to conventional tableting, and its implementation and regulatory approval may require more resources.

5. Conclusions

In conclusion, the preparation of the double-needle electrospun PVP NF was optimized, DS was loaded into the fibers, and the formulation was investigated. In order to achieve an orodisperse dosage form, the salt form of the drug and a water-soluble polymer were selected. Many properties facilitated the rapid disintegration and the fast dissolution, i.e., the large specific surface area, the structure of an amorphous solid dispersion, the good wettability, and the porous structure of the NF matrix. Both the rapid disintegration and the immediate dissolution were proved by two different methods. The formulation was biocompatible and stable for 6 months.

Author Contributions: Conceptualization, L.É.U. and R.A.; methodology, L.É.U., A.M.-N. and M.C.; validation, L.É.U., T.J. and C.H.; formal analysis, T.J.; investigation, L.É.U., T.J., C.H., O.J.-L. and K.B.; resources, M.C.; writing—original draft preparation, L.É.U.; writing—review and editing, R.A.; visualization, L.É.U.; supervision, R.A.; funding acquisition, R.A. All authors have read and agreed to the published version of the manuscript.

Funding: The publication was funded by The University of Szeged Open Access Fund (FundRef, No. 7569). The work was supported by the NKFIH OTKA K_146148 project. Project no TKP2021-EGA-32 has been implemented with the support provided by the Ministry of Culture and Innovation of Hungary from the National Research, Development and Innovation Fund, financed under the TKP2021-EGA funding scheme.

Institutional Review Board Statement: Not applicable.

Informed Consent Statement: Not applicable.

Data Availability Statement: No new data were created or analyzed in this study. Data sharing is not applicable to this article.

Acknowledgments: The authors would like to thank all the laboratory technicians who contributed to the project.

Conflicts of Interest: The authors declare no conflicts of interest.

Abbreviations

The following abbreviations are used in this manuscript:

ANOVA	analysis of variance
AS	artificial saliva
AS-to-FaSSGF	artificial saliva-to-Fasted State Simulating Gastric Fluid
CCD	charge-coupled device
DL%	drug loading
DS	diclofenac sodium
EE%	entrapment efficiency
ES	electrospinning

FaSSGF	Fasted State Simulating Gastric Fluid
FDA	U.S. Food and Drug Administration
HPMC	hydroxypropyl methylcelluloses
HPLC	High Performance Liquid Chromatography
MEM	Minimal Essential Medium
MTT	thiazolyl blue tetrazolium bromide
NF	nanofiber
ODF	orodispersible film
ODT	orally disintegrating tablet
PBS	phosphate-buffer solution
PEO	polyethylene oxide
PVA	polyvinyl alcohol
PVP	polyvinylpyrrolidone
SEM	scanning electron microscopy
USP	United States Pharmacopeia
UV	ultraviolet
UV-Vis	ultraviolet-visible
XRPD	X-ray powder diffraction

References

1. Cilurzo, F.; Musazzi, U.M.; Franzé, S.; Selmin, F.; Minghetti, P. Orodispersible Dosage Forms: Biopharmaceutical Improvements and Regulatory Requirements. *Drug Discov. Today* **2018**, *23*, 251–259. [CrossRef] [PubMed]
2. Haider, A.; Haider, S.; Kang, I.-K. A Comprehensive Review Summarizing the Effect of Electrospinning Parameters and Potential Applications of Nanofibers in Biomedical and Biotechnology. *Arab. J. Chem.* **2018**, *11*, 1165–1188. [CrossRef]
3. Hannan, P.A.; Khan, J.A.; Khan, A.; Safiullah, S. Oral Dispersible System: A New Approach in Drug Delivery System. *Indian J. Pharm. Sci.* **2016**, *78*, 2–7. [CrossRef] [PubMed]
4. Council of Europe. *European Pharmacopoeia*, 10th ed.; Council of Europe: Strasbourg, France, 2020.
5. Seager, H. Drug-Delivery Products and the Zydis Fast-Dissolving Dosage Form. *J. Pharm. Pharmacol.* **1998**, *50*, 375–382. [CrossRef]
6. Ning, T.; Zhou, Y.; Xu, H.; Guo, S.; Wang, K.; Yu, D.-G. Orodispersible Membranes from a Modified Coaxial Electrospinning for Fast Dissolution of Diclofenac Sodium. *Membranes* **2021**, *11*, 802. [CrossRef]
7. Balusamy, B.; Celebioglu, A.; Senthamizhan, A.; Uyar, T. Progress in the Design and Development of "Fast-Dissolving" Electrospun Nanofibers Based Drug Delivery Systems—A Systematic Review. *J. Control. Release* **2020**, *326*, 482–509. [CrossRef]
8. Yu, D.-G.; Shen, X.-X.; Branford-White, C.; White, K.; Zhu, L.-M.; Annie Bligh, S.W. Oral Fast-Dissolving Drug Delivery Membranes Prepared from Electrospun Polyvinylpyrrolidone Ultrafine Fibers. *Nanotechnology* **2009**, *20*, 055104. [CrossRef]
9. Zhang, Q.; Lin, Z.; Zhang, W.; Huang, T.; Jiang, J.; Ren, Y.; Zhang, R.; Li, W.; Zhang, X.; Tu, Q. Fabrication of Green Poly(Vinyl Alcohol) Nanofibers Using Natural Deep Eutectic Solvent for Fast-Dissolving Drug Delivery. *RSC Adv.* **2020**, *11*, 1012–1021. [CrossRef]
10. Asatiani, N.; Filipová, B.; Pechočiaková, M.; Kluk, A.; Kunc, Š.; Mikeš, P. Electrospun HPMC/PEO-Blend Orodispersible Films: How Slight Batch Differences Affect the Crucial Mechanical Properties. *Cellulose* **2023**, *30*, 4527–4541. [CrossRef]
11. Hennig, M. Oral Fast-Dissolving Risperidone Loaded Electrospun Nanofiber Drug Delivery Systems for Antipsychotic Therapy. *J. Drug Deliv. Sci. Technol.* **2023**, *92*, 105262.
12. Aytac, Z.; Ipek, S.; Erol, I.; Durgun, E.; Uyar, T. Fast-Dissolving Electrospun Gelatin Nanofibers Encapsulating Ciprofloxacin/Cyclodextrin Inclusion Complex. *Colloids Surf. B Biointerfaces* **2019**, *178*, 129–136. [CrossRef] [PubMed]
13. Celebioglu, A.; Uyar, T. Electrospun Formulation of Acyclovir/Cyclodextrin Nanofibers for Fast-Dissolving Antiviral Drug Delivery. *Mater. Sci. Eng. C* **2021**, *118*, 13. [CrossRef] [PubMed]
14. Guler, E.; Yekeler, H.B.; Ozdemir Kumral, Z.N.; Parviz, G.; Ozcan, G.S.; Uner, B.; Demirbas, S.G.; Ayyildiz, S.; Yazir, Y.; Kalaskar, D.; et al. Fabrication of Oro-Dispersible Sodium Valproate-Loaded Nanofibrous Patches for Immediate Epileptic Innervation. *ACS Biomater. Sci. Eng.* **2025**, *11*, 1523–1538. [CrossRef] [PubMed]
15. Uhljar, L.É.; Ambrus, R. Electrospinning of Potential Medical Devices (Wound Dressings, Tissue Engineering Scaffolds, Face Masks) and Their Regulatory Approach. *Pharmaceutics* **2023**, *15*, 417. [CrossRef]
16. Cimini, A.; Imperi, E.; Picano, A.; Rossi, M. Electrospun Nanofibers for Medical Face Mask with Protection Capabilities against Viruses: State of the Art and Perspective for Industrial Scale-Up. *Appl. Mater. Today* **2023**, *32*, 101833. [CrossRef]
17. Li, Y.; Zhu, J.; Cheng, H.; Li, G.; Cho, H.; Jiang, M.; Gao, Q.; Zhang, X. Developments of Advanced Electrospinning Techniques: A Critical Review. *Adv. Mater. Technol.* **2021**, *6*, 2100410. [CrossRef]

18. Beaudoin, É.J.; Kubaski, M.M.; Samara, M.; Zednik, R.J.; Demarquette, N.R. Scaled-Up Multi-Needle Electrospinning Process Using Parallel Plate Auxiliary Electrodes. *Nanomaterials* **2022**, *12*, 1356. [CrossRef]
19. Bi, Y.; Sunada, H.; Yonezawa, Y.; Danjo, K.; Otsuka, A.; Iida, K. Preparation and Evaluation of a Compressed Tablet Rapidly Disintegrating in the Oral Cavity. *Chem. Pharm. Bull.* **1996**, *44*, 2121–2127. [CrossRef]
20. Celebioglu, A.; Uyar, T. Fast Dissolving Oral Drug Delivery System Based on Electrospun Nanofibrous Webs of Cyclodextrin/Ibuprofen Inclusion Complex Nanofibers. *Mol. Pharm.* **2019**, *16*, 4387–4398. [CrossRef]
21. Banat, H.; Csóka, I.; Paróczai, D.; Burian, K.; Farkas, Á.; Ambrus, R. A Novel Combined Dry Powder Inhaler Comprising Nanosized Ketoprofen-Embedded Mannitol-Coated Microparticles for Pulmonary Inflammations: Development, In Vitro–In Silico Characterization, and Cell Line Evaluation. *Pharmaceuticals* **2024**, *17*, 75. [CrossRef]
22. Can-Herrera, L.A.; Oliva, A.I.; Dzul-Cervantes, M.A.A.; Pacheco-Salazar, O.F.; Cervantes-Uc, J.M. Morphological and Mechanical Properties of Electrospun Polycaprolactone Scaffolds: Effect of Applied Voltage. *Polymers* **2021**, *13*, 662. [CrossRef] [PubMed]
23. Fini, A.; Cavallari, C.; Ospitali, F. Diclofenac Salts, V. Examples of Polymorphism among Diclofenac Salts with Alkyl-Hydroxy Amines Studied by DSC and HSM. *Pharmaceutics* **2010**, *2*, 136–158. [CrossRef] [PubMed]
24. U.S. Department of Health and Human Services, Food and Drug Administration, Center for Drug Evaluation and Research (CDER). *Guidance for Industry—Orally Disintegrating Tablets*; 2008.
25. Kincl, M.; Meleh, M.; Veber, M.; Vrečer, F. Study of Physicochemical Parameters Affecting the Release of Diclofenac Sodium from Lipophilic Matrix Tablets. *Acta Chim. Slov.* **2004**, *51*, 409–425.
26. U.S. Department of Health and Human Services, Food and Drug Administration, Center for Drug Evaluation and Research (CDER). *Dissolution Testing and Acceptance Criteria for Immediate-Release Solid Oral Dosage Form Drug Products Containing High Solubility Drug Substances Guidance for Industry*; 2018; p. 8.
27. Uhljar, L.É.; Kan, S.Y.; Radacsi, N.; Koutsos, V.; Szabó-Révész, P.; Ambrus, R. In Vitro Drug Release, Permeability, and Structural Test of Ciprofloxacin-Loaded Nanofibers. *Pharmaceutics* **2021**, *13*, 556. [CrossRef]
28. Amelian, A.; Wasilewska, K.; Wesoły, M.; Ciosek-Skibińska, P.; Winnicka, K. Taste-Masking Assessment of Orally Disintegrating Tablets and Lyophilisates with Cetirizine Dihydrochloride Microparticles. *Saudi Pharm. J.* **2017**, *25*, 1144–1150. [CrossRef]
29. Tawfik, E.A.; Scarpa, M.; Abdelhakim, H.E.; Bukhary, H.A.; Craig, D.Q.M.; Barker, S.A.; Orlu, M. A Potential Alternative Orodispersible Formulation to Prednisolone Sodium Phosphate Orally Disintegrating Tablets. *Pharmaceutics* **2021**, *13*, 120. [CrossRef]

polymers
MDPI

Article

Novel PVA–Hyaluronan–Siloxane Hybrid Nanofiber Mats for Bone Tissue Engineering

Daniela Anahí Sánchez-Téllez [1], Shantal Lizbeth Baltierra-Uribe [2], Mónica Araceli Vidales-Hurtado [3], Alejandra Valdivia-Flores [2], Blanca Estela García-Pérez [2,*] and Lucía Téllez-Jurado [1,*]

1 Department of Engineering in Metalurgy and Materials, Escuela Superior de Ingeniería Química e Industrias Extractivas, Instituto Politécnico Nacional, Unidad Profesional Adolfo López Mateos (UPALM), Av. Instituto Politécnico Nacional S/N, Zacatenco, Mexico City 07738, Mexico
2 Department of Microbiology, Escuela Nacional de Ciencias Biológicas, Instituto Politécnico Nacional, Prolongación de Carpio y Plan de Ayala S/N, Casco de Santo Tomás, Mexico City 11340, Mexico
3 Centro de Investigación en Ciencia Aplicada y Tecnología Avanzada, Unidad Querétaro, Instituto Politécnico Nacional, Cerro Blanco 141, Colinas del Cimatario, Santiago de Querétaro 76090, Mexico
* Correspondence: blgarciap@ipn.mx or abrilestela@hotmail.com (B.E.G.-P.); ltellezj@ipn.mx (L.T.-J.)

Abstract: Hyaluronan (HA) is a natural biodegradable biopolymer; its biological functions include cell adhesion, cell proliferation, and differentiation as well as decreasing inflammation, angiogenesis, and regeneration of damaged tissue. This makes it a suitable candidate for fabricating nanomaterials with potential use in tissue engineering. However, HA nanofiber production is restricted due to the high viscosity, low evaporation rate, and high surface tension of HA solutions. Here, hybrids in the form of continuous and randomly aligned polyvinyl alcohol (PVA)–(HA)–siloxane nanofibers were obtained using an electrospinning process. PVA–HA fibers were crosslinked by a 3D siloxane organic–inorganic matrix via sol-gel that restricts natural hydrophilicity and stiffens the structure. The hybrid nanofiber mats were characterized by FT-IR, micro-Raman spectroscopy, SEM, and biological properties. The PVA/HA ratio influenced the morphology of the hybrid nanofibers. Nanofibers with high PVA content (10PVA-8 and 10PVA-10) form mats with few beaded nanofibers, while those with high HA content (5PVA-8 and 5PVA-10) exhibit mats with mound patterns formed by "ribbon-like" nanofibers. The hybrid nanofibers were used as mats to support osteoblast growth, and they showed outstanding biological properties supporting cell adhesion, cell proliferation, and cell differentiation. Importantly, the 5PVA-8 mats show 3D spherical osteoblast morphology; this suggests the formation of tissue growth. These novel HA-based nanomaterials represent a relevant advance in designing nanofibers with unique properties for potential tissue regeneration.

Keywords: nanofibers; electrospinning; hyaluronic acid; polyvinyl alcohol; siloxanes; hybrids; osteoblasts

1. Introduction

Hyaluronic acid (HA) is an important component of the extracellular matrix (ECM) in the human body [1]. Some properties of HA such as viscoelasticity, high moisture absorption capability and hygroscopic nature, make HA an excellent biopolymer for maintaining the viscoelasticity of the ECM, supporting cellular structure, and functioning as a lubricant [2]. Recently, HA has gained attention in the design of novel devices for application in the bioengineering and medical fields due to its exceptional physicochemical and biological attributes [3]. HA has been widely used in bone regeneration therapies, serving as either hyaluronic acid-incorporated scaffolds [3] or as carriers for cell seeding and/or bioactive components [4]. Even though hyaluronic acid itself has a limited effect on osteogenesis enhancement, HA derivatives and HA-incorporated hybrid scaffolds have shown the potential to improve osteogenesis and mineralization [5]. Based on current research, it is observed that HA is becoming the key biopolymer component in HA-based hybrid scaffolds [6,7] as it improves their biological osteoinductive properties.

The design of scaffolds from biocompatible nanofibers represents an excellent advantage, because they have a porous and 3D structure, and the high surface-to-volume ratio creates a more suitable microenvironment for cell growth [8]. Several in vitro studies have shown that these structures are ideal for cell adhesion, proliferation, and migration [9–12]. Moreover, this type of 3D microenvironment is necessary for nutrient transport, cell ingrowth, and vascularization [13,14].

Along with the composition of the scaffolds, the process of manufacturing fiber materials is important. One of the techniques used to produce micro/nanofiber mats in a continuous process is the electrospinning technique. This process allows for the design of fibers with diameters adjusted from nanometers to microns. The fiber morphologies and sizes can be regulated by tuning the following parameters: (1) solution parameters including viscosity, solvent, polymer concentration, molecular weight of the polymer, conductivity, and surface tension; (2) processing parameters including applied voltage, distance between tip and collector, needle diameter, feed rate/flow rate, and collector type; (3) ambient parameters including humidity and temperature [15,16].

Unfortunately, electrospinning of nanofibrous mats made up purely of biopolymers is difficult to achieve because very low polymer concentrations reach high solution viscosities [17]. However, to take advantage of the HA viscoelastic and biological properties, HA can be chemically modified and/or physically blended with other polymer solutions (PVA, PCL, PLA) to make it electrospinnable. Moreover, the need to use organic solvents rather than water or alcohol to obtain the critical chain entanglement concentration necessary to being electrospun reduces the nanofibers' biological and biomedical applicability. Sun et al. [18] successfully electrospun nanofibers from an aqueous complex coacervate solution composed of chitosan and hyaluronic acid. They eliminated the need for toxic solvent systems in the electrospun biopolymer solution and the risk of degradation effects.

Due to the excellent biocompatibility of polyvinyl alcohol (PVA), PVA-based hydrogels have been tried out as tissue regeneration materials such as artificial articular cartilage, intervertebral discs, and meniscus [19]. In addition, PVA has been shown to form easily electrospun fibers [19]. However, like other natural and synthetic polymers, its degradation rates can hardly be designed or adjusted to match the regeneration rates of bone tissue. Thus, to improve degradation resistance, mechanical properties, and osteoconductivity, polymers can be hybridized with inorganic components (hydroxyapatite [20], biomimetic calcium phosphate, SiO_2 [21]) or crosslinked with silicone (PDMS)-modified silica (SiO_2) 3D networks. Organic–inorganic materials are classified as hybrids or nanocomposites because organic and inorganic components are bonded in a single system at the molecular or nano level. Under optimal conditions, hybrid materials show the synergy of the properties of both components. Although this type of crosslinking has produced excellent results, the hybridization of biopolymers (sodium hyaluronate, chondroitin sulfate, chitosan, alginate) [22,23] or synthetic polymers (PVA, PLA, PCL, PEG) is rarely explored. Sánchez-Téllez et al. [6] proved in their study that the crosslinking of polysaccharide chains by a 3D siloxane organic–inorganic matrix via the sol-gel method gave sodium hyaluronate matrix strength, improving rheological properties and increasing thermal degradation temperatures.

Because the degradative and biological properties of hybrids can be modulated, the properties of electrospun nanofiber mats built up from an organic (sodium hyaluronate, PVA)–inorganic (siloxane) hybrid solution are magnified due to their high surface areas that enable the deposition of cell adhesive molecules (peptides) on the surface. This increases cell adhesion, promotes cell proliferation, and maintains cell phenotype. Therefore, to increase the desirable physicochemical and biological properties of hybrid materials, the aim of this work was to design 3D cell scaffolds based on hybrid electrospun nanofiber mats composed of PVA–HA hydrogels inorganically crosslinked by a PDMS-modified SiO_2 organic–inorganic network. The fabricated mats showed that they promote cell adhesion, proliferation, and mineralization of human osteoblasts; this highlights their potential use in bone tissue engineering.

2. Materials and Methods

Techniques and equipment used for the processing, characterization, and determination of chemical and biological properties of the hybrid hydrogels are described.

2.1. Materials

Electrospun PVA–HA hybrid nanofiber mats were prepared by chemical modification and hybridization of PVA–HA and then inorganically crosslinked via the sol-gel method. The starting reagents were: (1) sodium hyaluronate oral grade with a Mw = 150–300 kDa, PDI = 1.3 [24] (HA, Sigma-Aldrich, St. Louis, MO, USA); (2) N-(3-Dimethylaminopropil)-N′ethylcarbodiimide hydrochloride (EDC, $C_8H_{17}N_3 \cdot HCl$, Sigma-Aldrich); (3) N-hydroxysuccinimide (NHS, $C_4H_5NO_3$, Sigma-Aldrich); (4) 2-(N-Morpholino) ethanesulfonic acid hydrate (MES, $C_6H_{13}NO_4S \cdot H_2O$, Sigma-Aldrich); (5) 3-aminopropyltriethoxysilane (APTES, $H_2N(CH_2)_3Si(OCH_2CH_3)_3$, Sigma-Aldrich); (6) deionized water; (7) hydrolyzed polyvinyl alcohol (PVA, Mw = 77,000–79,000 Da, J.T. Baker, Mallinckrodt Baker, Inc., Phillipsburg, NJ, USA), PDI = 2–2.5 [25]; (8) tetraethylorthosilicate (TEOS, $Si(OCH_2CH_3)_4$, Sigma-Aldrich); (9) hydroxyl-terminated polydimethylsiloxane (PDMS, OH-$[Si(CH_3)_2$-O]nH, Mw = 550 gmol, viscosity = 25 cSk, Sigma-Aldrich), PDI = 1.17 [26].

2.2. Synthesis of the PVA–HA–Siloxane Hybrid Sol

For the synthesis process of PVA–HA hybrid nanofiber mats, two starting solutions were needed. For the first solution, an adequate amount of PVA (7% w/v) is dissolved into deionized water at 80 °C under constant stirring. The PVA aqueous solution was allowed to cool down until it reached room temperature. For the second solution [6], an adequate amount of HA (3% w/v) was dissolved into deionized water at room temperature. Then, a chemical modification of the HA was made by the amidation of its carboxyl groups (R-COO-Na) with the amine groups of the inorganic modifier APTES. Thus, EDC and NHS were added to the HA aqueous solution that acts as a carboxylic acid activating agent. After 30 min, the accurate amount of MES and APTES were also added to the solution [27]. Molar ratios of the HA/APTES/EDC/NHS/MES used are 1/1/1.5/1/2.5, respectively. Inorganic crosslinking of HA occurred using the sol-gel method by adding TEOS and PDMS to the solution; this led to a hybrid solution with a 3D siloxane organic–inorganic matrix being obtained. The weight ratios of the components APTES/TEOS/PDMS used are 0.1/0.1/0.8, respectively. Finally, both solutions were mixed using the following relations: PVA:HA = 10:1 and 5:1. The hydrolysis–condensation (crosslinking) reactions took place among PVA, TEOS, APTES, and PDMS [28–31] at room temperature under stirring conditions for 1 h. After that, the solution was dialyzed (MWCO of 6000–8000, Spectrum® Laboratories, Piscataway, NJ, USA) using ultrapure water to eliminate the excess of EDC/NHS. The resulting sol was poured into a plastic syringe for electrospinning injection.

2.3. Preparation of the Electrospun PVA–HA–Siloxane Nanofiber Mats

The prepared sols were electrospun to form organic–inorganic crosslinked nanofibers. The electrospinning equipment (Basic Electrospinning Series) consists of a high-voltage power supply, a syringe infusion pump, and a ground electrode (aluminum foil on a stainless-steel collector). The sols were loaded into a plastic syringe (10 mL). The experimental conditions such as needle–collector distance, electric field strength, and flow rate were optimized to achieve the best nanofiber formation. The voltage used was 7.5 kV. The solution was constantly supplied at a flow rate of 2.5 mL/h with a needle (27 G) 0.4 mm × 12 mm. The spinning distance between the needle tip and the ground electrode was 8 and 10 cm. Under these conditions, the obtained fiber network is 0.1–0.5 µm thick. The sample code of each hybrid mat is assigned according to (1) PVA–HA ratio and (2) spinning distance, e.g., sample 10PVA-8 that indicates a ratio of PVA:HA = 10:1 and obtained at a spinning distance of 8 cm.

2.4. Characterization of PVA–HA–Siloxane Nanofiber Mats

Structural characterization of PVA–HA hybrid nanofiber mats was performed by FT-IR, SEM, and micro-Raman spectroscopy. FT-IR spectra were obtained at room temperature in a Perkin Elmer Spectrum Two spectrometer coupled with an Attenuated Total Reflection, Perkin Elmer (Hong Kong) (ATR-FTIR). Spectra were recorded in transmittance mode in the range from 4000 to 600 cm^{-1} by 16 scans and with a resolution of 4 cm^{-1}. Morphological characterization of PVA–HA hybrid mats was performed in a high-resolution JEOL Scanning Electron Microscope, (JEOL, Peabody, MA, USA). The Raman spectra were obtained in a micro-Raman confocal LabRam (Horiba Jobin Yvon, Oberursel, Germany) coupled with microscopy.

2.5. Cell Culture

Human osteoblast cell line MG-63 (CRL-1427) was obtained from the American Type Culture Collection (ATCC, Rockville, MD, USA). Osteoblasts were cultured in α-Minimum Essential Medium Eagle (α-MEM, Sigma Aldrich, St. Louis, MO, USA) supplemented with 10% heat-inactivated fetal bovine serum (FBS, Gibco, Thermo Fisher Scientific, Waltham, MA, USA) under a 5% CO_2 atmosphere at 37 °C. Media was renewed each third day. When cells reached confluence, they were detached from culture flasks using 0.05 g/L trypsin and 0.05 g/L EDTA solution (Sigma Aldrich).

2.5.1. Cell Culture on PVA–HA–Siloxane Nanofiber Mats

Mat samples of 0.5 cm^2 were sterilized first by immersion into a 70% ethylic alcohol solution for 15 min and then by exposure under ultraviolet light for 30 min. Then mats were incubated with FBS-supplemented α-MEM media for 2 days. Media was changed each day. MG-63 osteoblast cells were cultured on the mats at a density of 5×10^4 cells per sample under a 5% CO_2 atmosphere at 37 °C. The evaluation of the cell viability, cytoskeletal arrangement, and cell attachment was performed at 24 h, 7, 14, and 21 days of culture. The media was changed every day until the experiment finished. Three replicates of each sample were tested.

2.5.2. Cell Viability on PVA–HA–Siloxane Nanofiber Mats

MG-63 osteoblast cells grown onto the nanofiber mats for 1, 7, 14, and 21 days were labeled with a mix of nucleic acid dyes (SYTO9; 5 µmol/L; Invitrogen, Carlsbad, CA, USA) and propidium iodide (PI; 30 µmol/L; Sigma Aldrich) to allow visualization of living and dead cells [32]. The samples were observed with a confocal laser scanning microscopy system (LSM5 Pascal, Zeiss, Jena, Germany).

2.5.3. Stain of Actin Filaments in Cells Adhered onto PVA–HA–Siloxane Nanofiber Mats

MG-63 osteoblast cells grown onto the nanofiber mats for 1, 7, 14, and 21 days were fixed with 4% paraformaldehyde solution (Sigma Aldrich) and then washed three times with PBS. Actin filaments were stained with phalloidin-tetramethylrhodamine B isothiocyanate (phalloidin-TRITC; 10 ng/µL; Sigma Aldrich) for 20 min at room temperature. The excess of the phalloidin-TRITC was eliminated by washing the nanofiber mats five times with PBS. Nuclei of cells were counterstained with DAPI (Vector Laboratories, Burlingame, CA, USA). The samples were observed with a confocal laser scanning microscopy system (LSM5 Pascal, Zeiss, Germany).

2.5.4. Scanning Electron Microscopy of Cells Adhered on PVA–HA Hybrid Nanofiber Mats

MG-63 osteoblast cells were grown onto the nanofiber mats for 1, 7, 14, and 21 days and were fixed with 2.5% glutaraldehyde solution and then post-fixed with 1% aqueous osmium tetroxide solution. Samples were washed three times and then dehydrated with increasing concentrations of ethyl alcohol solutions and critical point drying with liquid CO_2. Finally, for observation through scanning electron microscopy (SEM, 5800LV GenTech

Scientific, Arcade, NY, USA), preparations were coated with gold in a sputter coater (Denton Vacuum-Desk II, Moorestown, NJ, USA).

2.5.5. Calcium Deposits on PVA–HA Hybrid Nanofiber Mats

MG-63 osteoblast cells were grown onto the nanofiber mats for 1, 7, 14, and 21 days and were fixed with 4% paraformaldehyde solution for 15 min at room temperature. Then samples were washed with PBS solution to eliminate the excess of paraformaldehyde. The mineralization process was evidenced by the alizarin assay; the is considered the gold standard for quantification of osteoblast mineralization [33]. In brief, 200 μL of 40 mM alizarin red was added to each sample and incubated for 20 min at room temperature. Samples were washed 5 times with PBS to remove the colorant. Finally, samples were analyzed under an inverted microscope (Primovert, Zeiss, Jena, Germany).

3. Results and Discussion

3.1. Synthesis of the PVA–HA–Siloxane Hybrid Precursor Sol

The hybrid sol was prepared by the hybridization and crosslinking of PVA-modified HA polymers with organic–inorganic siloxane compounds: 3-aminopropyltriethoxysilane (APTES), tetraethylorthosilicate (TEOS), and polydimethylsiloxane (PDMS). The preparation worked as follows: (1) the amidation of sodium hyaluronate (HA) with the inorganic modifier APTES to form amide functional groups: HA-APTES modified molecules; (2) the hydrolysis of siloxane organic–inorganic precursors: HA-APTES modified molecules–silica alkoxide [TEOS]–silicone [PDMS] [6]; (3) the condensation of previously hydrolyzed organic–inorganic precursors among them and/or with PVA to form the 3D PVA-modified HA siloxane organic–inorganic hybrid sol [34–36].

The ethoxy groups ($-OC_2H_5$) from APTES and TEOS are hydrolyzed by water molecules to form silanol groups ($Si-OH$). Simultaneously, an ethanol molecule (C_2H_5-OH) is formed per each hydrolyzed radical. Then, silanol groups ($Si-OH$) either from hydrolyzed APTES-TEOS molecules or hydroxyl-terminated PDMS chains condense among them or with -OH groups from PVA; this results in a hybrid sol (Figure 1) that will become the precursor of a PVA–HA polymeric matrix crosslinked with a siloxane organic–inorganic 3D network. Because the rate of hydrolysis reactions is faster than the reaction rate of condensation reactions, the condensation process is finished during electrospinning when the nanofibers are formed.

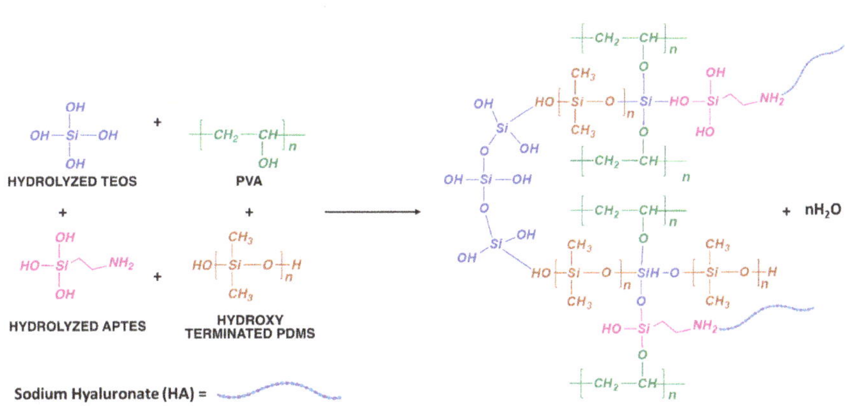

Figure 1. Scheme of the possible PVA-modified HA polymeric matrix crosslinked with a siloxane organic–inorganic 3D network.

3.2. Characterization of the PVA–HA–Siloxane Hybrid Nanofiber Mats

Hybrid nanofiber mats were obtained by electrospinning of a PVA–HA–siloxane solution. The electrospinning process allowed fibers to be shaped by the coaxial stretching of the

PVA–HA–siloxane colloidal solution, in which sol-gel reactions took place to form the 3D hybrid organic–inorganic network. Bands related to the intermolecular interactions of PVA, sodium hyaluronate, APTES, and siloxane network formed during hydrolysis-condensation reactions are observed in the infrared spectra (Figure 2) of the hybrid nanofiber membranes. The assignation of bands was performed according to the literature data. Bands related to PVA are observed: (1) the wide band with a maximum at 3320 cm^{-1} due to stretching vibrations (ϑ_{O-H}) of –OH bonds from intermolecular and intramolecular hydrogen bonds; (2) the band at 2920 cm^{-1} attributed to asymmetric stretching ($\vartheta_{a\ C-H}$) of $C-H$ bonds from alkyl groups; (3) the band at 1430 cm^{-1} corresponding to the deformation vibration of –OH; (4) the band at 1315 cm^{-1} related to the scissoring vibration of $-CH_2$ in PVA backbone; (5) the band at 1084 cm^{-1} due to the stretching vibrations (ϑ_{C-OH}) of alcohol in PVA; (6) the band at 836 cm^{-1} attributed to the out-of-plane vibration of $-CH$ bonds [21]. Other relevant bands demonstrating the chemical bonding between macromolecules of PVA and HA polymers and the siloxane organic–inorganic network to form the nanofiber mats can be observed in the FT-IR spectra: (1) the band at 1560 cm^{-1} is assigned to bending vibrations (δ_{N-H}) of amide II due to $N-H$ groups [37,38]. This results from the chemical modification of sodium hyaluronate with APTES [6,27] and confirms the presence of HA-APTES bonding within the nanofiber chemical composition; (2) from hydrolysis reactions, the bands at 920 cm^{-1} and 895 cm^{-1} are assigned to $Si-OH$ stretching due to silanol groups in hydrolyzed TEOS and to asymmetric stretching $\vartheta_a(Si-O)$ in $Si-OH$ bonds from hydroxyl-terminated PDMS, respectively; (3) the formation of siloxane structures (Q-D units) due to condensation is observed at 850 cm^{-1}, corresponding to (TEOS) $Si-O-Si$ (PDMS) bonds; (4) the band at 1100 cm^{-1} can be assigned to both asymmetric stretching ($\vartheta_{a\ Si-O}$) in $Si-O-Si$ bonds from siloxanes and asymmetric stretching ($\vartheta_{a\ Si-O-C}$) in $Si-O-C$ due to condensation of PVA with siloxane molecules [21]; (5) the band at 790 cm^{-1} assigned to $Si-O$ bonds due to silicon tetrahedron $[SiO_4]$ and $Si-C$ bonds due to PDMS and APTES molecules [6,28,30].

The composition of each PVA–HA–siloxane hybrid sol precursor influenced the chemical bonding between the polymer macromolecules (PVA, HA) and the siloxane (PDMS-TEOS) organic–inorganic hybrid matrix in the nanofiber mat. Condensation between different polymers competes with the condensation between molecules of the same polymer and/or the condensation with silanol groups from hydrolyzed TEOS and the hydroxyl-terminated PDMS.

Nanofiber mats with lower HA content and, therefore, more PVA content, (10PVA-8, 10PVA-10) showed more intense bands related to PDMS (1259, 895, 790 cm^{-1}) in comparison to mats with more HA content (5PVA-8, 5PVA-10). The band at 895 cm^{-1} attributed to $Si-OH$ from hydroxyl-terminated PDMS is more intense in mats with higher PVA content, meaning that a larger amount of PDMS chains did not condense with $-OH$ groups from TEOS and/or APTES in the presence of PVA. This suggests a higher affinity of PVA with PDMS. That may be confirmed by the band at 790 cm^{-1} showing freer $[SiO_4]$ tetrahedra. However, in mats with lower PVA content, as the band at 895 cm^{-1} decreased in intensity, the band at 850 cm^{-1} attributed to siloxane D-Q units increased, showing higher (PDMS) $Si-O-Si$ (TEOS) siloxane bonding, and thus a more condensed network. Moreover, bands related to HA (1640, 1560, 1220 cm^{-1}) appear to be more intense in mats with more HA content, indicating that more HA chains reacted with the hybrid siloxane system. Since the bands at 1640 cm^{-1} and 1220 cm^{-1} attributed to C=O and COO^- groups from HA, respectively, show higher intensity in comparison to the band at 1560 cm^{-1} attributed to amide II, HA chains were partially modified and hybridized by APTES. Thus, unhybridized HA macromolecules could have formed two types of hydrogen bonds with PVA polymer macromolecules [39]; that is, between two hydroxyl groups (PVA) $OH-OH$ (HA) and/or between the amide groups and the oxygen of the hydroxyl groups (HA) $NH-OH$ (PVA). Therefore, the broadening or shifting of characteristic bands attributed to HA and PVA may be related to the interaction between them [21]. Finally, unreacted polymeric chains from PVA, HA, and PDMS may coexist in equilibrium by electrostatic forces.

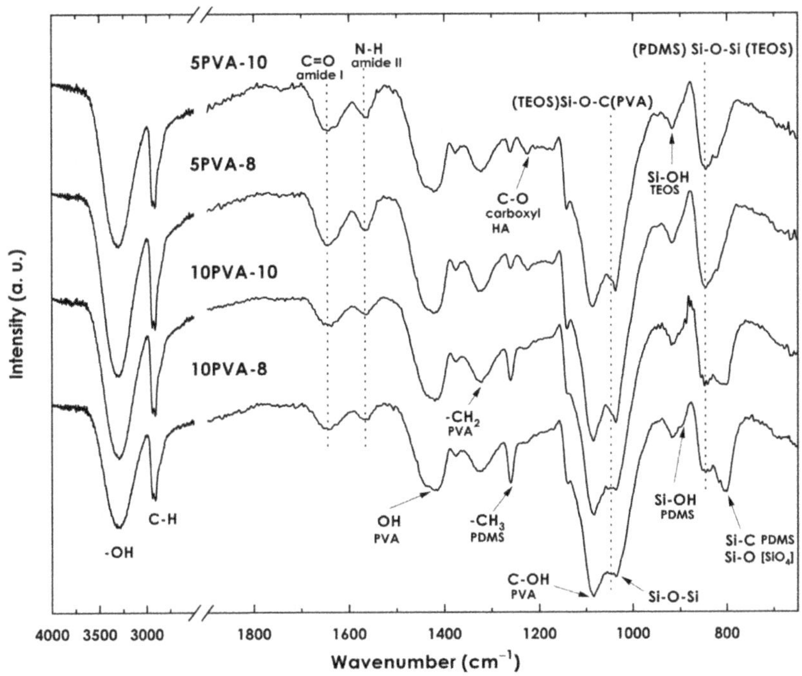

Figure 2. FT-IR spectra of electrospun PVA–HA–siloxane hybrid nanofibers. Nanofibers were obtained from relation 10:1 and 5:1 PVA:HA and spinning distances of 10 cm (10PVA-10) and 8 cm (10PVA-8). The principal bands associated with obtaining the PVA–HA–siloxane hybrid material are shown.

Figure 3a shows Raman spectra of electrospun PVA–HA–siloxane hybrid nanofibers in the range of 450 to 3200 cm^{-1}. In this region, there is a set of bands that, like the results shown in IR, present evidence of the transformation process of the precursors present in the chemical mixture for the formation of an organic–inorganic hybrid network. In general, it can be observed that the bands in the spectrum of sample 5PVA-8 are sharper and more intense, and this denotes a higher structural order. The band centered at 3000 cm^{-1} is the contribution of the vibrational modes of CH_2 and CH_3 from PVA and PDMS, respectively, and $-$OH groups of silanol (Si $-$ OH) [40].

Details of the region where most of the bands are found, from 450 to 1500 cm^{-1}, are shown in Figure 3b. The hydrolysis and condensation reactions of TEOS are evidenced by the two intense bands at 480 and 1037 cm^{-1} (the latter related to O $-$ Si $-$ O bonds), as well as two bands of very low intensity, one between 1122 and 1135 cm^{-1} (assigned to O $-$ Si $-$ O bonds bending) and the other centered on 980 cm^{-1} from Si $-$ O stretching vibration of silanol groups (Si $-$ OH). It is possible that the band in 1020 cm^{-1} can be associated with Si $-$ O $-$ C bond due to the integration of PVA into the network. The band at 480 cm^{-1} has been assigned to symmetric bending vibrations of the Si $-$ O $-$ Si linkage and/or to the O $-$ Si $-$ O deformation of the coupled 'tetrahedra' [SiO_4] groups [41] due to condensation of hydrolyzed TEOS and PDMS. This band is generally intense and broad in SiO_2 systems obtained by sol-gel; the full width at half maximum (FWHM) is associated with the wide distribution of Si $-$ O $-$ Si inter-tetrahedral angles within the structure. The decrease in its intensity indicates the lowering of network dimensionality by breaking

Si − O − Si bridges within the glass structure [41]. In the spectra of all samples, this band is relatively narrow and of low intensity due to the interconnection with functional groups such as methyl and methylene of PDMS and PVA; these prevent the formation of the long-range Si − O − Si bridge network due to the formation of the hybrid compound.

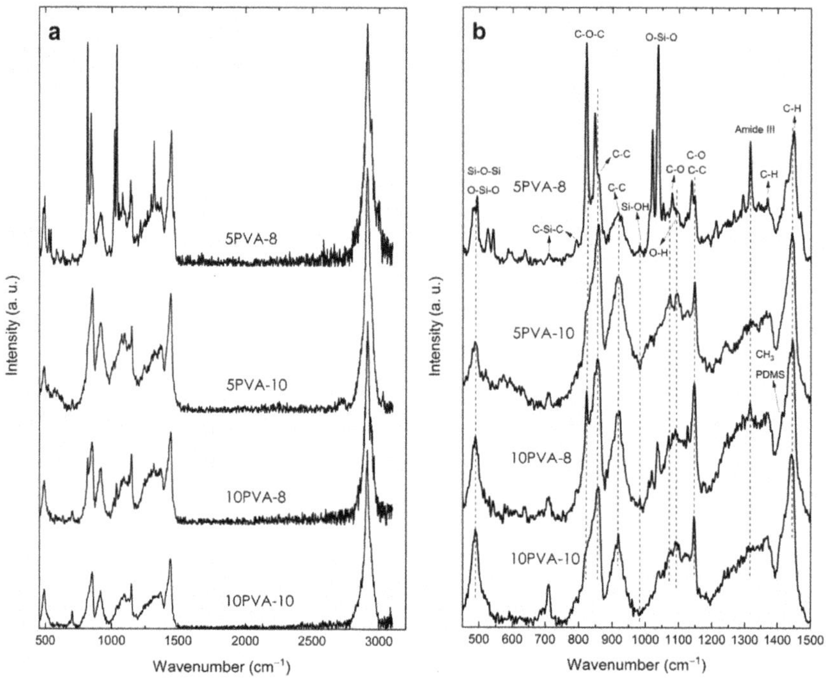

Figure 3. Micro-Raman spectra of electrospun PVA–HA–siloxane hybrid nanofibers. Hybrid nanofibers were obtained from relations 10:1 and 5:1 PVA:HA at spinning distances of 10 (10PVA-10 and 5PVA-10) and 8 cm (10PVA-8 and 5PVA-8). (**a**) 450–3200 cm^{-1} range. (**b**) Region below 1500 cm^{-1}. The sharper and more intense bands of the sample 5PVA-8 denote a higher structural order.

The low-intensity bands at 706 and 790 cm^{-1} are attributed to symmetric and asymmetric stretching vibrations, respectively, of C − Si − C bonds in PDMS [42]. The band at 821 cm^{-1} is related to stretching vibrations of C − O − C bonds from the hyaluronic acid repetitive units [6]. This structural part of the hybrid material is kept attached to the three-dimensional network through the APTES nitrogen atom by an amidation process, and this is confirmed by the amide III mode related to the band at 1314 cm^{-1} [43]. This band is sharper and more intense in samples 5PVA-8 and 10PVA-8, which is related to a higher structural order.

The bands related to PVA are those located at 856, 918, and 1147 cm^{-1}; the first two are related to C − C stretching vibrations of the polymer backbone. In particular, the band at 918 cm^{-1} is associated with a non-crystalline conformation of PVA carbon backbone, so an increase in its intensity may represent a decrease in crystallinity. The band at 1147 cm^{-1} is assigned to C − C stretching vibration in the crystalline phase of the PVA matrix [17,44]. The intensity of the band at 918 cm^{-1} in the sample 5PVA-8 is lower than the rest of the samples, which confirms a higher structural order. Low-intensity bands are also seen at 1068 and 1092 cm^{-1} attributed to C − O stretching and O − H bending vibrations [45]. The low-intensity bands at 1365 cm^{-1} and the intense band at 1442 cm^{-1} are due to C − H and –OH bending vibrations from groups in PVA, PDMS, and hydrolyzed APTES [6,41,45].

The PVA–HA–siloxane hybrid nanofiber mats have nanometric organic–inorganic interpenetrating networks. Figure 4 displays SEM images, at different magnifications, of the morphology of the electrospun mats. Mats with lower HA content (10PVA-8 and 10PVA-10) exhibit homogeneous hybrid mats with few beaded nanofibers. In contrast, mats with more HA content (5PVA-8 and 5PVA-10) exhibit hybrid mats with mound patterns formed by beaded or "ribbon-like" nanofibers. Furthermore, the mats obtained with a shorter distance between the needle tip and the collector (10PVA-8 and 5PVA-8) resulted in nanofibers with thinner diameters.

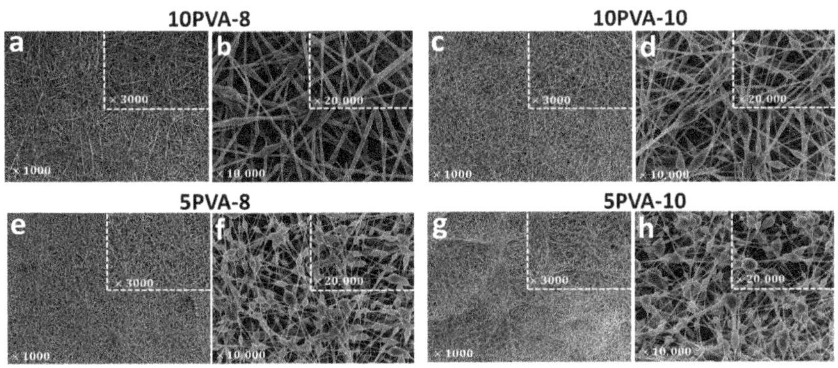

Figure 4. Ultrastructural morphology of electrospun PVA–HA–siloxane hybrid nanofiber mats. Images show the morphology difference between nanofibers. (**a**,**b**) Mats with hybrid nanofibers 10:1 PVA:HA and spinning distance of 8 cm. Mats show homogeneous nanofibers (**c**,**d**) Mats with hybrid nanofibers 10:1 PVA:HA and spinning distance of 10 cm. Mats show few beaded nanofibers. (**e**,**f**) Mats with hybrid nanofibers 5:1 PVA:HA and spinning distance of 8 cm. These mats show nanofibers beaded enrichment. (**g**,**h**) Mats with hybrid nanofibers 5:1 PVA:HA and spinning distance of 10 cm. These mats show a similar pattern to 5PVA-8 mats, with many beaded nanofibers. In all images, the square shows a magnification of the same sample. Magnification is indicated in each image.

The formation of beadles nanofibers is attributed to (1) the increase of PVA concentration and the decrease of HA concentration (higher PVA concentration enhances the electrospinning behavior into the polymeric solution by shaping fibers with increased diameters [31,46]) within the polymeric blending gave the solution the adequate composition and thus a good polymer chain entanglement (at the established electrospinning parameters). It was, therefore, able to overcome surface tension and this resulted in the formation of almost uniform beadles electrospun nanofibers, as observed in 10PVA-8 and 10PVA-10 mats; (2) the stretching of the polymeric solution and the increase in distance between the needle tip and the collector both tailored smaller-diameter fibers, as observed in the 10PVA-10 mat compared to the 10PVA-8 mat.

In contrast, the formation of beaded nanofibers (solvent–containing nanofibers) in mats with more HA content (5PVA-8, 5PVA-10) is attributed to the incomplete evaporation of the solvent (water/ethanol solution due to hydrolysis reactions in the sol-gel process) and the low-stretching of the polymeric blending in the flight between the needle and the metallic collector. Polymeric blending with a higher HA concentration (5PVA) appeared to be challenging to be electrospun [47] due to (1) the HA concentration within the polymeric blending increasing and changing miscibility and solution behavior [47,48]; (2) the decrease of PVA concentration allowed a higher concentration of colloidal particles within the polymeric solution when APTES–TEOS–PDMS started to be hydrolyzed and condensed during the sol-gel process. This resulted in faster gelation of PVA and HA polymers as viscosity increased in the sol, and this hindered electrospinning; (3) the decrease of PVA concentration might result in the reduction and breaking of entangled polymer chains before reaching the collector as the electric field is applied; (4) drying due to gelation of the polymeric solution at the tip hindering its flow through the needle tip and causing blockage; resulting in the formation of beaded nanofibers [15].

Bead morphology in all mats appear as elongated droplets or stretched droplets. This morphology is characteristic of viscous polymeric solutions [49]. However, slight changes in bead shapes are observed: in 10PVA-8 and 10PVA-10 mats, the few beaded nanofibers have stretched shapes, while in 5PVA-8 and 5PVA-10 mats, the abundant beaded nanofibers have more elongated shapes [15]. These variations are caused by concentration changes in the polymeric solution as PVA concentration decreases. Therefore, the concentration/viscosity of the polymeric solution determines the nanofibers' morphology. Likewise, since the interactions between two macromolecules depend on the presence of a solvent [21], different evaporation rates of the solvent (water/ethanol) may lead to phase separation. This results in the fabrication of highly porous electrospun nanofibers, as observed in the four studied PVA–HA–siloxane hybrid nanofiber mats [15].

PVA-based nanofiber mats have limited applicability due to their high solubility in aqueous media. However, to take advantage of PVA's electrospinning and biological properties [50], it is desirable to control and balance its hydrophilicity to modulate degradation rates and improved mechanical properties after being electrospun, shaping mats with structured pores (highly porous materials with a range of pore size) and interconnected structures with a morphology like ECM capable of supporting cell growth [39]. PVA has been chemically blended with HA and other natural polymers to improve its similarity with ECM. This results in interesting tailor-made materials [51–53]. Since HA has a pronounced hydrophilic character that causes fast degradation rates and weak mechanical properties, it has been suggested that HA must be used after chemical modification and/or crosslinked with organic or inorganic polymers [6,22]. Unfortunately, few studies have been conducted on electrospun pristine HA materials [47] since the high viscosity and surface tension at relatively low concentrations make it a polymer that is very difficult to electrospun [48].

To prove that PVA–HA–siloxane hybrid nanofibers exhibit good resistance to degradation, Figure 5 displays SEM images of the PVA–HA–siloxane mats after 21 days of cell culture in supplemented α-Minimum Essential Medium Eagle, which is used for the growth of osteoblasts.

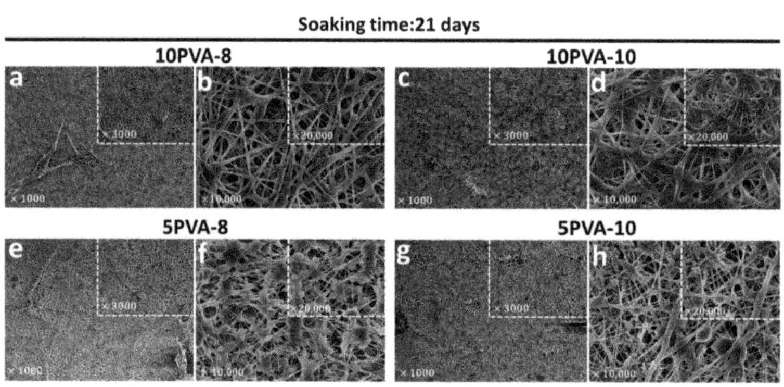

Figure 5. Electrospun PVA–HA–siloxane hybrid nanofibers morphology after 21 days of soaking into the culture medium. (**a**,**b**) Mats with hybrid nanofibers 10:1 PVA:HA and spinning distance of 8 cm. These mats show more resistance to degradation in the culture medium, and their morphology remained. (**c**,**d**) Mats with hybrid nanofibers 10:1 PVA:HA and spinning distance of 10 cm. The nanofibers of these mats lost their cylindrical shape after soaking. (**e**,**f**) Mats with hybrid nanofibers 5:1 PVA:HA and spinning distance of 8 cm. The beaded pattern of these nanofibers was lost and mats showed fractures after soaking. (**g**,**h**) Mats with hybrid nanofibers 5:1 PVA:HA and spinning distance of 10 cm. The fibers of these mats lost their shape and were thinner compared to mats before immersion. In all images, the small squares show a magnification of the same sample.

After 21 days of immersion, the PVA–HA–siloxane hybrid nanofiber mats maintained their morphological structure, as previously described in Figure 4. Most of the fibers in the mats with higher PVA content (10PVA-8 and 10PVA-10) maintain their small diameters. However, some of them are thicker and flatter; have lost their cylindrical shape and appear to blend into each other. Shape loss and blending seem to be higher in the 10PVA-10 mats. On the other hand, mats with lower PVA content (5PVA-8 and 5PVA-10) show fractures. In addition, their "ribbon-like" fibers are thinner compared to the fibers before immersion. This suggests hydrolytic degradation. The bead structures also experienced changes, and their elongated droplet shapes turned into several flat shapes and sizes. Hydrolytic degradation appears to be greater in the 10PVA-8 mats.

According to the observed changes, the variation in PVA:HA ratios affects the physicochemical characteristics such as swelling behavior, hydrolytic degradation, mechanical flexibility, and protein adsorption of the hybrid mats. Mats with lower HA content (10PVA-8, 10PVA-10) exhibited less hydrolytic degradation on their nanofibers due to (1) the PVA affinity with PDMS silicone which boosts their hydrophobic behavior and (2) that PVA contributes to the reduction of HA degradation rate, avoiding its rapid dissolution [54]. On the other hand, mats with higher HA content (5PVA-8, 5PVA-10) showed more hydrolytic degradation on their nanofibers since the PVA ratio decreased and HA concentration increased within the hybrid system, tailoring degradation rates in unhybridized HA macromolecules. Siloxane–inorganic crosslinking in all mats provides excellent physical and chemical support to allow cell growth.

3.3. Cell Viability on PVA–HA–Siloxane Nanofiber Mats

Biological responses to osteoblast cells, such as enhanced cell adhesion, cell growth stimulation, and calcium deposition, were studied in PVA–HA–siloxane hybrid nanofiber mats. The viability of osteoblast cells grown on nanofiber mats was evaluated by immunofluorescence using SYTO9 and PI. Confocal images showed viable cells for all fiber mats since the first day. However, the 5PVA-10 mats exhibited the maximum number of cells. On day 14, 10PVA-8 and 10PVA-10 mats showed a smaller number of dead cells. On day 21, all materials efficiently supported the viability and proliferation of osteoblasts. Decreasing the ratio of PVA in the PVA–HA–siloxane nanofiber mats had a positive effect for the first time on the adhesion and cell proliferation behavior since the HA ratio increased, boosting their biological properties. These results confirm the role of HA as a base component in scaffolds for tissue regeneration. Because this polymer is a natural component in ECM, it offers a similar extracellular environment to seeded cells. This results in cell adhesion, proliferation, and differentiation enhancement. Several studies have observed osteogenic-enhancing properties such as cell proliferation and differentiation in hyaluronic acid-based scaffolds [55,56]. These osteogenic properties of hyaluronic acid are molecular weight- and concentration-dependent [56]. Other studies have indicated that HA favors osteogenesis [57,58]. Having HA as part of the scaffold components can give the material the ability to bind water to it [59]. However, having chemically modified HA with APTES and then inorganically crosslinking it with siloxane molecules makes the HA less hydrophilic and more resistant to hydrolytic-degradation. The chemical interaction of organic and inorganic components gives the electrospun scaffolds the power of mechanical support and chemical stability to boost their intrinsic osteogenic properties (cell proliferation), as observed in Figure 6.

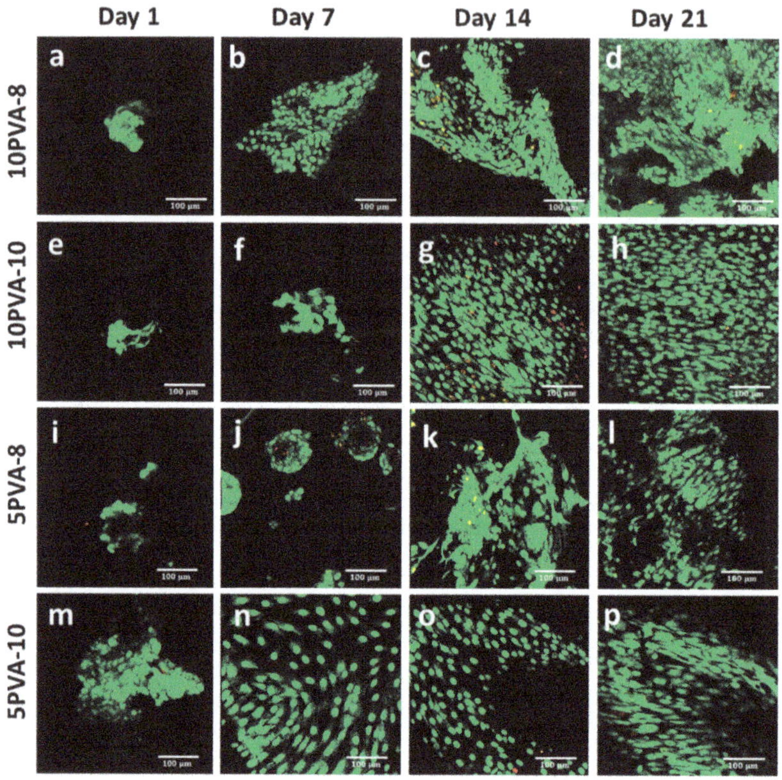

Figure 6. Viability of osteoblasts grown on PVA–HA–siloxane nanofiber mats. Osteoblasts were incubated on nanofiber mats for 1, 7, 14, and 21 days. After incubation, the cells were stained with SYTO9-PI solution. Live cells were dyed green, and dead cells were red. (**a–d**) Representative images of 10PVA-8 mats; the cell viability was maintained in these mats only on day 14, and some cells lost their viability. (**e–h**) Representative images of 10PVA-10 mats; cells on these mats conserved their viability until day 21, and a few cells lost their viability at day 14. (**i–l**) Representative images of 5PVA-8; cells on these mats maintained their viability until day 21, and at day 7, spheroid-like 3D structures were evident. (**m–p**) Representative images of 5PVA-10 mats; cell viability was maintained in these mats, and cells formed a circular growth pattern. In all samples, the increase in number of cells became evident over time. The images show the cells in a representative focal plane of the 3D nanofiber mats. The bar in all images represents 100 μm.

3.4. Cytoskeletal Arrangement of Cells Adhered onto PVA–HA–Siloxane Nanofiber Mats

The actin cytoskeleton is highly implicated in several cellular processes, such as cell survival, proliferation, and motility. These processes are related to the formation of tissues. According to this, the morphological analysis of MG-63 osteoblasts seeded on PVA–HA–siloxane biomaterials was evaluated by staining actin filaments. On the 7th day, cells developed an adherent monolayer on the 10PVA-10 mats. The actin filaments were organized as parallel bundles alongside cells, and they kept this pattern till day 21. Interestingly, on the rest of the materials (10PVA-8, 5PVA-8, and 5PVA-10) cells grew up with the shape of spheroid-like 3D structures, which have been appreciated since the 7th day. Cultured cells on 10PVA-8 and 5PVA-8 mats maintained this spheroid-like morphology until day 14, and 5PVA-8 mats until day 21. Finally, cells matured in all mats and they were found to be in close interaction on day 21. Their actin cytoskeleton was found to be well organized and to support the phenotypic morphology of osteoblasts (Figure 7).

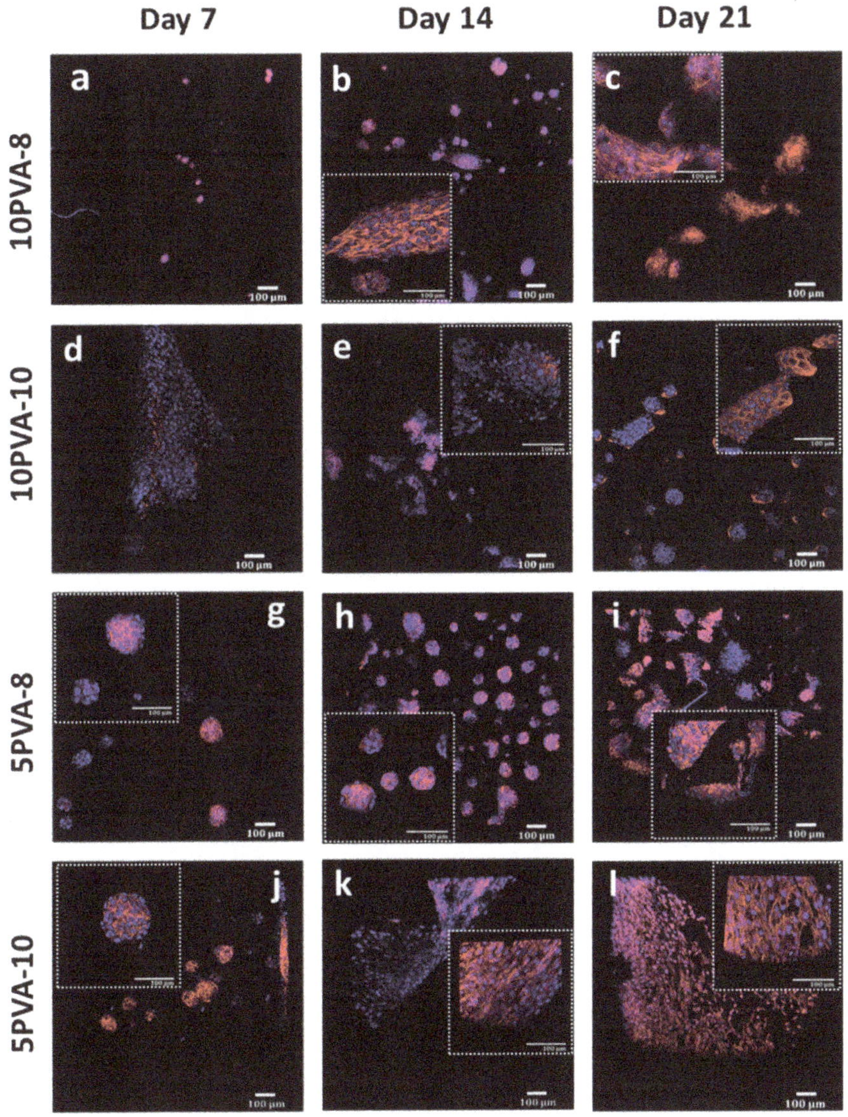

Figure 7. Actin cytoskeleton of osteoblasts grown on PVA–HA–siloxane nanofiber mats. (**a–c**) Representative images of cell growth on 10PVA-8 mats. The actin cytoskeleton was more evident at 14 and 21 days, and some spheroid structures were evident at 14 days. (**d–f**) Representative images of actin stain of the cells grown on 10PVA-10 mats. Actin filaments distribution was evident from 7 days and maintained until 21 days. (**g–i**) Images of cells grown on 5PVA-8 mats. The formation of spheroid-like 3D structures was evident for 7 days, and this pattern was conserved until 21 days. Actin filaments are present in these structures. (**j–l**) Representative images of osteoblasts grown on 5PVA-10 mats. The formation of 3D structures was visible for 7 days. The homogenous distribution of actin filaments was evident in cells at 14 and 21 days. Actin filaments were stained with rhodamine-phalloidin (red), and cell nuclei were labeled with DAPI (blue). The images represent only one focal plane of materials and represent two experiments.

The assembly of spheroid-like structures on HA-based composite materials has been previously reported [60]. These structures reproduce the 3D architecture of tumors, which is why their use as models for screening therapeutic drugs against cancer has been highlighted [60,61]. The cell proliferation and its successful growth on nanofibers depend on the establishment of the interaction of HA with the cellular surface marker CD44 receptor. CD44 is a transmembrane receptor expressed by many cells and binding ligands such as hyaluronic acid [62]. This interaction promotes the production of autocrine growth factors favoring cell proliferation [63]. It is also directly related to proteins that participate in the cytoskeleton rearrangement, such as Rac and Rho proteins [64]. Thus, CD44 interacts with the cytoskeleton, and confers motility to cells and further migration by (1) opening up tissue spaces through which cells move and (2) enabling cells to move [59]. The PVA–HA–siloxane electrospun scaffold is (1) the surface on which osteoblasts adhere and (2) the extracellular microenvironment. Having HA as a component in the PVA–HA–siloxane mats influences cell–cell and cell–scaffold interactions by enhancing cell motility and cell migration across the nanofiber mat. This results in granulation tissue formation [62]. Figure 7 shows the actin filaments of osteoblasts and how they are organized during the assay. It is observed that osteoblasts migrate and proliferate across the mats, but they settle down in the form of parallel cell-bundles; these are well organized and in close interaction to simulate granulation tissue formations. Mats with a higher HA ratio, especially the 5PVA-8 mat, showed more spheroid-like cell-bundles. This highlights the role of HA in the formation of these structures as part of the scaffold component.

3.5. Cell Attachment onto PVA–HA Hybrid Nanofiber Mats

To confirm osteoblasts' attachment onto nanofiber mats and observe their morphology, they were analyzed by scanning electron microscopy, as shown in Figure 8. In all mats, cells showed the typical spindle-shaped morphology of osteoblasts. In some cases, cells presented large projections such as filopodia, and they grew up covering the total surface of the nanofiber mats. It is notable that on the 5PVA-8 mat, cells proliferated and spread in the form of multilayers resembling 3D structures. The formation of multilayer-cell 3D structures suggests the beginning of tissue growth, supporting granulation tissue formation and confirming the outstanding biological properties of the PVA–HA–siloxane hybrid scaffolds.

Osteoblasts adhere to the electrospun mat scaffold, and become to be the extracellular microenvironment. Since HA is a natural ECM component, it provides a similar extracellular environment, helping osteoblasts to adhere and proliferate, as observed in Figure 8. The novel 3D organic–inorganic siloxane network crosslinking the sodium hyaluronate provided the polysaccharide chains with strong chemical bonds that restricted their natural hydrophilicity and gave stiffness to the nanofiber mats [6]; this supports cell attachment, cell proliferation, and growth of new tissue. Other studies have reported similar results in HA-based nanofiber mats. For example, the addition of hydroxyapatite and chitosan in electrospun PVA/HA nanofiber mats increases the WI38 cell adherence and proliferation, and this supports its properties for bone regeneration [65]. In addition, Pangon and colleagues reported the functionality of chitin whiskers to enhance the mechanical properties of chitosan/PVA nanofibers and improve osteoblasts´ viability and proliferation [66]. In another PVA-based scaffold, cell adhesion was improved by adding HA, and the increase in cell proliferation was due to the enrichment of hydroxyapatite [67]. Interestingly, the addition of hydroxyapatite was not required for cell adhesion and proliferation of osteoblasts in mats fabricated in this work. Moreover, in electrospun HA-based nanofiber with the incorporation of glycosaminoglycans (GAGs), cells such as human keratinocytes, fibroblasts, and mesenchymal stem cells firmly adhered to the nanofiber showed high cell viability and cellular proliferation [68].

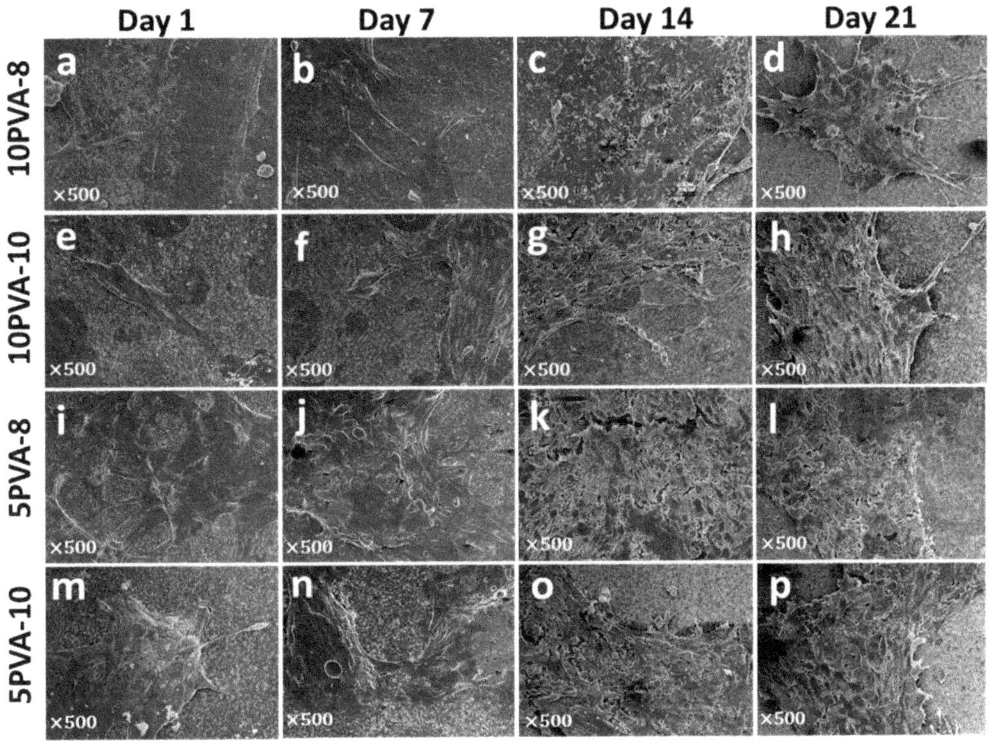

Figure 8. Ultrastructure of osteoblasts spread on PVA–HA–siloxane nanofiber mats. Osteoblasts attached to the surface of the fiber mats were observed using the conventional SEM technique. (**a–d**) Representative images of osteoblasts growth on 10PVA-8 mats; cells are visibly adhered on these mats. At the first days (1 to 7 days) few cells are found, and cellular density is more apparent at 14 and 21 days. Membranal filopodia are present at 21 days. (**e–h**) Ultrastructure morphology of cell growth on 10PVA-10 mats. On these mats, osteoblasts are more spread out, and cellular density is evident from 7 days. (**i–l**) Growth osteoblasts on 5PVA-8 mats; a greater cell density is evident in these mats with typical osteoblasts morphology. (**m–p**) Ultrastructure morphology of osteoblasts grown on 5PVA-10 mats; clusters of cells are visible on these mats, and cellular contact is more closed from cell to cell.

3.6. *Differentiation of Osteoblasts Grown on PVA–HA Hybrid Nanofiber Mats*

During cell maturation, osteoblasts differentially expressed several proteins. Some of these proteins induce mineralization in the extracellular matrix, and calcium deposits are accumulated to promote the formation of hydroxyapatite crystals. To evaluate the effect of PVA–HA hybrid nanofiber mats on the adhered osteoblasts′ ability to reach a mature stage, mineralization nodules were evaluated by Alizarin-staining, as shown in Figure 9. Alizarin-stained calcium deposits were found in all nanofiber mats. Their initial formation was more evident on the seventh day, and their abundant presence remained until the 21st day, except for the 10PVA-10 mat, in which fewer mineralization nodules were found on days 14 and 21. The favorable mineralization in HA-based nanofibers has been previously reported, for example, in poly(l-lactic acid)-co-poly(ε-caprolactone)-silk fibroin-hydroxyapatite-hyaluronic acid (PLACL–HaP–SF–HA) scaffolds, the osteoblastic bone formation was attributed to hyaluronic acid and the increase of mineralization was due to incorporation of silk fibroin on days 10 and 15 [69]. The role of HA in supporting the differentiation and calcium mineralization has also been highlighted in HA + FmocFRGD/PCL electrospun fibers [70].

The calcium deposits resulting from osteoblast maturation and differentiation inducing mineralization in the ECM confirm the outstanding biological properties of the PVA–HA–siloxane hybrid mats. HA working as a fundamental component within the nanofiber mats enhances the osteogenic effects of the tested hybrid scaffolds. It can be observed that HA acts as the vector for osteoinductive properties in the hybrid membranes.

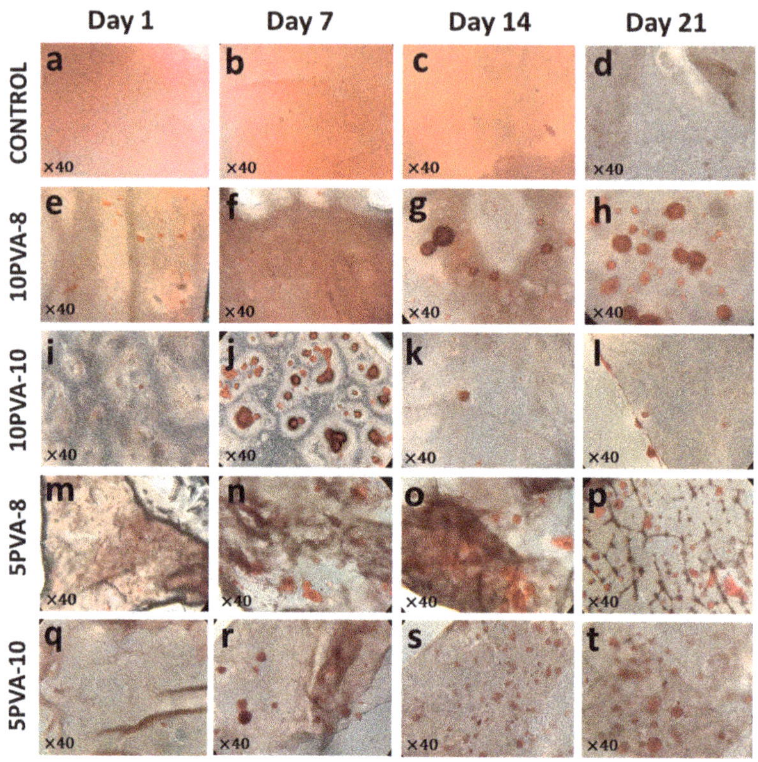

Figure 9. Calcium deposits produced from osteoblasts grown on PVA–HA–siloxane hybrid nanofiber mats. Calcium deposits were stained with the alizarin red solution. (**a–d**) Representative images of mats without cells; no alizarin red stain was visible on control mats. (**e–h**) Representative images of calcium deposit formation on 10PVA-8 mats. The mineralization increased over time; on days 14 and 21, the calcium deposits were more evident. (**i–l**) Formation of calcium deposits on 10PVA-10 mats; remarked at 7 days on. (**m–p**) Mineralization on 5PVA-8 mats was evidenced by alizarin stain, the calcium deposits were more apparent from 7 days to 21 days. (**q–t**) Mineralization induced on 5PVA-10 mats; the increase of alizarin stain was prominent from 7 until 21 days.

4. Conclusions

PVA–HA–siloxane hybrid nanofiber mats with outstanding biological properties were successfully synthesized. The improved biological properties were enhanced by the inorganic–siloxane chemical crosslinking of PVA–HA polymeric chains; this prevented their hydrolytic degradation and metabolism, provided support to cells, and boosted HA intrinsic osteogenic properties. The PVA, as a synthetic polymer with good biological properties, helped the electrospinning of HA. Hybrid nanofiber mats with a higher PVA ratio (10PVA-8 and 10PVA-10) show less cell growth during the seven days of culture but as time passes, cell growth significantly increases. On the other hand, hybrid nanofiber mats with lower PVA ratio (5PVA-8 and 5PVA-10) show higher cell growth from the first day of culture. However, these mats also presented a higher degradation at the end of the assay, which is attributed to the higher HA ratio within the fiber mats. All the hybrid mats support cell adhesion, cell proliferation, and cell differentiation (cell maturation and calcium deposits) during the 21 days of the assay. It is noteworthy to state two relevant facts: (1) the 5PVA-8 mat shows 3D spherical osteoblast morphology, suggesting 3D tissue growth formation, and (2) the functionality of the cells grown on the nanofiber mats was independent of the addition of both factors and stimulating-cell-growth cytokines, making these materials excellent candidates for their use in bone tissue repair.

Author Contributions: Conceptualization, L.T.-J. and B.E.G.-P.; methodology, D.A.S.-T., S.L.B.-U. and A.V.-F.; formal analysis, D.A.S.-T., L.T.-J., B.E.G.-P. and M.A.V.-H.; investigation, D.A.S.-T., S.L.B.-U. and A.V.-F.; resources, L.T.-J., B.E.G.-P. and M.A.V.-H.; writing—original draft preparation, D.A.S.-T.; writing—review and editing, L.T.-J. and B.E.G.-P.; funding acquisition, L.T.-J. and B.E.G.-P. All authors have read and agreed to the published version of the manuscript.

Funding: This research was funded by SECRETARIA DE INVESTIGACION Y POSGRADO, INSTITUTO POLITECNICO NACIONAL (SIP-IPN) grant numbers 20226980, 20231103, 20230384, 20240624, 20220580, 20230345 and the APC was funded by SIP-IPN.

Institutional Review Board Statement: Not applicable.

Data Availability Statement: The data presented in this study are available on request from the corresponding authors.

Acknowledgments: L.T.-J. and B.E.G.-P. received fellowships from COFAA, EDI and SNII. M.A.V.-H. receive fellowship from EDI and SNII. D.A.S.-T. and S.L.B.-U. would like to acknowledge CONACYT and SNII. The authors would also like to acknowledge the Electron Microscopy Central of ENCB/IPN and Centro de Nanociencias y Micro y Nanotecnologías of IPN for technical assistance.

Conflicts of Interest: The authors declare no conflicts of interest.

References

1. Almond, A. Hyaluronan. *Cell. Mol. Life Sci.* **2007**, *64*, 1591–1596. [CrossRef]
2. Lin, W.; Liu, Z.; Kampf, N.; Klein, J. The Role of Hyaluronic Acid in Cartilage Boundary Lubrication. *Cells* **2020**, *9*, 1606. [CrossRef]
3. Collins, M.N.; Birkinshaw, C. Hyaluronic acid based scaffolds for tissue engineering—A review. *Carbohydr. Polym.* **2013**, *92*, 1262–1279. [CrossRef]
4. Hosseini, H.; Shahraky, M.K.; Amani, A.; Landi, F.S. Electrospinning of polyvinyl alcohol/chitosan/hyaluronic acid nanofiber containing growth hormone and its release investigations. *Polym. Adv. Technol.* **2021**, *32*, 574–581. [CrossRef]
5. Zhai, P.; Peng, X.; Li, B.; Liu, Y.; Sun, H.; Li, X. The application of hyaluronic acid in bone regeneration. *Int. J. Biol. Macromol.* **2020**, *151*, 1224–1239. [CrossRef] [PubMed]
6. Sánchez-Téllez, D.A.; Rodríguez-Lorenzo, L.M.; Téllez-Jurado, L. Siloxane-inorganic chemical crosslinking of hyaluronic acid-based hybrid hydrogels: Structural characterization. *Carbohydr. Polym.* **2020**, *230*, 115590. [CrossRef]
7. Flegeau, K.; Toquet, C.; Rethore, G.; d'Arros, C.; Messager, L.; Halgand, B.; Dupont, D.; Autrusseau, F.; Lesoeur, J.; Veziers, J.; et al. In Situ Forming, Silanized Hyaluronic Acid Hydrogels with Fine Control Over Mechanical Properties and In Vivo Degradation for Tissue Engineering Applications. *Adv. Healthc. Mater.* **2020**, *19*, e2000981. [CrossRef] [PubMed]
8. Wang, S.; Hu, F.; Li, J.; Zhang, S.; Shen, M.; Huang, M.; Shi, X. Design of electrospun nanofibrous mats for osteogenic differentiation of mesenchymal stem cells. *Nanomedicine* **2018**, *7*, 2505–2520. [CrossRef] [PubMed]
9. Leong, M.F.; Chian, K.S.; Mhaisalkar, P.S.; Ong, W.F.; Ratner, B.D. Effect of electrospun poly(D,L-lactide) fibrous scaffold with nanoporous surface on attachment of porcine esophageal epithelial cells and protein adsorption. *J. Biomed. Mater. Res. A* **2009**, *89*, 1040–1048. [CrossRef]
10. Yu, D.; Wang, J.; Qian, K.J.; Yu, J.; Zhu, H.Y. Effects of nanofibers on mesenchymal stem cells: Environmental factors affecting cell adhesion and osteogenic differentiation and their mechanisms. *J. Zhejiang Univ. Sci. B* **2020**, *21*, 871–884. [CrossRef]
11. Canha-Gouveia, A.; Rita Costa-Pinto, A.; Martins, A.M.; Silva, N.A.; Faria, S.; Sousa, R.A.; Salgado, A.J.; Sousa, N.; Reis, R.L.; Neves, N.M. Hierarchical scaffolds enhance osteogenic differentiation of human Wharton's jelly derived stem cells. *Biofabrication* **2015**, *7*, 035009. [CrossRef]
12. Wu, T.; Xue, J.; Li, H.; Zhu, C.; Mo, X.; Xia, Y. General Method for Generating Circular Gradients of Active Proteins on Nanofiber Scaffolds Sought for Wound Closure and Related Applications. *ACS Appl. Mater. Interfaces* **2018**, *10*, 8536–8545. [CrossRef]
13. Gupte, M.J.; Swanson, W.B.; Hu, J.; Jin, X.; Ma, H.; Zhang, Z.; Liu, Z.; Feng, K.; Feng, G.; Xiao, G.; et al. Pore size directs bone marrow stromal cell fate and tissue regeneration in nanofibrous macroporous scaffolds by mediating vascularization. *Acta Biomater.* **2018**, *82*, 1–11. [CrossRef]
14. Zhang, X.; Meng, Y.; Gong, B.; Wang, T.; Lu, Y.; Zhang, L.; Xue, J. Electrospun nanofibers for manipulating soft tissue regeneration. *J. Mater. Chem. B* **2022**, *10*, 7281–7308. [CrossRef]

15. Haider, A.; Haider, S.; Kang, I. A comprehensive review summarizing the effect of electrospinning parameters and potential applications of nanofibers in biomedical and biotechnology. *Arab. J. Chem.* **2018**, *11*, 1165–1188. [CrossRef]
16. Ding, J.; Zhang, J.; Li, J.; Li, D.; Xiao, C.; Xiao, H.; Yang, H.; Zhuang, X.; Chen, X. Electrospun polymer biomaterials. *Prog. Polym. Sci.* **2019**, *90*, 1–34. [CrossRef]
17. Brenner, E.K.; Schiffman, J.D.; Thompson, E.A.; Toth, L.J.; Schauer, C.L. Electrospinning of hyaluronic acid nanofibers from aqueous ammonium solutions. *Carbohydr. Polym.* **2012**, *87*, 926–929. [CrossRef] [PubMed]
18. Sun, J.; Perry, S.L.; Schiffman, J.D. Electrospinning Nanofibers from Chitosan/Hyaluronic Acid Complex Coacervates. *Biomacromolecules* **2019**, *20*, 4191–4198. [CrossRef] [PubMed]
19. Kobayashi, M.; Hyu, H.S. Development and Evaluation of Polyvinyl Alcohol-Hydrogels as an Artificial Atrticular Cartilage for Orthopedic Implants. *Materials* **2010**, *3*, 2753–2771. [CrossRef]
20. Shirzaei, S.I.; Rezaei, M.; Baradar, K.A.; Razzaghi, D. Preparation and characterization of polycaprolactone/chitosan-g-polycaprolactone/hydroxyapatite electrospun nanocomposite scaffolds for bone tissue engineering. *Int. J. Biol. Macromol.* **2021**, *182*, 1638–1649. [CrossRef] [PubMed]
21. Mehrasa, M.; Anarkoli, A.O.; Rafienia, M.; Ghasemi, N.; Davary, N.; Bonakdar, S.; Naeimi, M.; Agheb, M.; Salamat, M.R. Incorporation of zeolite and silica nanoparticles into electrospun PVA/collagen nanofibrous scaffolds: The influence on the physical, chemical properties and cell behavior. *Int. J. Polym. Mater.* **2016**, *65*, 457–465. [CrossRef]
22. Sánchez-Téllez, D.A.; Téllez-Jurado, L.; Rodríguez-Lorenzo, L.M. Hydrogels for Cartilage Regeneration, from Polysaccharides to Hybrids. *Polymers* **2017**, *9*, 671. [CrossRef]
23. Hernández-González, A.C.; Téllez-Jurado, L.; Rodríguez-Lorenzo, L.M. Alginate hydrogels for bone tissue engineering, from injectables to bioprinting: A review. *Carbohydr. Polym.* **2020**, *229*, 115514. [CrossRef]
24. Lee, B.M.; Park, S.J.; Noh, I.; Kim, C.H. The effects of the molecular weights of hyaluronic acid on the immune responses. *Biomater. Res.* **2021**, *25*, 27. [CrossRef] [PubMed]
25. Marin, E.; Rojas, J.; Ciro, Y. A review of polyvinyl alcohol derivatives: Promising materials for pharmaceutical and biomedical applications. *Afr. J. Pharm. Pharmacol.* **2013**, *8*, 674–684.
26. Yang, L.H.; Park, K.E.; Yoon, S. Synthesis and Hydrophilicity Analysis of bis(propane-1,2-diol) Terminated Polydimethylsiloxanes (PDMSs). *Materials* **2022**, *15*, 753. [CrossRef]
27. Hu, X.; Li, D.; Zhou, F.; Gao, C. Biological hydrogel synthesized from hyaluronic acid, gelatin and chondroitin sulfate by click chemistry. *Acta Biomater.* **2011**, *7*, 1618–1626. [CrossRef]
28. Sánchez-Téllez, D.A.; Téllez-Jurado, L.; Rodríguez-Lorenzo, L.M. Optimization of the CaO and P_2O_5 contents on PDMS–SiO_2–CaO–P_2O_5 hybrids intended for bone regeneration. *J. Mater. Sci.* **2015**, *50*, 5993–6006. [CrossRef]
29. Peña-Alonso, R.; Rubio, F.; Rubio, J. The Role of γ-Aminopropyltriethoxysilane (γ-APS) on Thermal Stability of TEOS-PDMS Ormosils. *J. Sol-Gel Sci. Technol.* **2005**, *36*, 77–85. [CrossRef]
30. Téllez, L.; Rubio, J.; Rubio, F.; Morales, E.; Oteo, J.L. FT-IR Study of the Hydrolysis and Polymerization of Tetraethyl Orthosilicate and Polydimethyl Siloxane in the Presence of Tetrabutyl Orthotitanate. *Spectrosc. Lett.* **2004**, *37*, 11–31. [CrossRef]
31. Perween, S.; Khan, Z.; Singh, S.; Ranjan, A. PVA-PDMS-Stearic acid composite nanofibrous mats with improved mechanical behavior for selective filtering applications. *Sci. Rep.* **2018**, *8*, 16038. [CrossRef]
32. Arrieta-Oliva, H.I.; Gutiérrez-Ventura, R.I.; Sánchez-Téllez, D.A.; Téllez-Jurado, L.; García-Pérez, B.E. Osteoblast response to zirconia modified-ORMOSILs. *Mater. Sci. Eng. C Mater. Biol. Appl.* **2020**, *109*, 110546. [CrossRef]
33. Puchtler, H.; Meloan, S.N.; Terry, M.S. On the history and mechanism of alizarin and alizarin red S stains for calcium. *J. Histochem. Cytochem.* **1969**, *17*, 110–124. [CrossRef] [PubMed]
34. Pirzada, T.; Arvidson, S.A.; Saquing, C.D.; Shah, S.S.; Khan, S.A. Hybrid silica-PVA nanofibers via sol-gel electrospinning. *Langmuir* **2012**, *28*, 5834–5844. [CrossRef] [PubMed]
35. Sabetghadam, A.; Mohammadi, T. PVA–APTEOS/TEOS Hybrid Sol–Gel Pervaporation Membrane for Dehydration of Ethanol. *Compos. Interfaces* **2010**, *17*, 223–228. [CrossRef]
36. Yu, L.; Li, C.M.; Zhou, Q.; Luong, J.H. Poly(vinyl alcohol) functionalized poly(dimethylsiloxane) solid surface for immunoassay. *Bioconj. Chem.* **2007**, *18*, 281–284. [CrossRef]
37. Eslahi, N.; Simchi, A.; Mehrojoo, M.; Shokrgozar, M.A.; Bonakdar, S. Hybrid cross-linked hydrogels based on fibrous protein/block copolymers and layered silicate nanoparticles tunable thermosensitivity, biodegradability and mechanical durability. *RSC Adv.* **2016**, *6*, 62944–62957. [CrossRef]
38. Haxaire, K.; Maréchal, Y.; Milas, M.; Rinaudo, M. Hydration of polysaccharide hyaluronan observed by IR spectrometry. I. Preliminary experiments and band assignments. *Biopolymers* **2003**, *72*, 10–20. [CrossRef] [PubMed]
39. Teodorescu, M.; Bercea, M.; Morariu, S. Biomaterials of Poly(vinyl alcohol) and Natural Polymers. *Polym. Rev.* **2018**, *58*, 247–287. [CrossRef]
40. de Oliveira, A.A.R.; Ciminelli, V.; Dantas, M.S.S.; Mansur, H.S.; Pereira, M.M. Acid character control of bioactive glass/polyvinyl alcohol hybrid foams produced by sol-gel. *J. Sol-Gel Sci. Technol.* **2008**, *47*, 335–346. [CrossRef]
41. Kalampounias, A.G. IR and Raman spectroscopic studies of sol–gel derived alkaline-earth silicate glasses. *Bull. Mater. Sci.* **2011**, *34*, 299–303. [CrossRef]

42. Borjanović, V.; Bistričić, L.; Vlasov, I.; Furić, K.; Zamboni, I.; Jakšić, M.; Shenderova, O. Influence of proton irradiation on the structure and stability of poly(dimethylsiloxane) and poly(dimethylsiloxane)-nanodiamond composite. *J. Vac. Sci. Technol. B* **2009**, *7*, 2396–2403. [CrossRef]
43. La Verde, G.; Sasso, A.; Rusciano, G.; Capaccio, A.; Fusco, S.; Mayol, L.; Biondi, M.; Silvestri, T.; Netti, P.A.; La Commara, M.; et al. Characterization of Hyaluronic Acid-Coated PLGA Nanoparticles by Surface-Enhanced Raman Spectroscopy. *Int. J. Mol. Sci.* **2022**, *24*, 601. [CrossRef] [PubMed]
44. Fischer, R.L.; McCoy, M.G.; Grant, S.A. Electrospinning collagen and hyaluronic acid nanofiber meshes. *J. Mater. Sci. Mater. Med.* **2012**, *7*, 1645–1654. [CrossRef] [PubMed]
45. Dodda, J.M.; Bělský, P.; Chmelař, J.; Remiš, T.; Smolná, K.; Tomáš, M.; Kullová, L.; Kadlec, J. Comparative study of PVA/SiO_2 and PVA/SiO_2/glutaraldehyde (GA) nanocomposite membranes prepared by single-step solution casting method. *J. Mater. Sci.* **2015**, *50*, 6477–6490. [CrossRef]
46. Koski, A.; Yim, K.; Shivkumar, S. Effect of molecular weight on fibrous PVA produced by electrospinning. *Mater. Lett.* **2004**, *58*, 493–497. [CrossRef]
47. Um, I.C.; Fang, D.; Hsiao, B.S.; Okamoto, A.; Chu, B. Electro-spinning and electro-blowing of hyaluronic acid. *Biomacromolecules* **2004**, *5*, 1428–1436. [CrossRef] [PubMed]
48. Snetkov, P.; Morozkina, S.; Uspenskaya, M.; Olekhnovich, R. Hyaluronan-Based Nanofibers: Fabrication, Characterization and Application. *Polymers* **2019**, *11*, 2036. [CrossRef]
49. Fong, H.; Chun, I.; Reneker, D.H. Beaded nanofibers formed during electrospinning. *Polymer* **1999**, *40*, 4585–4592. [CrossRef]
50. Stocco, E.; Barbon, S.; Dalzoppo, D.; Lora, S.; Sartore, L.; Folin, M.; Parnigotto, P.P.; Grandi, C. Tailored PVA/ECM scaffolds for cartilage regeneration. *Biomed. Res. Int.* **2014**, *2014*, 762189. [CrossRef]
51. Agostino, A.D.; La Gatta, A.; Busico, T.; De Rosa, M.; Schirald, C. Semi-interpenetrated Hydrogels Composed of PVA and Hyaluronan or Chondroitin Sulphate: Chemico-Physical and Biological Characterization. *J. Biotechnol. Biomater.* **2012**, *2*, 1000140. [CrossRef]
52. Kodavaty, J.; Deshpande, A.P. Mechanical and Swelling Properties of Poly (vinyl alcohol) and Hyaluronic Acid Gels used in Biomaterial Systems—A comparative Study. *Def. Sci. J.* **2014**, *64*, 222–229. [CrossRef]
53. Fahmy, A.; Kamoun, E.A.; El-Eisawy, R.; El-Fakharany, E.M.; Taha, T.H.; El-Damhougy, B.K.; Abdelhai, F. Poly(vinyl alcohol)-hyaluronic Acid Membranes for Wound Dressing Applications: Synthesis and in vitro Bio-Evaluations. *J. Braz. Chem. Soc.* **2015**, *26*, 1466–1474. [CrossRef]
54. Ding, J.; He, R.; Zhou, G.; Tang, C.; Yin, C. Multilayered Mucoadhesive Hydrogel Films Based on Thiolated Hyaluronic Acid and Polyvinylalcohol for Insulin Delivery. *Acta Biomater.* **2012**, *8*, 3643–3651. [CrossRef] [PubMed]
55. Pavasant, P.; Shizari, T.M.; Underhill, C.B. Distribution of hyaluronan in the epiphysial growth plate: Turnover by CD44-expressing osteoprogenitor cells. *J. Cell Sci.* **1994**, *107*, 2669–2677. [CrossRef] [PubMed]
56. Huang, L.; Cheng, Y.Y.; Koo, P.L.; Lee, K.M.; Qin, L.; Cheng, J.C.Y.; Kumta, S.M. The effect of hyaluronan on osteoblast proliferation and differentiation in rat calvarial-derived cell cultures. *J. Biomed. Mater. Res.—Part A* **2003**, *66*, 880–884. [CrossRef] [PubMed]
57. Teixeira, E.; Boeckel, D.; Sesterheim, P.; Fulginiti, R.; Shinkai, R.; Machado, D. Mesenchymal stem cells and hyaluronic acid for bone grafting. *Clin. Oral Implant. Res.* **2018**, *29*, 12724. [CrossRef]
58. Liu, L.S.; Thompson, A.Y.; Heidaran, M.A.; Poser, J.W.; Spiro, R.C. An osteoconductive collagen/hyaluronate matrix for bone regeneration. *Biomaterials* **1999**, *20*, 1097–1108. [CrossRef]
59. Kogan, G.; Šoltés, L.; Stern, R.; Schiller, J.; Mendichi, R. Hyaluronic Acid: Its Function and Degradation in in vivo Systems. *Stud. Nat. Prod. Chem.* **2008**, *34*, 789–882. [CrossRef]
60. Carvalho, M.P.; Costa, E.C.; Miguel, S.P.; Correia, I.J. Tumor spheroid assembly on hyaluronic acid-based structures: A review. *Carbohydr. Polym.* **2016**, *150*, 139–148. [CrossRef] [PubMed]
61. Kondapaneni, R.V.; Shevde, L.A.; Rao, S.S. A Biomimetic Hyaluronic Acid Hydrogel Models Mass Dormancy in Brain Metastatic Breast Cancer Spheroids. *Adv. Biol.* **2023**, *7*, e2200114. [CrossRef]
62. Chen, W.Y.; Abatangelo, G. Functions of hyaluronan in wound repair. *Wound Repair. Regen.* **1999**, *7*, 79–89. [CrossRef] [PubMed]
63. Hamann, K.J.; Dowling, T.L.; Neeley, S.P.; Grant, J.A.; Leff, A.R. Hyaluronic acid enhances cell proliferation during eosinopoiesis through the CD44 surface antigen. *J. Immunol.* **1995**, *154*, 4073–4080. [CrossRef] [PubMed]
64. Oliferenko, S.; Kaverina, I.; Small, J.V.; Huber, L.A. Hyaluronic acid (HA) binding to CD44 activates Rac1 and induces lamellipodia outgrowth. *J. Cell Biol.* **2000**, *148*, 1159–1164. [CrossRef]
65. Salim, S.A.; Loutfy, S.A.; El-Fakharany, E.M.; Taha, T.H.; Hussein, Y.; Kamoun, E.A. Influence of chitosan and hydroxyapatite incorporation on properties of electrospun PVA/HA nanofibrous mats for bone tissue regeneration: Nanofibers optimization and in-vitro assessment. *J. Drug Deliv. Sci. Technol.* **2021**, *62*, 102417. [CrossRef]
66. Pangon, A.; Saesoo, S.; Saengkrit, N.; Ruktanonchai, U.; Intasanta, V. Hydroxyapatite-hybridized chitosan/chitin whisker bionanocomposite fibers for bone tissue engineering applications. *Carbohydr. Polym.* **2016**, *144*, 419–427. [CrossRef] [PubMed]
67. Chocholata, P.; Kulda, V.; Dvorakova, J.; Kolaja Dobra, J.; Babuska, V. Biological Evaluation of Polyvinyl Alcohol Hydrogels Enriched by Hyaluronic Acid and Hydroxyapatite. *Int. J. Mol. Sci.* **2020**, *21*, 5719. [CrossRef]
68. Bhowmick, S.; Rother, S.; Zimmermann, H.; Lee, P.S.; Moeller, S.; Schnabelrauch, M.; Koul, V.; Jordan, R.; Hintze, V.; Scharnweber, D. Biomimetic electrospun scaffolds from main extracellular matrix components for skin tissue engineering application—The role of chondroitin sulfate and sulfated hyaluronan. *Mater. Sci. Eng. C Mater. Biol. Appl.* **2017**, *79*, 15–22. [CrossRef]

69. Sujana, A.; Venugopal, J.R.; Velmurugan, B.; Góra, A.; Salla, M.; Ramakrishna, S. Hydroxyapatite-intertwined hybrid nanofibres for the mineralization of osteoblasts. *J. Tissue Eng. Regen. Med.* **2017**, *11*, 1853–1864. [CrossRef]
70. Rachmiel, D.; Anconina, I.; Rudnick-Glick, S.; Halperin-Sternfeld, M.; Adler-Abramovich, L.; Sitt, A. Hyaluronic Acid and a Short Peptide Improve the Performance of a PCL Electrospun Fibrous Scaffold Designed for Bone Tissue Engineering Applications. *Int. J. Mol. Sci.* **2021**, *22*, 2425. [CrossRef]

polymers

MDPI

Article

Development and Characterization of Antimicrobial Chitosan/Polyethylene Oxide/Bacterial Cellulose Nanofibers

Fatma Sude Cetin [1,2], Tubanur Avci [1,2], Emre Uygur [1,2], Elif Ilhan [1], Elif Kaya [3], Gulgun Bosgelmez Tinaz [4], Liviu Duta [5,*], Canan Dogan [1,2,*] and Oguzhan Gunduz [1,2]

1 Center for Nanotechnology and Biomaterials Application and Research (NBUAM), Marmara University, Istanbul 34722, Turkey; sudegoktug@hotmail.com (F.S.C.); avcitubanur@gmail.com (T.A.); emreygr@gmail.com (E.U.); eliffguven@gmail.com (E.I.); oguzhan@marmara.edu.tr (O.G.)
2 Department of Metallurgical and Materials Engineering, Faculty of Technology, Marmara University, Istanbul 34722, Turkey
3 Department of Biochemistry, Health Sciences Institute, Marmara University, Istanbul 34854, Turkey; ekaya14@marun.edu.tr
4 Department of Basic Pharmaceutical Sciences, Faculty of Pharmacy, Marmara University, Istanbul 34668, Turkey; gulgun.tinaz@marmara.edu.tr
5 National Institute for Laser, Plasma and Radiation Physics, 409 Atomistilor Street, 077125 Magurele, Romania
* Correspondence: liviu.duta@inflpr.ro (L.D.); canan.dogan@marmara.edu.tr (C.D.)

Abstract: This study introduces novel chitosan (CS) and polyethylene oxide (PEO) copolymers reinforced with bacterial cellulose (BC) to fabricate nanofibers using the electrospinning method. SEM analysis confirmed uniform nanofiber formation, with CS/PEO/BC nanofibers (~240 nm) exhibiting a larger diameter than CS/PEO ones (~190 nm). FTIR spectroscopy confirmed BC integration, while Differential scanning calorimetry analysis indicated minimal impact on glass transition temperature. Notably, as compared to CS/PEO nanofibers, the CS/PEO/BC ones demonstrated superior swelling capacity, accelerated biodegradation, and enhanced mechanical (i.e., tensile) properties, with maximum stress and strain values of ~3.41 MPa and ~0.01% vs. ~2.14 MPa and ~0.01%. Antimicrobial assays confirmed activity against bacterial strains, and biocompatibility tests showed high cell viability at day seven (99.26% for CS/PEO/BC nanofibers). These findings highlight the potential of CS/PEO/BC nanofibers as promising candidates for tissue engineering, offering improved strength, biodegradability, and antimicrobial properties.

Keywords: bacterial cellulose; electrospinning; biomaterial; antimicrobial efficiency; biocompatibility testing

1. Introduction

A biomaterial is defined as any natural or synthetic substance, or a combination thereof, that can be used for an extended period as a whole or part of a system to treat, augment, or replace any tissue, organ, or function of the body [1]. The utilization of biomaterials is crucial due to their biocompatibility, ability to support tissue repair, and potential to enhance the efficacy of medical treatments.

Biomaterials have a wide range of applications across various medical fields. In orthopedics and traumatology, biomaterials are employed in bone and joint prosthetics, as well as in screws and nails designed to restore skeletal structures. In dentistry, they are utilized in dental implants, filling materials, and protective coatings. Within the cardiovascular system, biomaterials are integral components of stents and artificial heart valves, used to manage vascular disorders and heart diseases. For skin and soft tissue applications, they

 https://doi.org/10.3390/polym17050693

serve as wound dressings, particularly in the treatment of burns. Additionally, they play an essential role in ophthalmology, especially in the development of contact lenses and corneal implants. These diverse applications underscore the versatility and critical importance of biomaterials in modern medicine [2–4].

Several key parameters are essential in the development of biomaterials, including biocompatibility, biodegradability, mechanical strength, biomimetic properties, and machinability. Biocompatibility ensures that the material does not provoke an adverse immune response from body tissues and remains non-toxic [5]. Biodegradability refers to the material's capacity to degrade or decompose naturally within the biological environment over time [6], minimizing long-term residue. Mechanical strength is essential to meet the structural and mechanical demands specific to the tissue or organ being treated [7]. Biomimetic properties enable the material to support cell adhesion, proliferation, and differentiation, thereby mimicking the native characteristics of biological tissues [8]. Lastly, machinability ensures that the material can be readily shaped and processed to meet practical and clinical requirements.

Nanofibers, as a promising biomaterial, have garnered substantial attention in recent years due to their potential applications in tissue engineering. These nanofibers can be fabricated from natural polymers, such as collagen, chitosan (CS), gelatin, casein, and cellulose, as well as synthetic polymers, including polyethylene oxide (PEO), polypropylene, polyvinyl alcohol, polymethyl methacrylate, polylactic acid, and poly(ε-caprolactone) [9,10]. In this study, CS, and bacterial cellulose (BC) were selected as natural polymer sources, while PEO served as the synthetic polymer. CS, derived from the exoskeletons of crustaceans like shrimp and crab, is renowned for its antimicrobial activity against bacteria, fungi, and certain viruses [11,12]. It is also biocompatible, biodegradable, and exhibits low toxicity [13]. BC, a highly pure form of cellulose synthesized by bacteria, is distinguished by its exceptional mechanical strength, high purity, nanofibril structure, and a large surface area, making it a valuable candidate for biomedical applications [14,15]. Additionally, its microporous structure enhances its functionality [16]. PEO, a petroleum-derived synthetic polymer, is water-soluble and biocompatible. Its ability to enhance viscosity [17,18] and its widespread application in medical contexts, including drug delivery and tissue scaffolding, further underscore its significance in this study.

Electrospinning is a versatile technique for fabricating polymer fibers, particularly nanofibers, characterized by sub-micron dimensions and high surface-to-volume ratios. The process involves an electrostatic field to transform a polymer solution into fine fibers, which are subsequently deposited as solid nanostructures on a grounded collector. This method enables precise control over fiber morphology, which is influenced by key parameters such as applied voltage, flow rate, collector distance, and the physical-chemical properties of the polymer solution [19–22]. Electrospun nanofibers closely mimic the hierarchical structure of the extracellular matrix, making them particularly suitable for various biomedical applications, including tissue engineering, drug delivery systems, and wound dressings. Beyond the biomedical field, electrospinning has garnered attention for its innovative applications in the food and packaging industries. Electrospun fibers enhance the mechanical properties of packaging materials, extend food shelf life, and act as carriers for bioactive compounds such as essential oils and antioxidants. Additionally, they serve as scaffolds for bacterial cultures during fermentation processes and can be integrated with pH-sensitive sensors for smart packaging solutions [21,23].

In this study, nanofibers combining CS and PEO copolymers with BC were fabricated using electrospinning. Unlike previous studies, this approach employs a distinctive combination of materials that enhances processability and structural integrity while leveraging the synergistic effects of BC, CS, and PEO. The integration of these components significantly improves the mechanical properties, antimicrobial activity, and biocompatibility of the resulting nanofibers, offering a novel contribution to the field. By addressing existing challenges in nanofiber production, this study establishes a foundation for the development of eco-friendly, functional biomaterials tailored for advanced biomedical applications.

2. Materials and Methods

2.1. Materials

PEO (M_w = 6000 g/mol) and CS (M_w = 100.00 g/mol) were procured from Sigma Aldrich (Saint Louis, MO, USA). Glucose (M_w = 180.16 g/mol), citric acid (M_w = 192.12 g/mol), and di-sodium hydrogen phosphate (M_w = 141.96 g/mol) were purchased from Merck (Darmstadt, Germany). Casein peptone Type I and yeast extract were obtained from Neogen (Lexington, KY, USA), while glacial acetic acid (AA, M_w = 60.05 g/mol, ≥100%), dimethylacetamide (DMAc, M_w = 87.12 g/mol, ≥99.0%), and lithium chloride (LiCl, M_w = 42.39 g/mol) were supplied by ISOLAB (Windsor, ON, Canada).

2.2. Bacterial Cellulose Production

Gluconacetobacter xylinus (ATCC 23767, Virginia, USA) was used for the production of BC pellicles. The bacteria were cultured in Hestrin and Schramm (HS) medium, composed of 2% (*w*/*v*) glucose, 0.5% (*w*/*v*) yeast extract, 0.5% (*w*/*v*) peptone, 0.27% (*w*/*v*) di-sodium hydrogen phosphate, and 0.115% (*w*/*v*) citric acid per liter of distilled water. The pH of the medium was adjusted to 5.5 with AA, as this value was demonstrated to be optimal for cellulose production [24]. The medium was sterilized by autoclaving at 121 °C for 20 min and cooled to room temperature (RT) before use.

Cells were initially precultured in test tubes and then inoculated into a 100 mL Erlenmeyer flask containing 40 mL of HS medium. The flask was incubated statically at RT for 3 days. Upon the appearance of a thin cellulose layer on the surface, 1% of the cells and medium were transferred to a new container and subsequently to Petri dishes, where they were incubated at RT for 7 days.

To purify the cellulose pellicles, they were immersed in a 0.1 M sodium hydroxide solution for 24 h at RT to remove cells and residual components of the culture medium. The pellicles were then neutralized overnight by washing with deionized water. For sterilization, the pellicles were autoclaved again and stored at −20 °C until further use. Before use, frozen pellicles were thawed and prepared by freeze-drying.

As illustrated in Figure 1, the BC production process involves culturing and purification steps, culminating in the electrospinning setup, where the prepared BC is utilized for nanofiber fabrication.

2.3. Dissolution and Acetylation of Bacterial Cellulose

A 100:1 (*v*/*v*) mixture of DMAc and BC was stirred and heated at 170 °C for 1 h using a sand bath in a condenser system. Subsequently, LiCl (0.4%) was added to the mixture, which was then stirred at 100 °C for an additional hour. The mixture was further stirred at RT for 24 h, ultimately resulting in the formation of a viscous substance.

2.4. Electrospinning Process

CS was prepared by dissolving a 10% (*v*/*v*) AA-water solution at a concentration of 3 wt.%. BC was dissolved at concentrations of 5, 7, and 10 wt.% in DMAc, while PEO was dissolved at 2 and 5 wt.% in distilled water, each at RT with continuous stirring for 4 h. Subsequently, 3% CS, 5% BC, and 5% PEO solutions were mixed in a 1:1:1 ratio and stirred overnight at RT.

The prepared solutions were immediately electrospun using a system consisting of a 10 mL syringe with an 18-gauge needle (inner diameter: 0.84 mm), a ground electrode (21.5 cm diameter stainless steel sheet mounted on a rotating drum with variable speed), and a high-voltage source. The electrospinning process was conducted at a voltage of 25 kV, a working distance of 15 cm, and a solution flow rate of 0.5–0.7 mL/h. All electrospinning procedures were performed at RT, and the resulting nanofibers were stored at RT for subsequent characterization tests.

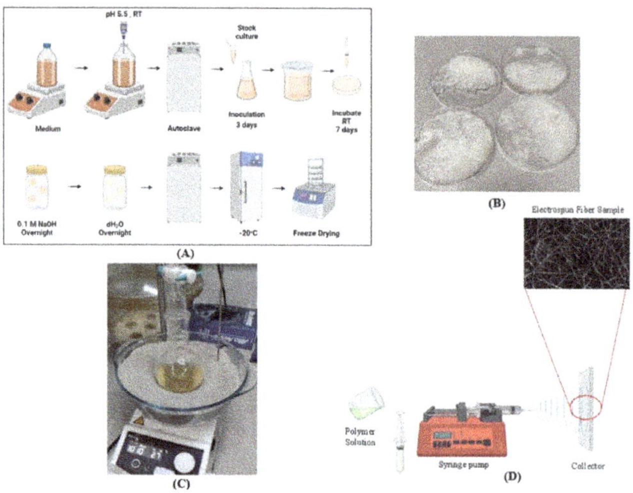

Figure 1. (**A**) Process of bacterial cellulose production, (**B**) Freeze-dried bacterial cellulose, (**C**) Bacterial cellulose dissolved using a sand bath, (**D**) Schematic representation of the electrospinning process.

2.5. Scanning Electron Microscopy (SEM)

The morphological microstructure of the nanofibers was analyzed by SEM (ZEISS, Jena, Germany). To prevent electrostatic charging and enhance image resolution, the nanofiber samples were coated with a gold layer for 120 s using a sputtering technique. The diameters of the nanofibers were measured using ImageJ software (v1.47, NIH, Bethesda, MD, USA) for precise analysis.

2.6. Fourier Transform Infrared (FTIR) Spectroscopy

FTIR analysis was conducted using a FT/IR-4700 spectrometer (JASCO Corporation, Tokyo, Japan) to qualitatively identify the components of the nanofiber samples. The spectra were recorded in the range of 4000–400 cm^{-1}, with 32 scans performed at a resolution of 4 cm^{-1}.

2.7. Differential Scanning Calorimetry (DSC)

A DSC-60 Plus instrument (Shimadzu Corporation, Kyoto, Japan) was used to investigate the thermal properties of the nanofibers. Approximately 2–3 mg of nanofiber samples were placed in aluminum pans and positioned in the DSC sample holder. The thermal analysis was conducted by heating the samples from 25 °C to 300 °C at a rate of 10 °C/min.

2.8. Swelling-Degradation Test

The swelling-degradation test is commonly conducted to evaluate the response of a material when exposed to a specific liquid or medium. To assess the swelling and degradation properties of the produced nanofibers, the initial dry mass (W_d, Equation (1)) of each sample was recorded. The nanofibers were then immersed in 1 mL of phosphate-buffered saline (PBS) at pH 7.4 for the swelling test. The samples were incubated at RT in a thermal shaker for 24 h.

At 24 h intervals, the samples were removed from the solution, and excess liquid on the surface was gently blotted using filter paper. The swollen mass (W_w) of each sample was then measured (Equation (1)). This procedure was repeated at regular intervals for up to 336 h. The swelling ratio was calculated using the following formula [25]:

$$S = \frac{W_w - W_d}{W_d} \times 100 \quad (1)$$

For the degradation test, the initial mass (W_0, Equation (2)) of each sample was measured. The samples were then immersed in 1 mL of PBS at pH 7.4 and incubated in a thermal shaker at RT for 24 h. After incubation, the samples were removed from the PBS, dried in an oven at 37 °C for 24 h, and their dry mass (W_t) was recorded (Equation (2)). This process was repeated at 24 h intervals for up to 168 h. The degradation rate was calculated using the following formula [25]:

$$D = \frac{W_0 - W_t}{W_0} \times 100 \quad (2)$$

2.9. Tensile Test

To evaluate the mechanical properties of the nanofiber patches, tensile testing was performed using a Shimadzu EZ-LX device (Shimadzu Corporation, Kyoto, Japan). The nanofiber patches were prepared as rectangular specimens measuring 10 mm × 50 mm. Prior to testing, the thickness of each patch was measured using a digital micrometer (Mitutoyo MTI Co., Dusseldorf, Germany). The tensile test was conducted at a speed of 5 mm/min with an applied force of 0.1 N. Each measurement was performed in triplicate for each group to ensure accuracy and reproducibility.

2.10. Antimicrobial Testing

The microorganisms *E. coli* (ATCC 25922) and *S. aureus* (ATCC 23235) were inoculated onto 5% sheep blood Columbia Agar (COS; Biomerieux, Isère, France) and McConkey Agar (MCK; Biomerieux, Isère, France) using dilution seeding methods. The inoculated plates were incubated overnight at 35 °C. For further testing on Mueller Hinton Broth medium (MHB; Biomerieux, Isère, France), bacterial suspensions were prepared with a turbidity of 0.5 McFarland (1–5 × 10^8 CFU/mL) and spread onto Mueller Hinton E. Agar (MHE; Biomerieux, Isère, France).

For these tests, the disk diffusion method was employed, Thus, small circular disks (of ~5–6 mm) were cut from each scaffold, sterilized under UV light for 10 min, and placed in the Petri dish. As recommended by the protocol [26], a distance of 12–20 mm was maintained between the samples and from the edge of the Petri dish to prevent any potential interference of the samples. Inhibition zones were determined by measuring the areas where there was no microbial growth.

2.11. Cell Culture Assay

Human dermal fibroblast (HDF) cell lines (ATCC-PCS-201-012) were grown in a 5% CO_2 incubator at 37 °C (ThermoFisher, Waltham, MA, USA) using Dulbecco's Modified Eagle Medium (DMEM) (Gibco, Carlsbad, CA, USA). When the cells reached ~80% confluency, they were detached using trypsin/EDTA (Gibco, Carlsbad, CA, USA) and harvested. The detached cells were centrifuged at 2500 rpm for 5 min to form a pellet, which was subsequently lysed in an appropriate volume of DMEM. The total number of cells was quantified using a cell counting device (Anvajo Science, Dresden, Germany).

To assess cell viability, circular disks with a 6 mm diameter were cut from the nanofibers and sterilized under UV light for 2 h. Following sterilization, the disks were rinsed with PBS and transferred into a 96-well plate. Each well was subsequently inoculated with DMEM supplemented with 10% fetal bovine serum (FBS) and 1% penicillin/streptomycin, followed by a 24 h incubation period to stabilize the system.

HDFs were seeded onto the sterilized nanofibers at an initial density of 5×10^3 cells per well. As a control, an equivalent number of cells were seeded directly onto empty wells (2D control). Both the cell-seeded nanofibers and the control samples were incubated at 37 °C with 5% CO_2 for one week.

Cell proliferation was assessed using the 3-(4, 5-dimethylthiazolyl-2)-2, 5-diphenyltetrazolium bromide (MTT) assay (ElabScience, Houston, TX, USA). At 1, 3, and 7 days post-seeding, MTT solution was added to the wells and incubated for 3 h at 37 °C. Dimethyl sulfoxide was added to solubilize the formazan crystals formed by metabolically active cells. The absorbance of the resultant formazan product was measured at 570 nm using a microplate reader (Agilent BioTek, Epoch, Santa Clara, CA, USA). All experiments were conducted in triplicate to ensure reproducibility and statistical robustness.

2.12. Statistical Analysis

Statistical significance was evaluated using one-way ANOVA, followed by Turkey's post hoc test. A significance level of $p < 0.05$ was considered, with data labeled as follows: (*) for $p < 0.05$, (**) for $p < 0.01$, (***) for $p < 0.001$. Results are expressed as mean ± standard deviation (SD). ImageJ software (v1.47, NIH, Bethesda, MD, USA) was used to determine the nanofiber diameter from the SEM measurements.

3. Results and Discussion

3.1. SEM Investigation

The SEM images and corresponding histograms of fiber diameter distributions for CS/PEO/BC and CS/PEO nanofibers are shown in Figure 2. The SEM images clearly depict well-defined fibers with a uniform diameter distribution, homogeneous structure, and smooth morphology in both nanofiber architectures.

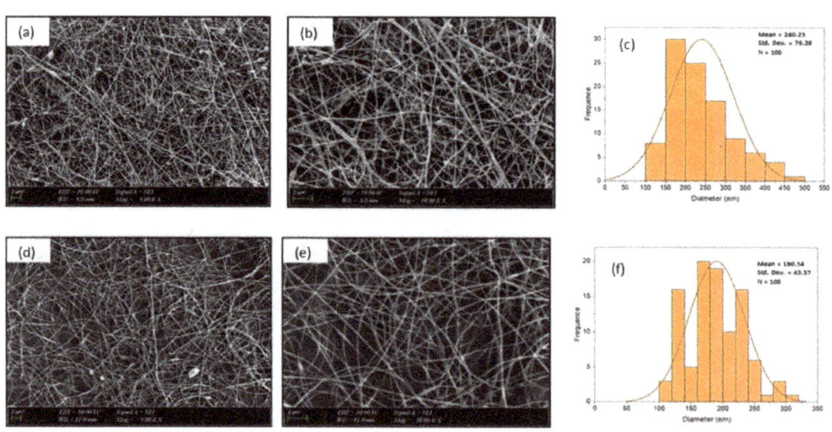

Figure 2. SEM images and diameter distributions of CS/PEO/BC and CS/PEO nanofibers: (**a**,**b**) CS/PEO/BC nanofibers at ×5000 and ×10,000 magnifications, respectively; (**c**) Diameter distribution of CS/PEO/BC nanofibers; (**d**,**e**) CS/PEO nanofibers at ×5000 and ×10,000 magnifications, respectively; (**f**) Diameter distribution of CS/PEO nanofibers.

The average diameter of CS/PEO nanofibers was measured at 190 ± 44 nm, while CS/PEO/BC nanofibers exhibited a larger diameter of 240 ± 79 nm. As illustrated in the histogram graphs, the incorporation of BC led to a noticeable increase in fiber diameter. This effect can be attributed to the high crystallinity and fibrillar network of BC, which contribute to nanofiber diameter expansion.

The presence of BC enhances the viscosity of the polymer solution, thereby facilitating the formation of thicker fibers. Additionally, BC's high surface area and water retention capacity promote the development of a more organized fiber structure. These structural characteristics enable BC to function as a reinforcing component within the polymer matrix, ultimately resulting in an increased fiber diameter [27].

3.2. FTIR Spectroscopy Analysis

FTIR spectroscopy confirmed the chemical composition of the CS/PEO/BC nanofiber structure, as demonstrated by the spectra presented in Figure 3. The spectra distinctly exhibit the characteristic absorption bands of BC, CS, and PEO, validating their presence within the nanofiber matrix.

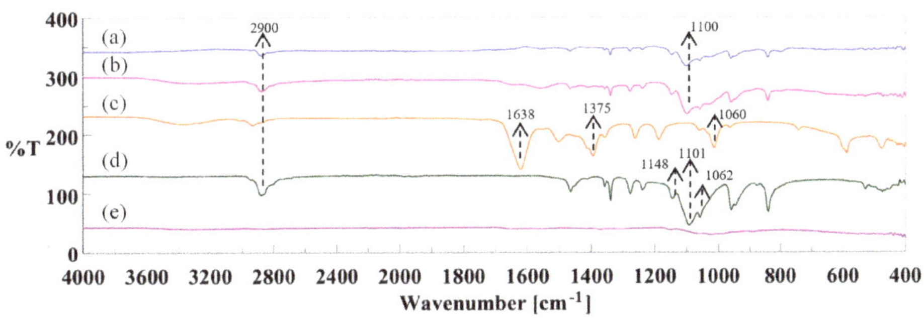

Figure 3. FTIR spectra of: (a) CS/PEO/BC nanofibers, (b) CS/PEO nanofibers, (c) BC, (d) PEO, and (e) CS.

A prominent aliphatic C-H stretching vibration peak observed around 2900 cm^{-1} is particularly noteworthy, as it corresponds to the C–H stretching vibrations in both PEO and cellulose, confirming the presence of PEO [28]. This observation is further supported by the sharp peak at approximately 1100 cm^{-1}, attributed to the stretching vibrations of characteristic C–O–C ether bonds in the CS/PEO/BC composite.

In pristine BC, distinct absorption bands were identified at 1638 cm^{-1} and 1375 cm^{-1}, corresponding to carbonyl groups and CH_2 bending vibrations, respectively. Additionally, a broad peak at approximately 1060 cm^{-1}, associated with glycosidic bonds, was observed [29].

For CS, characteristic absorption peaks were identified at 1680 cm^{-1} and 1538 cm^{-1}, corresponding to the carbonyl group (C=O–NHR) and amine ($-NH_2$) groups, respectively. Additionally, bands associated with the amide I, amide II, and amide III regions further highlight the structural features of CS.

The presence of PEO was confirmed by the characteristic C–O–C stretching vibrations observed at 1148, 1101, and 1062 cm^{-1}, indicative of its polyether chain structure [30]. The incorporation of BC was confirmed by the appearance of characteristic peaks, supporting the formation of a new composite material.

In conclusion, the FTIR spectrum of the CS/PEO/BC composite exhibits the characteristic vibration bands of all three components, confirming the successful integration of their chemical structures.

3.3. DSC Analysis

The DSC curves of CS/PEO/BC and CS/PEO nanofibers are presented in Figure 4. The glass transition temperatures (T_g) of both nanofibers are remarkably similar, with CS/PEO/BC nanofibers exhibiting a T_g of 62.67 °C, while CS/PEO nanofibers show a T_g of 61.67 °C.

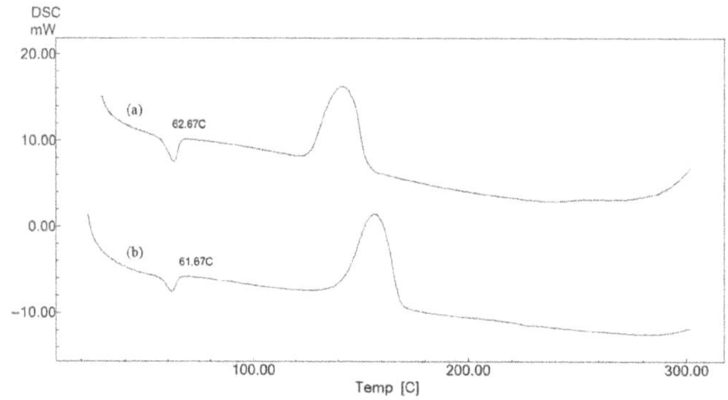

Figure 4. Differential scanning calorimetry curves of: (a) CS/PEO/BC nanofibers, and (b) CS/PEO nanofibers.

This small difference in T_g indicates that the thermal properties of both compositions are highly comparable. The incorporation of BC resulted in only a 1 °C increase in the glass transition temperature, suggesting that BC has a minimal effect on the glass transition behavior when integrated into the polymer matrix.

The T_g of BC-incorporated fibers is significantly higher than that of native BC (13.94 °C) and NaOH-treated BC membranes (41.41 °C), as reported by George et al. [31]. This elevated T_g reduces material aging at storage temperatures, thereby enhancing the longevity and stability of the composite material [32].

3.4. Swelling-Degradation Tests

The swelling ratio for both investigated group samples exhibited high values within the first 24 h, followed by a slight increase in the CS/PEO/BC group between 48 and 72 h. Over time, a gradual decrease in the swelling ratio was observed in both groups. The reason the swelling rate decreases over time is that the material reaches water saturation and enters the structural stabilization process. Initially, the polymer network structure expands by rapidly absorbing water, but over time, the water uptake is balanced and the polymer chains enter the relaxation process by rearranging. In addition, polymer degradation processes or water evaporation may also contribute to this decrease in swelling ratio [33]. However, the CS/PEO/BC group consistently demonstrates a higher swelling capacity compared to the CS/PEO group, as illustrated in Figure 5.

The incorporation of BC appears to enhance the swelling ratio and strengthen the material's interaction with water, likely due to an increase in hydrophilicity [34]. Over time, both groups exhibit mass loss, with the CS/PEO/BC group generally experiencing a slightly greater mass loss compared to the CS/PEO group (Figure 6).

Mass loss occurs most rapidly within the first 24–48 h, after which it stabilizes at a more constant rate. The higher mass loss observed in the CS/PEO/BC group may be attributed to the enhanced biodegradation rate introduced by BC incorporation. This characteristic could be advantageous for biomedical applications, particularly in controlled drug release systems [35].

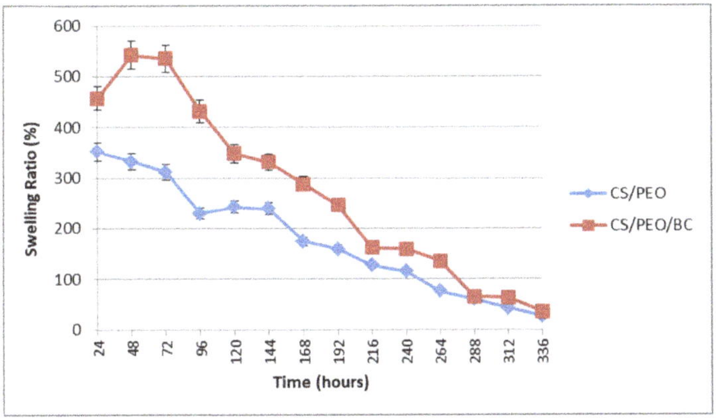

Figure 5. Swelling behavior of nanofiber samples.

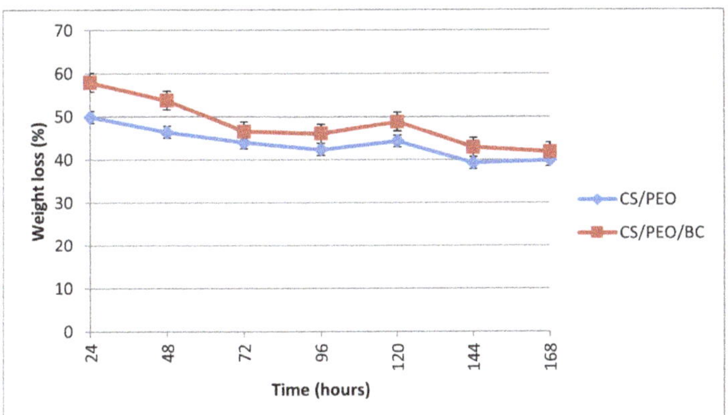

Figure 6. Degradation behavior of nanofiber samples.

3.5. Tensile Tests

The mechanical behavior of nanofibrous meshes, including tensile strength and strain at break, has been comprehensively analyzed to assess their applicability across various fields. Key factors such as material composition and intrinsic chemical interactions play a pivotal role in determining these properties, ultimately influencing the durability and performance of the nanofibers.

Tensile testing is particularly significant for biomaterials, as it provides critical insights into their strength and flexibility, which are critical for ensuring functionality and reliability in biomedical applications, such as tissue engineering and drug delivery systems. The tensile testing results for both electrospun nanofiber types are presented in Figure 7.

In this study, the control group of CS/PEO nanofibers exhibited a maximum stress of 2.139 ± 0.532 MPa and a maximum strain of 0.012 ± 0.003%. In a study conducted by Erdem et al., the tensile strength of CS/PEO nanofibers was reported as 4.2 MPa, with a tensile strain of 4.7% [36]. The higher strain values reported in their study can be attributed to the smaller fiber diameters, as the CS/PEO (60/40%) nanofibers had an average diameter of 72 ± 15 nm. In contrast, the CS/PEO nanofibers in our study exhibited and average diameter of 190 ± 44 nm, which likely contributed to the lower strain values observed in our results. Notably, the incorporation of BC into the nanofiber matrix resulted in an increase in both maximum stress (3.407 ± 0.824 MPa) and maximum strain (0.013 ± 0.004%), highlighting its significant impact on enhancing the mechanical properties of the nanofibers.

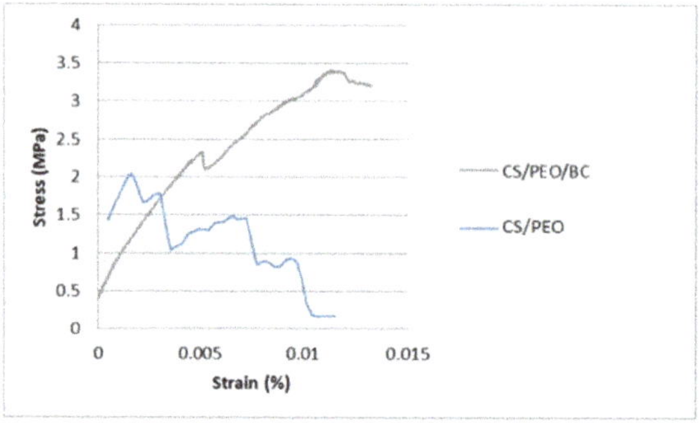

Figure 7. Representation of strain vs. stress in the case of CS/PEO and CS/PEO/BC nanofiber samples.

The superior mechanical performance can be attributed to the high crystallinity of BC and its strong internal and external hydrogen bonding, which contribute to an increased fiber diameter and the formation of a robust internal network. These structural characteristics enhance the mechanical durability and resilience of the nanofibers [23,37].

3.6. Antimicrobial Activity

The antibacterial effects of the nanofibers against both microorganisms are illustrated in Figure 8, with the corresponding values summarized in Table 1.

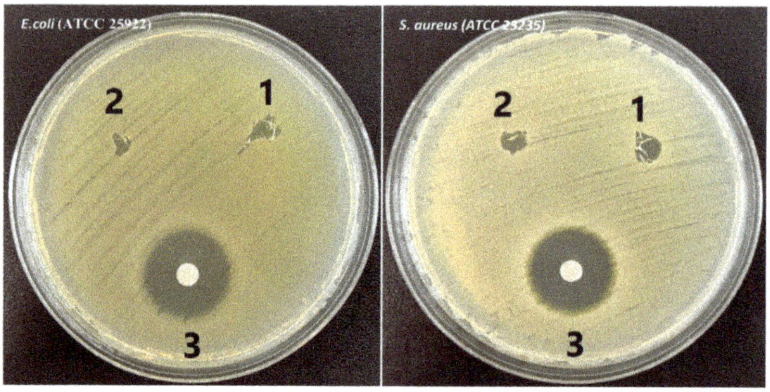

Figure 8. Antimicrobial effects of (**1**) CS/PEO nanofibers, (**2**) CS/PEO/BC nanofibers, and (**3**) Control (Gentamicin) against *E. coli* and *S. aureus*.

Table 1. Antimicrobial activity of the nanofibers, presented as the mean inhibition zone diameters (mm) against *E. coli* and *S. aureus*.

Type of Bacteria	Zone of Inhibition—Mean Diameter [mm]		
	CS/PEO	CS/PEO/BC	Control (Gentamicin/10 μg)
E. coli	6	4	20
S. aureus	7	6	19

The diameters of the inhibition zones for all nanofiber samples against both microorganisms were measured according to the CLSI guidelines, which define the zone boundary—including the disk diameter—as a distinct, visible, non-growing area detectable by the naked eye. The results confirmed the antimicrobial activity of both investigates samples, with CS/PEO nanofibers exhibiting slightly higher efficacy compared to CS/PEO/BC ones. Specifically, the maximum inhibition zone against *S. aureus* was 7 mm for CS/PEO nanofibers and 6 mm for CS/PEO/BC ones. Similarly, the inhibition zones against *E. coli* were 6 mm and 4 mm for CS/PEO and CS/PEO/BC nanofibers, respectively.

Possible explanations for the irregular inhibition zones observed in samples 1 and 2 include: (i) uneven diffusion of the tested compounds, potentially caused by differences in their solubility or viscosity, which may have led to an asymmetric distribution within the agar medium; and/or (ii) interaction with the medium, which could have influenced the diffusion dynamics of the tested compounds, resulting in a non-uniform distribution. It should be emphasized that the control zone (i.e., zone 3) was created using a standard antibiotic disk, which is known to diffuse uniformly within the agar medium, resulting in a well-defined, circular inhibition zone. In contrast, both CS/PEO and CS/PEO/BC nanofibers, unlike the control disk, possess a very fine structure, making it challenging to cut them into a smooth, circular shape. Consequently, the inhibition zones they formed were irregular. Furthermore, the differences observed in samples 1 and 2 are likely attributed to their distinct chemical or physical properties, which may have influenced their diffusion behavior compared to the control.

The literature highlights the broad-spectrum antimicrobial properties of CS, which can disrupt bacterial cell walls and compromise their structural integrity. This characteristic makes CS highly promising for applications in food safety and agriculture.

Numerous studies have demonstrated that CS-based biomaterials effectively inhibit microbial growth, exhibiting antimicrobial efficacy against both Gram-positive and Gram-negative bacteria. This effectiveness has been validated across various pH conditions and concentrations [38,39], further supporting the potential of CS-based materials in antimicrobial applications.

In contrast, the high hydrophilicity of BC can hinder the penetration of antimicrobial agents, thereby reducing their effectiveness. Additionally, the crystalline structure of BC may restrict the diffusion of antibacterial agents within the matrix, further contributing to its relatively lower antimicrobial activity.

These factors provide valuable insights into the subtle differences observed in the antimicrobial effects between CS and BC [40,41].

3.7. Cell Viability Tests

The proliferation of human fibroblast cell lines cultured on the fabricated material was assessed at 1, 4, and 7 days post-seeding using the MTT assay (Figure 9). By the seventh day, the cell viability of both CS/PEO/BC and CS/PEO nanofibers was determined to be 99.26% and 89.32%, respectively.

The seven-day MTT study demonstrated that cell viability in all groups exceeded 70% compared to the control group. Across both groups, cell viability presented reduced values on days one and three but increased significantly by day seven. This trend is consistent with the typical cellular response to a new culture environment, where cells experience initial stress during the adaptation phase, temporarily reducing viability. Over time, cells acclimate to their new conditions and begin to proliferate, leading to a notable recovery and increased viability [42]. The observed decrease in cell viability on days one and three for both scaffolds can be attributed to this initial adaptation phase, followed by a recovery to satisfactory viability levels by day seven.

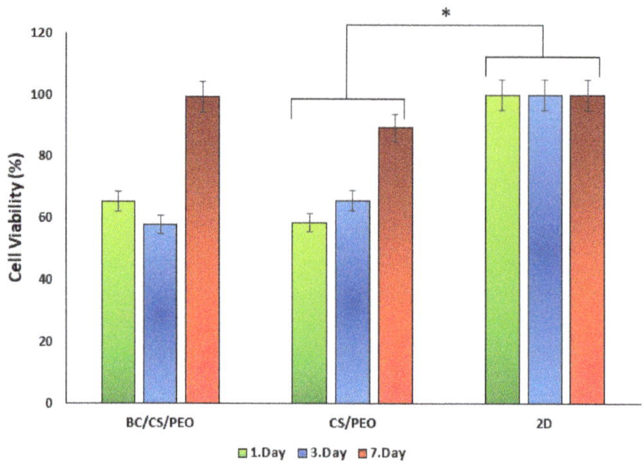

Figure 9. Cell viability of the human fibroblast cell line cultured on nanofiber samples over the designated time periods (* $p < 0.05$).

The CS used in nanofibers is a polymer with well-documented biocompatibility, as demonstrated in numerous previous studies. Its high biocompatibility is attributed to its structural and functional resemblance to glycosaminoglycans, which are integral components of the extracellular matrix in the human body [43,44].

The elevated biocompatibility values observed on day seven in both nanofiber groups can be primarily attributed to the CS content within the nanofiber composition. On day three, a slight decrease in BC/CS/PEO nanofibers is observed compared to CS/PEO nanofibers. Although BC exhibits excellent biocompatibility and water retention capacity, it may somewhat hinder cell adhesion. Additionally, its dense fibrous structure could limit cell penetration and spreading [45]. The decrease in cell viability on day three may be attributed to this factor. However, cells gradually adapt to their environment from day one to day seven. Although BC slightly inhibits cell viability, it exhibits an exceptional capacity for water absorption and retention, creating a moist microenvironment that supports cell viability, proliferation, and differentiation [46]. The slightly higher cell viability observed in the CS/PEO/BC scaffold compared to the CS/PEO scaffold on day seven is likely attributed to the enhanced biocompatibility conferred by BC.

4. Conclusions

In this study, nanofibers were successfully developed by integrating bacterial cellulose (BC), chitosan (CS), and polyethylene oxide (PEO) via electrospinning. The CS/PEO nanofibers exhibited a tensile strength of ~2.14 MPa and an elongation of ~0.01%. With the incorporation of BC, the tensile strength of CS/PEO/BC nanofibers increased by ~59%, reaching ~3.4 MPa, while maintaining a similar elongation (~0.01%). Additionally, the composite nanofibers demonstrated a water absorption capacity exceeding 200% within 24 h, highlighting their potential for tissue engineering and wound-healing applications. Cell viability assays indicated a 99.26% viability rate on day seven, confirming the excellent biocompatibility of CS/PEO/BC nanofibers and their ability to support cell proliferation. These findings align with previous reports demonstrating that BC incorporation significantly enhances the mechanical strength, water retention capacity, and biocompatibility of nanofibrous scaffolds.

In conclusion, CS/PEO/BC nanofibers, with their improved mechanical properties, superior water retention capacity, and excellent biocompatibility, hold significant potential for biomedical applications, particularly in tissue engineering.

Author Contributions: Conceptualization: C.D. and O.G.; methodology: F.S.C., T.A., E.U., E.I., E.K., G.B.T., C.D. and O.G.; software: F.S.C., L.D., C.D. and O.G.; validation: L.D., C.D. and O.G.; formal analysis: F.S.C., T.A., E.U., E.I., E.K., G.B.T. and L.D.; investigation: F.S.C., G.B.T., L.D., C.D. and O.G.; resources: L.D., C.D. and O.G.; data curation: L.D., C.D. and O.G.; writing—original draft preparation: F.S.C., T.A., E.U., E.I. and E.K.; writing—review and editing: F.S.C., T.A., E.U., E.I., E.K., G.B.T., C.D. and O.G.; visualization: L.D., C.D. and O.G.; supervision: L.D., C.D. and O.G.; project administration: L.D. and C.D.; funding acquisition: L.D. All authors have read and agreed to the published version of the manuscript.

Funding: This research received no external funding.

Institutional Review Board Statement: Not applicable.

Data Availability Statement: The original contributions presented in this study are included in the article. Further inquiries can be directed to the corresponding authors.

Acknowledgments: L.D. acknowledges with thanks the support of this work by the Romanian Ministry of Research, Innovation, and Digitization under Romanian National Nucleu Program LAPLAS VII—Contract No. 30N/2023.

Conflicts of Interest: The authors declare no conflicts of interest.

List of Abbreviations

AA	acetic acid
CS	Chitosan
CS/PEO/BC	Chitosan and polyethylene oxide combined with bacterial cellulose
BC	bacterial cellulose
CFU	colony forming units
DMAc	dimethylacetamide
DMEM	Dulbecco's Modified Eagle Medium
DSC	Differential scanning calorimetry
FBS	fetal bovine serum
FTIR	Fourier Transform Infrared spectroscopy
HDF	Human dermal fibroblast
HS	Hestrin and Schramm medium
LiCl	lithium chloride
MHB	Mueller Hinton Broth medium
MHE	Mueller Hinton E. agar
MTT	3-(4, 5-dimethylthiazolyl-2)-2, 5-diphenyl-238 tetrazolium bromide
NaOH	sodium hydroxide
PBS	phosphate-buffered saline
PEO	Polyethylene oxide
RT	room temperature
SEM	scanning electron microscopy
SD	standard deviation
T_g	glass transition temperature
W_0	initial mass
W_d	initial dry mass
W_t	dry mass
W_w	swollen mass

References

1. Gilmore, B.; Carson, L. 8-Bioactive biomaterials for controlling biofilms. In *Biomaterials and Medical Device-Associated Infections*; Elsevier eBooks: Amsterdam, The Netherlands, 2015; pp. 163–183. [CrossRef]
2. Parida, P.; Behera, A.; Mishra, S.C. Classification of Biomaterials used in Medicine. *Int. J. Adv. Appl. Sci.* **2012**, *1*, 31–35. [CrossRef]
3. Vlase, G.; Segneanu, A.; Bejenaru, L.E.; Bradu, I.A.; Sicoe, C.; Vlase, T.; Mogoşanu, G.D.; Buema, G.; Herea, D.; Ciocîlteu, M.V.; et al. Wild-Grown Romanian *Eupatorium cannabinum*: Advancing Phyto-Nanocarriers via Maltodextrin Micro-Spray Encapsulation—Metabolite Profiling, Antioxidant, Antimicrobial, and Cytotoxicity Insights. *Polymers* **2025**, *17*, 482. [CrossRef] [PubMed]
4. Hazar, E.; Hazar, A. Fracture Resistance of Glass-Fiber-Reinforced Direct Restorations on Endodontically Treated Molar Teeth with Furcal Perforation. *Polymers* **2025**, *17*, 370. [CrossRef]
5. Khan, M.Y.; Chen, M.-H. A Review on role of biomaterials in biomedical field. *Int. J. Bio-Pharma Res.* **2019**, *8*, 2788–2793.
6. Hench, L.L.; Polak, J.M. Third-Generation biomedical materials. *Science* **2002**, *295*, 1014–1017. [CrossRef]
7. Krishani, M.; Shin, W.Y.; Suhaimi, H.; Sambudi, N.S. Development of Scaffolds from Bio-Based Natural Materials for Tissue Regeneration Applications: A Review. *Gels* **2023**, *9*, 100. [CrossRef] [PubMed]
8. Liu, S.; Yu, J.; Gan, Y.; Qiu, X.; Gao, Z.; Wang, H.; Chen, S.; Xiong, Y.; Liu, G.; Lin, S.; et al. Biomimetic natural biomaterials for tissue engineering and regenerative medicine: New biosynthesis methods, recent advances, and emerging applications. *Mil. Med. Res.* **2023**, *10*, 16. [CrossRef]
9. Abdulhussain, R.; Adebisi, A.; Conway, B.R.; Asare-Addo, K. Electrospun nanofibers: Exploring process parameters, polymer selection, and recent applications in pharmaceuticals and drug delivery. *J. Drug Deliv. Sci. Tec.* **2023**, *90*, 105156. [CrossRef]
10. Hu, Y.; Ye, Y.; Wang, J.; Zhang, T.; Jiang, S.; Han, X. Functionalization of chitosan and its application in flame retardants: A review. *Int. J. Biol. Macromol.* **2025**, *295*, 139615. [CrossRef] [PubMed]
11. Desai, N.; Rana, D.; Salave, S.; Gupta, R.; Patel, P.; Karunakaran, B.; Sharma, A.; Giri, J.; Benival, D.; Kommineni, N. Chitosan: A potential biopolymer in drug delivery and biomedical applications. *Pharmaceutics* **2023**, *15*, 1313. [CrossRef]
12. Hemmami, H.; Amor, I.B.; Amor, A.B.; Zeghoud, S.; Ahmed, S.; Alhamad, A.A. Chitosan, its Derivatives, sources, preparation Methods, and applications: A review. *J. Turk. Chem. Soc. Sect. A Chem.* **2024**, *11*, 341–364. [CrossRef]
13. Jiménez-Gómez, C.P.; Cecilia, J.A. Chitosan: A Natural Biopolymer with a Wide and Varied Range of Applications. *Molecules* **2020**, *25*, 3981. [CrossRef] [PubMed]
14. Ahmed, J.; Gultekinoglu, M.; Edirisinghe, M. Bacterial cellulose micro-nano fibres for wound healing applications. *Biotechnol. Adv.* **2020**, *41*, 107549. [CrossRef]
15. Popa, L.; Ghica, M.V.; Tudoroiu, E.; Ionescu, D.; Dinu-Pîrvu, C. Bacterial Cellulose—A remarkable polymer as a source for biomaterials tailoring. *Materials* **2022**, *15*, 1054. [CrossRef] [PubMed]
16. Portela, R.; Leal, C.R.; Almeida, P.L.; Sobral, R.G. Bacterial cellulose: A versatile biopolymer for wound dressing applications. *Microb. Biotechnol.* **2019**, *12*, 586–610. [CrossRef]
17. Son, W.K.; Youk, J.H.; Lee, T.S.; Park, W.H. The effects of solution properties and polyelectrolyte on electrospinning of ultrafine poly(ethylene oxide) fibers. *Polymer* **2004**, *45*, 2959–2966. [CrossRef]
18. Pakravan, M.; Heuzey, M.; Ajji, A. A fundamental study of chitosan/PEO electrospinning. *Polymer* **2011**, *52*, 4813–4824. [CrossRef]
19. Wu, J.; Lorenz, H.P. Electrospinning of biomaterials and their applications in tissue engineering. *Nano Life* **2012**, *2*, 1230010. [CrossRef]
20. Zhang, X.; Shi, X.; Gautrot, J.E.; Peijs, T. Nanoengineered electrospun fibers and their biomedical applications: A review. *Nanocomposites* **2020**, *7*, 1–34. [CrossRef]
21. Vasconcelos, F.; Reis, R.L.; Martins, A.; Neves, N.M. Biomedical applications of fibers produced by electrospinning, microfluidic spinning and combinations of both. In *Electrospun Nanofibers*; Springer: Cham, Switzerland, 2022; pp. 251–295. [CrossRef]
22. Zulkifli, M.Z.A.; Nordin, D.; Shaari, N.; Kamarudin, S.K. Overview of electrospinning for tissue engineering applications. *Polymers* **2023**, *15*, 2418. [CrossRef]
23. Kumar, T.S.M.; Kumar, K.S.; Rajini, N.; Siengchin, S.; Ayrilmis, N.; Rajulu, A.V. A comprehensive review of electrospun nanofibers: Food and packaging perspective. *Compos. Part B-Eng.* **2019**, *175*, 107074. [CrossRef]
24. Tripathi, M.; Lal, B.; Syed, A.; Mishra, P.; Elgorban, A.M.; Verma, M.; Singh, R.; Mohammad, A.; Srivastava, N. Production of fermentable glucose from bioconversion of cellulose using efficient microbial cellulases produced from water hyacinth waste. *Int. J. Biol. Macromol.* **2023**, *252*, 126376. [CrossRef]
25. Ilhan, E.; Cesur, S.; Guler, E.; Topal, F.; Albayrak, D.; Guncu, M.M.; Cam, M.E.; Taskin, T.; Sasmazel, H.T.; Aksu, B.; et al. Development of Satureja cuneifolia-loaded sodium alginate/polyethylene glycol scaffolds produced by 3D-printing technology as a diabetic wound dressing material. *Int. J. Biol. Macromol.* **2020**, *161*, 1040–1054. [CrossRef] [PubMed]
26. Clinical and Laboratory Standards Institute (CLSI). Performance Standards for Antimicrobial Susceptibility Testing (30th ed., CLSI Supplement M100). 2020. Print ISBN 978-1-68440-066-9. Available online: https://clsi.org/media/3481/m100ed30_sample.pdf (accessed on 1 March 2025).

27. Eslahi, N.; Mahmoodi, A.; Mahmoudi, N.; Zandi, N.; Simchi, A. Processing and Properties of Nanofibrous Bacterial Cellulose-Containing Polymer Composites: A review of recent advances for biomedical applications. *Poly Rev.* **2019**, *60*, 144–170. [CrossRef]
28. Colom, X.; Carrillo, F. Crystallinity changes in lyocell and viscose-type fibres by caustic treatment. *Eur. Polym. J.* **2002**, *38*, 2225–2230. [CrossRef]
29. Azarniya, A.; Eslahi, N.; Mahmoudi, N.; Simchi, A. Effect of graphene oxide nanosheets on the physico-mechanical properties of chitosan/bacterial cellulose nanofibrous composites. *Compos. Part A Appl. Sci. Manuf.* **2016**, *85*, 113–122. [CrossRef]
30. Fuh, Y.K.; Chen, S.Z.; He, Z.Y. Direct-write, highly aligned chitosan-poly(ethylene oxide) nanofiber patterns for cell morphology and spreading control. *Nanoscale Res. Lett.* **2013**, *8*, 97. [CrossRef]
31. George, J.; Ramana, K.V.; Sabapathy, S.N.; Jagannath, J.H.; Bawa, A.S. Characterization of chemically treated bacterial (*Acetobacter xylinum*) biopolymer: Some thermo-mechanical properties. *Int. J. Biol. Macromol.* **2005**, *37*, 189–194. [CrossRef]
32. Rabek, C.L.; Van Stelle, R.; Dziubla, T.D.; Puleo, D.A. The effect of plasticizers on the erosion and mechanical properties of polymeric films. *J. Biomater. Appl.* **2013**, *28*, 779–789. [CrossRef]
33. Meyvis, T.K.L.; De Smedt, S.C.; Demeester, J.; Hennink, W.E. Influence of the Degradation Mechanism of Hydrogels on Their Elastic and Swelling Properties during Degradation. *Macromolecules* **2000**, *33*, 4717–4725. [CrossRef]
34. Girard, V.; Chaussé, J.; Vermette, P. Bacterial cellulose: A comprehensive review. *J. Appl. Polyme Sci.* **2024**, *141*, e55163. [CrossRef]
35. Huang, Y.; Zhu, C.; Yang, J.; Nie, Y.; Chen, C.; Sun, D. Recent advances in bacterial cellulose. *Cellulose* **2013**, *21*, 1–30. [CrossRef]
36. Erdem, R.; Sancak, E. Investigation of Morphological and Mechanical Properties of PVA/Aloe barbadensis and PEO/Chitosan Blended Nanofibrous Structures Fabricated by Dual Electrospinning System. *J. Sci. Eng.* **2014**, *14*, 1–6. [CrossRef]
37. Paul, S. Electrospinning of bacterial cellulose fibres: Challenges for an innovative bioengineered fibre production technique. *Trends Text. Eng. Fash. Technol.* **2024**, *10*, 1146–1149. [CrossRef]
38. Ke, C.; Deng, F.; Chuang, C.; Lin, C. Antimicrobial actions and applications of chitosan. *Polymers* **2021**, *13*, 904. [CrossRef]
39. Xing, K.; Zhu, X.; Peng, X.; Qin, S. Chitosan antimicrobial and eliciting properties for pest control in agriculture: A review. *Agron. Sustain. Dev.* **2014**, *35*, 569–588. [CrossRef]
40. Gorgieva, N.; Trček, N. Bacterial cellulose: Production, modification and perspectives in biomedical applications. *Nanomaterials* **2019**, *9*, 1352. [CrossRef]
41. Lahiri, D.; Nag, M.; Dutta, B.; Dey, A.; Sarkar, T.; Pati, S.; Edinur, H.A.; Kari, Z.A.; Noor, N.H.M.; Ray, R.R. Bacterial cellulose: Production, characterization, and application as antimicrobial agent. *Int. J. Mol. Sci.* **2021**, *22*, 12984. [CrossRef]
42. Halliwell, B. Oxidative stress in cell culture: An under-appreciated problem? *FEBS Lett.* **2003**, *540*, 3–6. [CrossRef]
43. Saylam, E.; Akkaya, Y.; Ilhan, E.; Cesur, S.; Guler, E.; Sahin, A.; Cam, M.E.; Ekren, N.; Oktar, F.N.; Gunduz, O.; et al. Levodopa-Loaded 3D-Printed poly (Lactic) Acid/Chitosan neural tissue scaffold as a promising drug delivery system for the treatment of Parkinson's disease. *Appl. Sci.* **2021**, *11*, 10727. [CrossRef]
44. Harugade, A.; Sherje, A.P.; Pethe, A. Chitosan: A review on properties, biological activities and recent progress in biomedical applications. *React. Funct. Polym.* **2023**, *191*, 105634. [CrossRef]
45. Helenius, G.; Bäckdahl, H.; Bodin, A.; Nannmark, U.; Gatenholm, P.; Risberg, B. In vivo biocompatibility of bacterial cellulose. *J. Biomed. Mater. Res. A* **2005**, *76A*, 431–438. [CrossRef] [PubMed]
46. Utoiu, E.; Manoiu, V.S.; Oprita, E.I.; Craciunescu, O. Bacterial cellulose: A sustainable source for hydrogels and 3D-Printed scaffolds for tissue engineering. *Gels* **2024**, *10*, 387. [CrossRef]

polymers

MDPI

Article

Antifungal Activity of Chitosan/Poly(Ethylene Oxide) Blend Electrospun Polymeric Fiber Mat Doped with Metallic Silver Nanoparticles

Leire Murillo [1], Pedro J. Rivero [1,2,*], Xabier Sandúa [1,2], Gumer Pérez [3], José F. Palacio [4] and Rafael J. Rodríguez [1,2]

1 Engineering Department, Public University of Navarre (UPNA), Campus Arrosadía S/N, 31006 Pamplona, Spain; murillo.127083@e.unavarra.es (L.M.); xabier.sandua@unavarra.es (X.S.); rafael.rodriguez@unavarra.es (R.J.R.)
2 Institute for Advanced Materials and Mathematics (INAMAT2), Public University of Navarre (UPNA), Campus Arrosadía S/N, 31006 Pamplona, Spain
3 Genetics, Genomics and Microbiology Research Group, Institute for Multidisciplinary Research in Applied Biology (IMAB), Public University of Navarre (UPNA), 31006 Pamplona, Spain; gumer.perez@unavarra.es
4 Centre of Advanced Surface Engineering, AIN, 31191 Cordovilla, Spain; jfpalacio@ain.es
* Correspondence: pedrojose.rivero@unavarra.es

Abstract: In this work, the implementation of advanced functional coatings based on the combination of two compatible nanofabrication techniques such as electrospinning and dip-coating technology have been successfully obtained for the design of antifungal surfaces. In a first step, uniform and beadless electrospun nanofibers of both polyethylene oxide (PEO) and polyethylene (PEO)/chitosan (CS) blend samples have been obtained. In a second step, the dip-coating process has been gradually performed in order to ensure an adequate distribution of silver nanoparticles (AgNPs) within the electrospun polymeric matrix (PEO/CS/AgNPs) by using a chemical reduction synthetic process, denoted as in situ synthesis (ISS). Scanning electron microscopy (SEM) has been used to evaluate the surface morphology of the samples, showing an evolution in average fiber diameter from 157 ± 43 nm (PEO), 124 ± 36 nm (PEO/CS) and 330 ± 106 nm (PEO/CS/AgNPs). Atomic force microscopy (AFM) has been used to evaluate the roughness profile of the samples, indicating that the ISS process induced a smooth roughness surface because a change in the average roughness Ra from 84.5 nm (PEO/CS) up to 38.9 nm (PEO/CS/AgNPs) was observed. The presence of AgNPs within the electrospun fiber mat has been corroborated by UV-Vis spectroscopy thanks to their characteristic optical properties (orange film coloration) associated to the Localized Surface Plasmon Resonance (LSPR) phenomenon by showing an intense absorption band in the visible region at 436 nm. Energy dispersive X-ray (EDX) profile also indicates the existence of a peak located at 3 keV associated to silver. In addition, after doping the electrospun nanofibers with AgNPs, an important change in the wettability with an intrinsic hydrophobic behavior was observed by showing an evolution in the water contact angle value from 23.4° ± 1.3 (PEO/CS) up to 97.7° ± 5.3 (PEO/CS/AgNPs). The evaluation of the antifungal activity of the nanofibrous mats against *Pleurotus ostreatus* clearly indicates that the presence of AgNPs in the outer surface of the nanofibers produced an important enhancement in the inhibition zone during mycelium growth as well as a better antifungal efficacy after a longer exposure time. Finally, these fabricated electrospun nanofibrous membranes can offer a wide range of potential uses in fields as diverse as biomedicine (antimicrobial against human or plant pathogen fungi) or even in the design of innovative packaging materials for food preservation.

Keywords: electrospinning; dip-coating; blend electrospun mat; chitosan; polyethylene oxide; LSPR; AgNPs; antifungal

1. Introduction

Among all the fabrication techniques, the electrospinning process is being shown great interest in the scientific community because it presents several important advantages such as uniform pore diameter, ease of fabrication, low cost, great flexibility in surface functionality and high degree of reproducibility [1–4]. The fundamental of this technique is based on the application of a high voltage between a syringe filled with a polymeric solution with a desired viscosity and a collector mounted at a fixed distance from the needle/syringe setup [5–8]. The diameter of the fibers can be reduced from micro- to nanometer scale, enabling a large increase in the surface area-to-volume ratio and, as a result, several properties can be directly improved for many industrial (photocatalysis, anticorrosion and anti-icing) [9–12] or biomedical applications (tissue engineering, drug delivery and wound healing dressing) [13–16]. According to this last field, the use of biopolymers such as chitosan has received increasing interest in the medical sector due to its natural origin (abundant in nature) combined with exceptional properties in terms of biodegradability, biocompatibility and being a good inhibitor against the growth of a wide variety of yeasts, fungi and bacteria, diminishing infection of pathogens [17–19]. However, the resultant electrospinability of chitosan is very difficult because of the cationic nature of the solution, its rigid chemical structure and interaction of inter- and intra-chain hydrogen bonding [20]. One of the most successful strategies for enhancing the electrospinability of the chitosan is blending with a second natural polymer (i.e., hyaluronic acid, gelatin, zein and silk fibroin) [21–24] or synthetic polymer (i.e., polyvinyl alcohol- PVA, polyethylene oxide-PEO and polyacrylamide-PAM) [25–29] in order to form a polymeric blend with better physico-chemical and mechanical properties in comparison with pure chitosan [30]. In this sense, the use of PEO with its intrinsic characteristics (hydrophilic, biocompatible, non-cytotoxic and biodegradable) is a good candidate in the biomedical field [31], enabling the formation of uniform fibers with additional functionalities due to the specific interactions between amino groups of chitosan and ether groups of PEO [27,32]. More specifically, the formation of hydrogen bonds renders to chitosan a better electrospinability with an increase in the solution chain entanglements [33].

The incorporation of AgNPs into the electrospun fiber mats is a very interesting approach for the design of inorganic–organic hybrid materials [34] that can provide a synergistic antibacterial effect achieved by combining both AgNPs and chitosan [35]. The antimicrobial properties of chitosan are enhanced due to the presence of polymer chain's positive charges and free amino groups [36], and it has been demonstrated that protonated amine groups of the chemical structure of the chitosan can bind negatively charged bacteria, breaking their reproduction [37]. These interactions lead to modified membrane permeability, interruption of the microbial membrane and subsequent leakage of proteins and other intracellular constituents that cause cell death [38–40]. In addition, thanks to the cationic-polyelectrolyte character and gel-forming ability related to the chitosan, this effect can be used as an effective binder of metal ions [31], the protonated amine and hydroxide sites being the main reactive sites for interactions with metal ions, whereas the use of chemical reduction methods is the most common synthetic route for the preparation of AgNPs into thin-films [41]. It is well known that AgNPs possess an inherent antibacterial activity against a broad spectrum of pathogenic microorganisms [42–44] on the benefit of their intrinsic optical colorimetric properties [45] associated to the Localized Surface Plasmon Resonance (LSPR) phenomenon, making possible the synthesis of nanoparticles with a desired shape, size and morphology [46]. Compared to bulk materials, the size, surface area and morphology of nanoparticles endow them with unique physical and chemical properties [47]. The experimental results also suggest that AgNPs can be used as effective growth inhibitors in various microorganisms, making them applicable to diverse medical devices and antimicrobial control systems [48,49]. There can be found in the bibliography more than one mechanism related to the antimicrobial activity of silver at the nanoscale [50–53].

In terms of antibacterial efficacy, a big number of publications are focused on the entrapment of the AgNPs into blended electrospun PEO/CS nanofibrous membranes in an only one-step fabrication process, showing promising results against Gram-positive bacteria (*Staphylococcus aureus* and *Propionibacterium acnes*) and Gram-negative bacteria (*Escherichia coli* and *Pseudonomas aeruginosa*) [54–59]. However, very few research publications have been found to be active against both bacteria and fungi by using this methodology of embedding AgNPs into blended electrospun fibers with the presence of chitosan [60,61]. The antimicrobial properties of the blended electrospun fiber mats with embedded AgNPs have been tested against a common skin pathogenic fungus such as *Candida albicans* by using the agar diffusion method.

The implementation of hybrid inorganic–organic electrospun fiber mat against other fungi is of great interest. One of them is *Pleurotus ostreatus*, an edible filamentous basidiomycete with chitin and lignin degrader with important biotechnological properties and resources [62]. Their dikaryotic mycelia are the result of the fusion of two compatible monokaryons, which are formed in basidia after karyogamy (final step in the process of fusing two nuclei to form a unique diploid nucleus) and immediately undergo meiosis [63]. Previous works have shown that chitosan has great potential as an antifungal agent against ascomycetes fungus such as *Candida albicans*, *Fusarium solani* or *Aspergillus niger*, among others [64]. This compound can be obtained from *Pleurotus* sp. Some chitosan resistance strains of *P. ostreatus* have shown an overexpression on chitin deacetylases enzymes, which catalyze the deacetylation of chitin, result in the formation of chitosan [65,66]. An interesting approach for increasing the antimicrobial activity of the chitosan against a wide variety of fungi strains can be derived for a higher amount and more homogeneous distribution of the synthesized AgNPs on the outer surface of the blended polymeric electrospun matrix. One method that can provide this possibility is the in situ *synthesis* (denoted as ISS) process [67] where a polymeric matrix is firstly doped with silver ions (Ag^+) and a further reduction step is performed in order to obtain AgNPs perfectly distributed along the thin-film surface [68]. The implementation nanofibers doped with AgNPs can be used in the design of active packaging to increase the shelf life of food or even to exert an inhibitory effect on fungal growth [69–71].

The novelty of this research work is the use of two different nanofabrication techniques such as electrospinning and dip-coating technology, which are totally compatible, enabling the development of highly effective antifungal surfaces, thanks to the combination of chitosan and AgNPs, respectively. For the fabrication of the functionalized thin films, a dual step is performed. Firstly, a PEO/CS blend electrospun nanofiber mat has been obtained and, secondly, the implementation of a chemical synthetic route (ISS process) is performed onto the electrospun mat with the aim of obtaining AgNPs, which are totally distributed in the outer surface. The resultant surface morphology (fiber diameter and roughness) of the electrospun nanofibers have been determined by SEM and AFM, whereas both UV-Vis spectra and EDX profile have been used for corroborating the presence of AgNPs within the polymeric fiber mat. Finally, notorious differences have been observed in terms of wettability and antifungal efficacy between blend nanofibers with and without AgNPs. To the best of our knowledge, this is the first time that the combination of both electrospinning and dip-coating techniques is presented in the bibliography for the design of functionalized surfaces against the fungus strain of *Pleurotus ostreatus*.

2. Materials and Methods

2.1. Materials

Chitosan (CS, medium molecular weight, ≥75% deacetylation degree), polyethylene oxide (PEO, 400 kDa), silver nitrate ($AgNO_3$, ≥99%), borane dimethylamine complex (DMAB, 97%), glacial acetic acid (≥99%) and deionized water (DW) were purchased from Sigma-Aldrich (St. Louis, MI, USA) and used without any further purification.

2.2. Thin-Film Fabrication Process

The fabrication of the functionalized thin films was divided into two main steps. Firstly, electrospinning was used for the preparation of PEO/CS blend electrospun nanofibers. Secondly, the dip-coating technology was performed on the previously electrospun fiber mat in order to obtain AgNPs by a chemical reduction synthetic process (ISS).

2.2.1. Preparation of PEO/CS Blend Structure

The resultant PEO/CS blend solution was fabricated by mixing CS-acetic acid and PEO aqueous solution in a 1:1 (*v*/*v*) ratio. Initially, 5 wt% PEO solution dissolved in ultrapure water was prepared and, then, 4 wt% CS solution was prepared by using 90% acetic acid as solvent. Both solutions were left under magnetic agitation in a sealed container at room temperature during 24 h, being mixed at a 1:1 (*v*/*v*) ratio. Finally, the blend solution was left at room condition under magnetic agitation during 24 h, respectively, previous to perform the electrospinning process.

2.2.2. Electrospinning of PEO/CS Blend

Electrospinning was performed using the "Electrospinning Professional Machine" from Doxa Microfluidics (Doxa Microfluidics, Malaga, Spain). The PEO/CS blend solution was introduced into a 5 mL syringe (BD Plastic, Franklin Lakes, NJ, USA) and pushed through a Teflon tube to a metal needle (inner diameter of 0.6 mm). The optimized experimental parameters were a flow rate of 1.3 mL/h, a distance of 20 cm from tip to collector and a specific voltage of 26 kV between the metal needle and the collector (V+ = 23 kV, V− = 3 kV), respectively. All the experiments were performed at room condition, showing a temperature of 23.7 °C and 32.45% of relative humidity (RH). Finally, all the coatings were exposed for a period of time of 90 min in the same experimental conditions.

2.2.3. In Situ Synthesis of Silver Nanoparticles (AgNPs)

In order to obtain the AgNPs, $AgNO_3$ was used as the loading metallic agent (Ag^+) and DMAB wasused as the reducing agent, which are necessary to perform the ISS process by using the dip-coating technique. The whole process is divided into the following steps. Firstly, the previously fabricated membrane was immersed in 0.01 M $AgNO_3$ solution for 10 min, obtaining a membrane loaded with silver ions (Ag^+). Then, this membrane was immersed in 0.1 M DMAB solution for 10 min, enabling the chemical reduction from silver ions (Ag^+) to metallic silver (Ag^0) nanoparticles. Between each loading and reduction step, the samples were immersed in ultrapure water (denoted as rinsing step). This same process was repeated to obtain a greater amount of AgNPs distributed in the polymeric electrospun membrane [66]. The dip-coating process was carried out by using ND-R Rotatory Dip-Coater (Nadetech Innovations, Navarra, Spain) [67]. Finally, a schematic representation of the whole procedure for the fabrication of the thin films is shown in Figure 1.

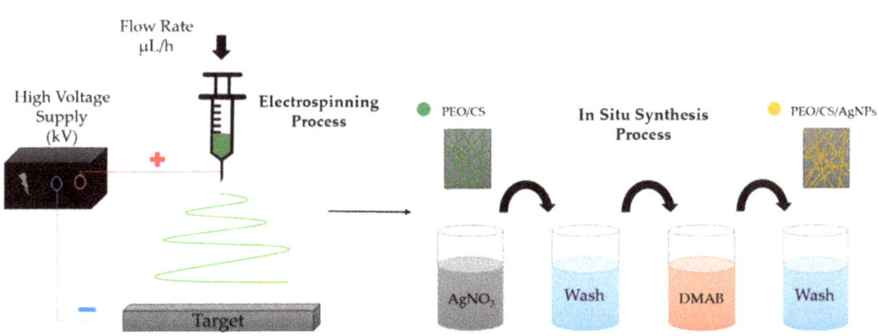

Figure 1. Schematic representation of the thin-film fabrication method used, separated into two steps: firstly, the production of PEO/CS electrospun nanofibers and, secondly, ISS process of AgNPs by using dip-coating technique onto the electrospun nanofiber mat.

2.3. Characterization Techniques

The morphology of the nanofibers was observed using a field-emission scanning electron microscope (FESEM, Hitachi S4800, Tokyo, Japan) and an atomic force microscope (AFM, Veeco Innova AFM, Veeco Instruments, Plainview, NY, USA). UV-Vis analysis was performed to confirm the formation of silver nanoparticles in the electrospun matrix. For this purpose, a Jasco V-730 (Agilent, Santa Clara, CA, USA) spectrophotometer was used. The analysis was carried out in a wavelength range of 350–900 nm. In addition, a thermogravimetric analysis (TGA) is performed by the TGA/DSC 3+ (METTLER-TOLEDO) in order to determine the amount of AgNPs in the electrospun samples. Three different process conditions can be distinguished during the thermal process. Firstly, it was measured in the temperature range from 35 °C to 775 °C with a nitrogen flow rate of 50 mL/min and a heating rate of 10 °C/min, respectively. Secondly, in the range from 775 °C to 800 °C, the resultant heating rate was decreased up to 5 °C/min, maintaining the same nitrogen flow rate. Thirdly, from 800 °C to 950 °C, an air flow rate of 50 mL/min was employed in order to eliminate the pyrolysis residues of the blend structure, maintaining the same heating rate. The chemical bonding and the presence of specific functional groups were analyzed by Fourier transform infrared spectroscopy using ATR mode (FTIR-ATR). The FTIR spectra were obtained by a Perkin Elmer Frontier spectrophotometer (Waltham, MA, USA) in the range of 650–4000 cm^{-1}. Finally, the wettability of the surface was studied by measuring the contact angle with an optical tensiometer (CAM 100 KSV Instruments, Burlington, VT, USA). The water contact angle values (WCA) were calculated as an average of three measurements performed at different locations in each sample.

2.4. Antifungal Test against "Pleurotus Ostreatus"

The *Pleurotus ostreatus dikaryotic* strain dkN001 (Spanish Type Culture Collection accession CECT20600) was used in this work. This strain was cultured on square Petri dishes 120 × 120 mm containing Malt Extract Solid Medium (MESM: malt extract, 20 g/L; and bacteriologic agar, 15 g/L). Three glass slides were coated with PEO, PEO plus chitosan polymer (PEO/CS), and AgNPs plus PEO/CS sample (PEO/CS/AgNPs), whereas an uncoated glass slide was used as a control reference substrate, respectively. A piece of fungus mycelium was placed in the center of the Petri dishes and the three coated electrospun samples and the uncoated slide were placed at the end of the four axes of the Petri dish (see scheme of the Figure 2). Fungus was incubated in darkness conditions at 24 °C and the growth (apical form) was performed until the mycelium colonized the square Petri dish.

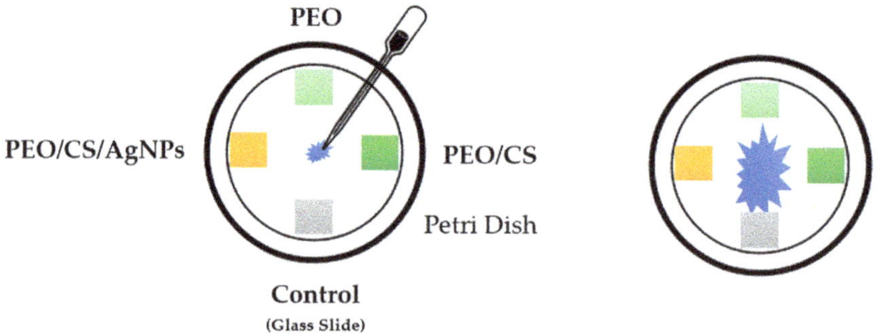

Figure 2. A schematic representation of the antifungal essay against the strain of *Pleurotus ostreatus* performed in this work.

The growth-inhibiting rate was calculated by the formula:

$$I = \frac{C - T}{C} \times 100\%$$

where "*I*" is the growth-inhibiting rate (%), while "*C*" and "*T*" represent the average colonization slides of the reference control (no electrospun fiber mat deposited) as well as the experimental electrospun coated glass slides (PEO, PEO/CS and PEO/CS/AgNPs samples), respectively. Data obtained were analyzed with IBM SPSS statistics 27.0 software (IBM Corp. Released 2020. IBM SPSS Statistics, Version 27.0. Armonk, New York: IBM Corp) using the Scheffe post hoc statistic to compare all possible simple and complex pairs of means between the different electrospun coated glass slides and the reference control slide. Finally, all the experiments have been evaluated in triplicate.

3. Results and Discussion

The FTIR spectra of PEO, PEO-CS and PEO-CS+AgNPs films are presented in Figure 3. Additionally, the characteristic peaks of the spectra are marked. Those in black correspond to functional groups of PEO and those marked in red correspond to functional groups of chitosan, respectively. It should be noted that the red color spectrum contains both compounds (PEO and CS) so that, in addition to the peaks marked in red, there are also the peaks marked in black. The same is true for the blue spectrum. PEO shows absorption peaks at 2885 cm^{-1} (-CH_2 stretching vibrations), at 1467 cm^{-1}, 1360 cm^{-1}, 1342 cm^{-1}, 1280 cm^{-1}, 1240 cm^{-1} and 842 cm^{-1} (all corresponding to CH bending vibrations) and, finally, at 1126 cm^{-1}, 1098 cm^{-1}, 1061 cm^{-1} and 961 cm^{-1} (C-O-C stretching vibrations) [72]. The characteristic absorption peaks of chitosan appear in the 3600–3200 cm^{-1} region, where a slight valley is observed (O-H and N-H stretching), at 1655 cm^{-1} (-C=O stretching) and at 1567 cm^{-1} (amide -NH_2 bond) [73]. These last two peaks are conditioned by the degree of deacetylation of the chitosan, since the higher the degree, the greater the amount of glucosamine units and, therefore, the greater the presence of amino groups. Finally, no significant differences are observed between the FTIR spectra of the PEO/CS and the PEO/CS doped with AgNPs films.

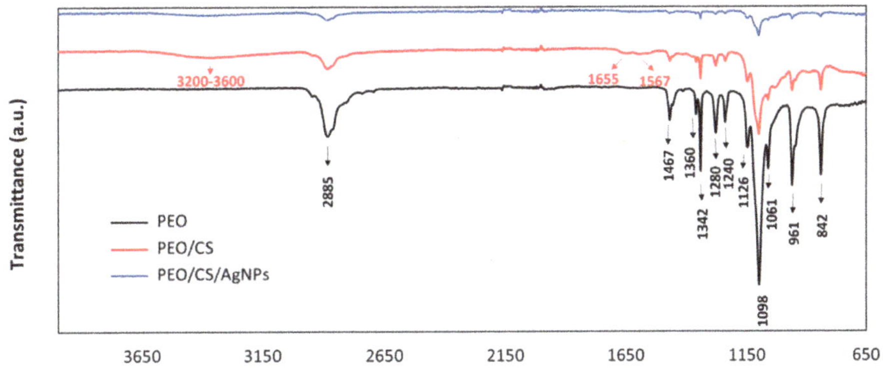

Figure 3. FTIR spectra of the electrospun samples: PEO (black plot), PEO/CS (red plot) and PEO/CS/AgNPs (blue plot), respectively.

The surface morphology of the electrospun nanofibrous fiber mats is presented in Figure 4 (scale bar 5 µm) by using SEM micrographs. As has been previously mentioned, neat chitosan shows a very poor electrospinability and only beads or droplets are formed. It has been demonstrated that the addition of PEO can greatly facilitate the electrospinning process of chitosan in the blend at room temperature with the formation of beadless nanofibers [26]. In addition, the fiber diameter in the blended polymeric mat tends to decrease slightly when the CS amount is increased in the PEO/CS ratio [30]. In this work, uniform and defect-free nanofibers with diameters of 100–300 nm can be observed for PEO sample (Figure 4a) and PEO/CS sample (Figure 4b), showing a higher diameter for

PEO/CS/AgNPs sample (Figure 4c). In order to clarify this result of the fiber diameter, Figure 5 presents SEM images (scale bar 2 μm) for all the samples of this study, showing different measurements of the nanofibers. The experimental results corroborate that an average fiber diameter of 157 ± 43 nm has been observed for PEO sample, whereas an average of 124 ± 36 nm has been obtained for PEO/CS sample. In the PEO/CS/AgNPs sample, the resultant fiber wall has become thicker, clearly indicating the integration of AgNPs within the nanofibers and, as a result, an increase in the fiber size has been clearly obtained, which is in concordance with the literature [56]. In this sense, the average in fiber diameter of 330 ± 106 nm has been observed for PEO/CS/AgNPs sample, showing a notable increase in comparison with the previous samples. According to this experimental result, Figure 6 shows the average corresponding to the diameter of the nanofibers as well as their corresponding distribution for each sample of this study [74]. The main reason for this phenomenon is associated to the hydrophilic behavior of both PEO and CS polymers, which show a high affinity to the aqueous media of the ISS process. As a result, a swelling behavior of the nanofibers can be appreciated with the corresponding increase in the fiber diameter. In addition, another relevant aspect is that the characteristic porosity associated to the electrospinning deposition process has been decreased because of this increase in the resultant fiber diameter after ISS process of the AgNPs. This result has been previously corroborated by performing the layer-by-layer deposition technique after the electrospinning process [12,75].

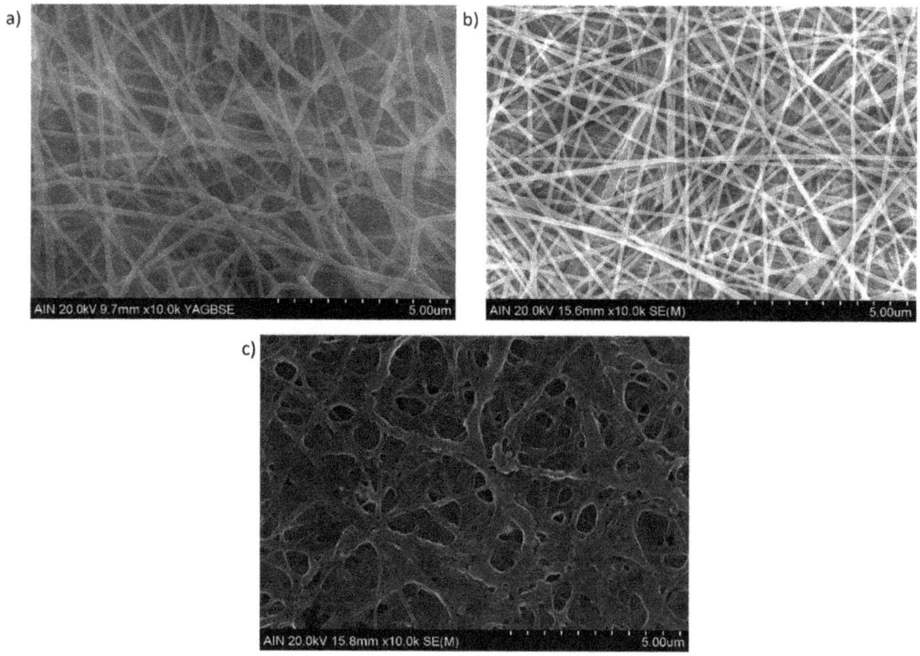

Figure 4. SEM images (scale bar 5 μm) of the blended electrospun fiber mats for PEO sample (**a**); PEO/CS sample (**b**) and PEO/CS/AgNPs sample (**c**), respectively.

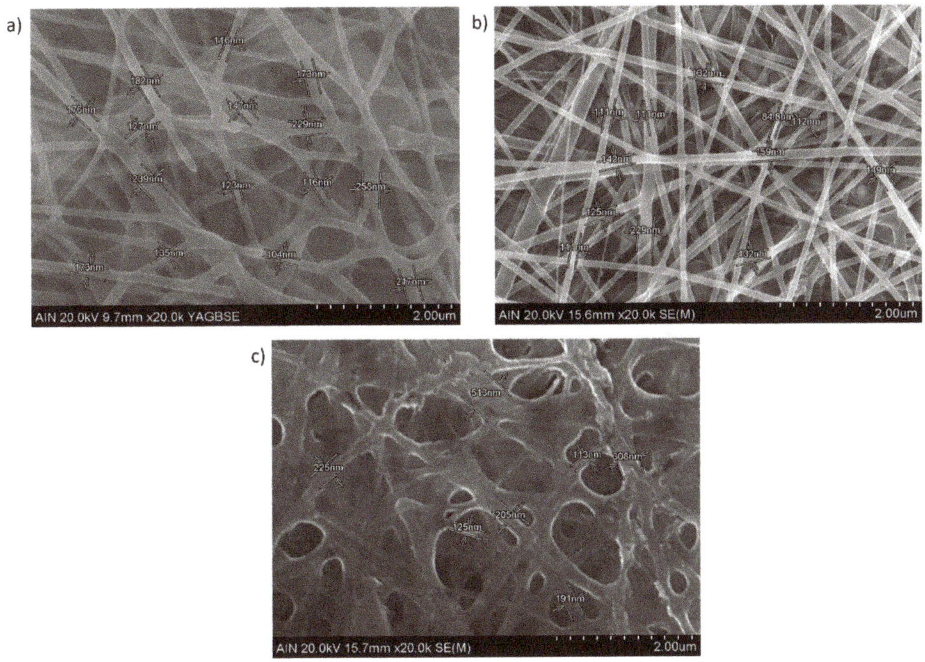

Figure 5. SEM images (scale bar 2 μm) of the blended electrospun fiber mats for PEO sample (**a**); PEO/CS sample (**b**) and PEO/CS/AgNPs sample (**c**) with the corresponding measurement of the fiber diameter.

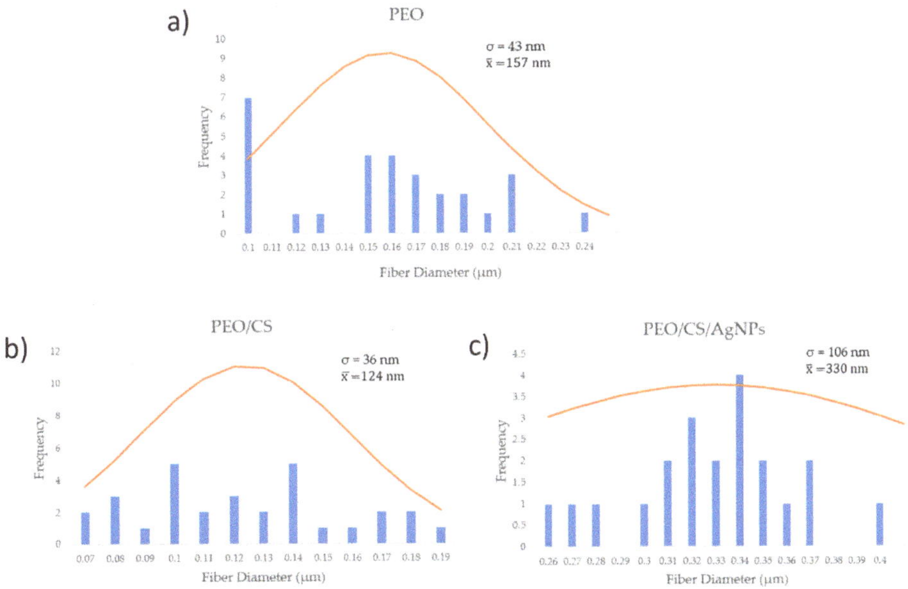

Figure 6. Nanofiber statistical diameter distribution of the electrospun mats for PEO sample (**a**), PEO/CS sample (**b**) and PEO/CS/AgNPs sample (**c**), respectively.

In order to corroborate the presence of AgNPs in the outer surface of the blended electrospun fiber mat, energy-dispersive X-Ray (EDX) is presented in Figure 7, where the location of the peak related to Ag in 3 keV is clearly observed (see Figure 7b). However, the presence of other chemical elements (i.e., Si, Mg, Ca and Na) have also been observed in the EDX profile, which are inherent to the reference glass slides used for the fabrication of the electrospun nanofibrous coatings. The resultant SEM image also shows a random distribution of AgNPs in the complete outer surface with a certain degree of aggregation due to the formation of clusters combined with smaller-sized AgNPs (bright white spots) along the whole electrospun nanofibers.

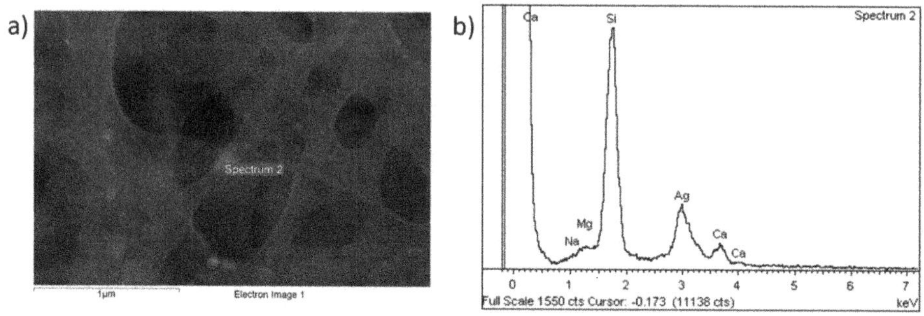

Figure 7. (**a**) SEM image of the blended electrospun fiber mat corresponding to the PEO/CS/AgNPs sample at scale bar of 1 μm; (**b**) EDX of the outer surface of the sample with the presence of Ag (bright spots).

In order to determine with a high accuracy the resultant amount of AgNPs within PEO/CS/AgNPs sample, TGA analysis was carried out. In Figure 8, the percentage of weight loss of the sample in a temperature range from 35 °C up to 950 °C, respectively, is shown. The first weight loss corresponds to residual solvent (i.e., acetic acid) at 35 °C, which is still present in the sample. Then, the second weight loss is due to the degradation of the CS/PEO blend, which starts disintegrating at 210 °C, showing two slopes of disintegration, and typical TGA graphic shape when both CS and PEO polymers are blended [76,77]. After the degradation of the blend structure in a nitrogen atmosphere, a black appearance residue appeared due to the pyrolysis of the CS/PEO compound. This residue of pyrolysis prevents us from obtaining the amount of silver that remains in the specimen. Thus, the heating rate velocity is reduced and air is introduced in the chamber. At 800 °C, air is introduced, and this fact leads to a total disintegration of the pyrolysis residues of the blend structure to CO_2, respectively. In this TGA figure, weight loss can be clearly observed as soon as the air is introduced in the chamber. Nevertheless, the remaining silver is oxidized, obtaining Ag_2O product. Finally, the amount of silver is calculated from the remaining weight of the sample, taking into account the amount of Ag presented in the Ag_2O sample. According to Equation (1), the percentage of silver presented in the Ag_2O is obtained. In this attempt, the initial and final weight of the specimen in the TGA are 8.197 mg and 0.901 mg, respectively. According to this, Equation (2) is used to calculate the overall amount of silver in the electrospun sample from the remaining Ag_2O product, obtaining a final value of 10.23% in the PEO/CS/AgNPs sample.

$$\mathrm{Ag}\% = \frac{\mathrm{MW_{Ag}}}{\mathrm{MW_O} + \mathrm{MW_{Ag}}} = \frac{2\cdot 107.868}{15.999 + 2\cdot 107.868} = 93.1\% \text{ of Ag in the } \mathrm{Ag_2O} \text{ molecule} \quad (1)$$

$$0.931\cdot\frac{\mathrm{Ag_2O\ Weight}}{\mathrm{Initial\ Weight}} = 0.931\cdot\frac{0.901}{8.197} = 10.23\% \text{ of Ag in electrospun sample} \quad (2)$$

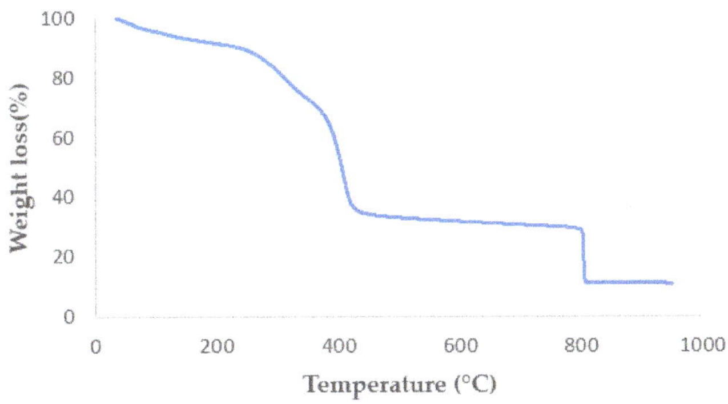

Figure 8. Thermogravimetric analytic curve for PEO/CS/AgNPs electrospun sample.

AFM imaging has been performed in order to examine the morphology and to determine the surface roughness of the blended electrospun nanofibers. In Figure 9, two-dimensional (2D) and three-dimensional (3D) AFM images for PEO/CS sample (Figure 9a) and for PEO/CS/AgNPs sample (Figure 9b) can be observed. In this last sample, light spots are also observed corresponding to the AgNPs distributed over the blended electrospun fibers. In addition, the surface roughness was investigated with AFM by measuring the line profile of both samples, the average roughness R_a being 84.5 nm for PEO/CS sample and 38.9 nm for PEO/CS/AgNPs sample, respectively. This result corroborates that ISS process over the blended electrospun nanofiber mat clearly induces a smooth roughness surface, which has been corroborated by the AFM profile images.

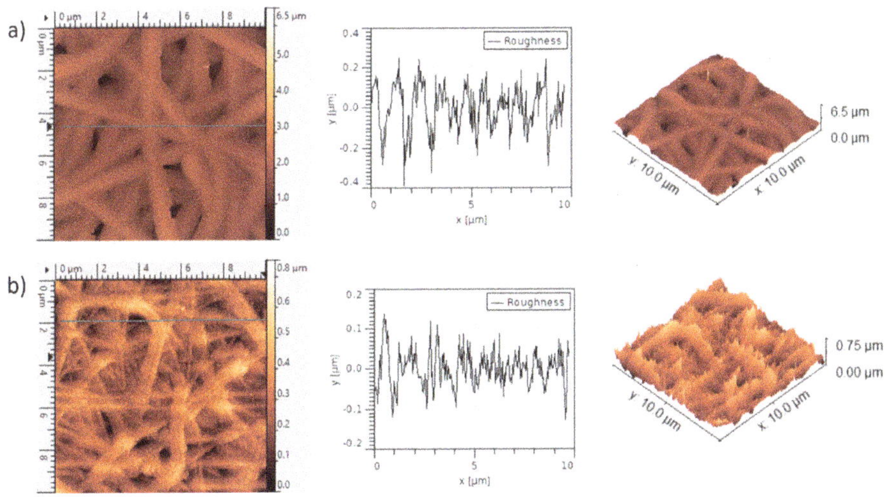

Figure 9. AFM images in 2D and 3D with their corresponding roughness profile for PEO/CS sample (**a**) and for PEO/CS/AgNPs sample (**b**), respectively.

Another way to corroborate the presence of the metallic AgNPs into the electrospun nanofiber mats is by using UV-Vis spectroscopy, and the UV-Vis spectra of all the samples of this study are shown in Figure 9. Significant differences can be extrapolated from these spectra data because, in one of them, an intense absorption band can be clearly observed in the visible region. First of all, an aspect indicative is that a dramatic color change from transparent (PEO or PEO/CS) to orange (PEO/CS/AgNPs) has been clearly observed after the two loading/reduction cycles of the ISS process (see inset of the Figure 10). This colored appearance is the result of the presence of a well-centered absorption band in the visible region inherent to the LSPR phenomenon, indicating the complete synthesis and incorporation of the AgNPs within the polymeric electrospun fiber mat [45,46].

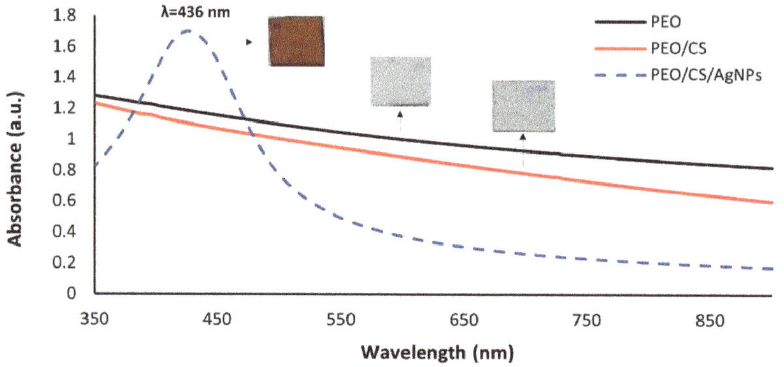

Figure 10. UV-Vis spectra of PEO sample (black plot), PEO/CS sample (red plot) and PEO/CS/AgNPs (dashed blue plot) with the presence of the LSPR absorption band centered at 436 nm inherent to the presence of AgNPs within the electrospun blended polymeric membrane.

Previous works have demonstrated that the amount of AgNPs is higher when the number of loading/reduction cycles is gradually increased, showing at the same time an increase in intensity of the LSPR absorption band, which is directly proportional to the amount of AgNPs trapped into the polymeric overlay [68,78]. The location of the LSPR absorption band is inherent to multiple factors such as shape, size, morphology, aggregation state, distribution and the resultant interparticle interactions of the nanoparticles, among others [79,80]. A previous work has demonstrated that LSPR absorption bands located at 600–700 nm are indicative of the presence of AgNPs with variable shape (i.e., rod, hexagonal and triangle), whereas LSPR absorption bands located at 400–500 nm indicate the presence of AgNPs with mostly spherical shape [81]. More specifically, the UV-Vis spectrum related to PEO/CS/AgNPs sample shows a maximum LSPR absorption band located at 436 nm, which is indicative of the synthesis of AgNPs with spherical shape with a variable size (polydispersity) and even associated to silver aggregates in the electrospun fiber mat, which has also been corroborated in the SEM image for EDX analysis [68].

In Figure 11, the static water contact angle evolution of the electrospun samples of this work is presented. First, it is important to remark that, as a function of the resultant chemical composition as well as surface energy, the materials can vary their corresponding wettability degree [82]. The wettability of a material surface, i.e., the hydrophobicity or hydrophilicity, is often used as a primary descriptor of surface biocompatibility [83]. In this sense, high surface energy materials are relatively easy to wet out with a high degree of affinity (hydrophilic behavior), whereas low surface energy materials present a very little attraction to any molecule (hydrophobic behavior), without showing any chemical affinity. In the top of each column of Figure 11, the picture of a water droplet deposited onto the outer surface of each sample is shown, and a great difference in the resultant shape of this droplet can be clearly observed, which is inherent to its wettability. In this sense, a more hydrophilic surface is observed when the resultant water droplet is more dispersed onto the electrospun fiber mat, which corresponds to a smaller water contact angle value. In contrast, a more hydrophobic surface is obtained when the water droplet presents a more intrinsic spherical shape, which corresponds to a bigger water contact angle value.

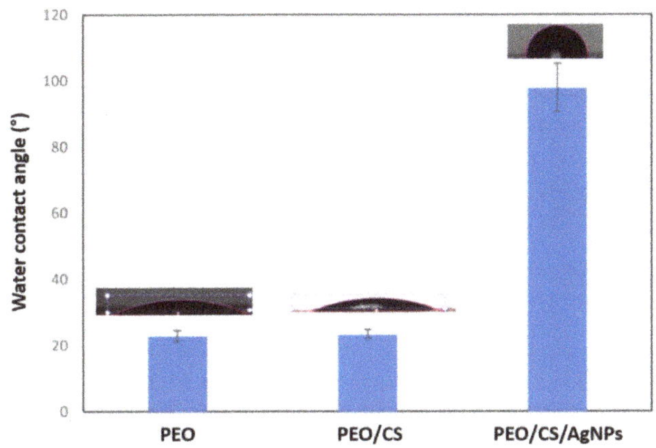

Figure 11. Water contact angle (WCA) values for all the samples of this study: PEO, PEO/CS and PEO/CS/AgNPs, showing an important difference in the resultant wettability from hydrophilic (PEO, PEO/CS) up to hydrophobic behavior (PEO/CS/AgNPs).

An initial interesting experimental result is that differences in the resultant wettability have not been observed for the PEO (22.9° ± 1.6) and PEO/CS (23.4° ± 1.3) samples, showing in both cases a hydrophilic nature, this being in concordance with the reported bibliography [84]. The main reason for this hydrophilic behavior can be associated to the presence of free polar groups in chitosan combined with the ether group of PEO, which show a great affinity to absorbing water molecules. The distribution of these functional groups on the outer surface of the electrospun nanofibers can reduce the interfacial tension and increase the affinity to water droplet, as the presence of this blend structure (PEO/CS) on polymeric matrices has been reported, even with hydrophobic nature [85]. However, performing the ISS of AgNPs onto the blend electrospun mat produced an important change in the resultant wettability by showing an important increase in the water contact angle. More specifically, the presence of the AgNPs confers a hydrophobic character to the PEO/CS blend electrospun mat, showing a water contact angle value of 97.7° ± 5.3, respectively. This experimental result of the PEO/CS/AgNPs can be associated to a dual effect. On one hand, previous research works have demonstrated that AgNPs have shown a hydrophobic behavior, indicating that surface properties can be tuned with AgNPs [86]. On the other hand, the increase in the resultant fiber diameter after doping with AgNPs by the ISS process (corroborated by SEM images) can contribute that the outer surface produces lower accessibility of the water droplets within the fiber mat with the corresponding increase in the water contact angle [87]. This increase in hydrophobicity is of interest for the design of a desired functionality in terms of anti-adherence activity [88] because it makes it difficult for microorganisms to colonize the surface [89].

The analysis of the antifungal effect of the different samples on the growth of dkN001 strain of *P. ostreatus* was performed during a period of time of 13 days until fungus had completely colonized the Petri dishes (see Figure 12). First of all, the experimental results clearly indicate that, until the first week, the fungus showed an apical growth in all directions without touching the coated glass slides. However, after 10 days of exposure, apical hypha (filamentous structure of mycelium) established contact with all the coated glass slides, being more notorious after 13 days of exposure.

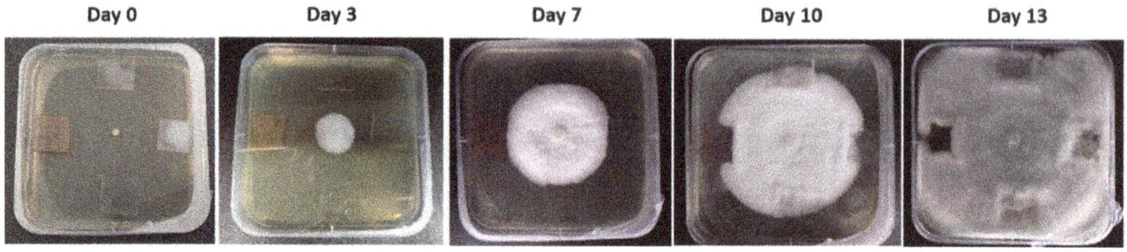

Figure 12. Evolution time related to the characteristic apical growth of the strain *Pleurotus ostreatus* from day 0 up to day 13 until a complete colonization of the Petri dish.

In order to appreciate significant differences between all the samples of the study, in Figure 13, the evolution is presented in more detail corresponding to days 10 and 13, respectively. It can be observed that the mycelium has shown growth in a normal form continuing without perturbations in the axes where PEO sample and reference control glass slide (no electrospun fiber mat deposited) have been placed (day 10). However, an interesting result with important differences is that PEO/CS and PEO/CS/AgNPs samples showed an inhibition zone during mycelium growth in this period of time of 10 days (see Figure 13a). Once the whole Petri dishes were completely colonized after 13 days, only PEO/CS/AgNPs electrospun sample showed antifungal activity on the dkN001 strain of *P. ostreatus* (see Figure 13b,c), observing a mean decrease in the growth rate of 39.7% compared to reference control.

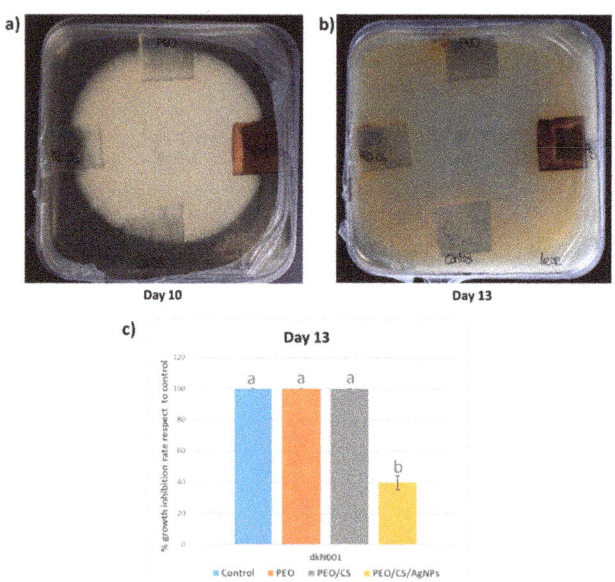

Figure 13. (**a**) Inhibition zone related to the PEO/CS sample (left axis) and PEO/CS/AgNPs sample (right axis) in comparison with PEO sample (superior axis) and control reference glass slide without coating (inferior axis) where the apical growth of the fungus strain after 10 days of exposure is clearly observed; (**b**) antifungal activity of the PEO/CS/AgNPs electrospun sample in comparison with PEO/CS sample, which is totally colonized after 13 days of exposure; (**c**) percentage inhibition of different electrospun samples in comparison to control after 13 days of exposure. Bars show the data average (n = 3) and SD (standard deviation). Lower case letters (a,b) indicate significant differences at p-value < 0.05 according to Scheffe's test (see Materials and Methods).

The antifungal effect of only chitosan polymer is well known [64], and it has been described that fungus such as *Cryptococcus* or *Pleurotus* can express chitin deacetylase genes during vegetative growth to provide integrity and proper rigidity to the cell wall and, in this way, cope with the effect of self-chitosan production [65,66]. In this work, the antifungal potency was clearly increased after doping the blended electrospun nanofibers with AgNPs in comparison with the other nanofiber formulation (PEO/CS) as per the zone of inhibition, which can be associated to their intrinsic hydrophobic nature and its antimicrobial behavior. According to the literature, AgNPs have emerged as effective antimicrobial agents by showing growth inhibition of different pathogen fungi, such as Aspergillus or *Candida*, among others. The antifungal mechanism related to the AgNPs can be associated to the production of reactive oxygen species (ROS) and free radicals, which can cause protein denaturation, nucleic acid and proton pump damage, lipid peroxidation, and cell wall damage. All these sequential steps enable an alteration in cell membrane permeability and, as a result, cell death is produced [90,91]. Thereby, it is also established that the stabilizing agents of AgNPs as well as the resultant surface charge can play an important role in the shaping of their fungicidal properties [92]. Finally, it is the first time that antifungal activity against *P. Ostreatus* has been demonstrated thanks to the combination of both AgNPs and chitosan by using this dual fabrication process of electrospinning and dip-coating techniques.

4. Conclusions

In this work, a novel strategy is presented based on a dual-step fabrication process for the design of functional surfaces in terms of antifungal activity. In a first step, the electrospinning process was selected for the fabrication of PEO/CS blend structure, showing uniform and defect-free nanofibers with an average in fiber diameter of 124 $\pm$ 36 nm. In a second step, the dip-coating technology was used for the ISS of AgNPs onto the electrospun fiber mat, and the resultant PEO/CS/AgNPs has shown an increase in fiber diameter up to a value of 330 $\pm$ 106 nm, corroborated by SEM images. In addition, AFM analysis was used to evaluate the roughness profile of the samples, indicating that the ISS process induced a smooth roughness surface because a change in the average roughness Ra from 84.5 nm (PEO/CS) up to 38.9 nm (PEO/CS/AgNPs) was observed. The presence of the AgNPs was clearly observed by the change in the optical properties of the electrospun films from transparent (PEO/CS) to orange coloration (PEO/CS/AgNPs), respectively. This aspect is associated to the presence of an intense absorption band in the visible region, which is associated to the LSPR phenomenon. In addition, the EDX profile also corroborates the presence of silver (peak at 3 keV), whereas SEM images indicate that a mixture of smaller AgNPs combined with silver aggregates was obtained. After performing the ISS process, an amount of 10.23% of silver was present within the electrospun fibers (corroborated by TGA). These AgNPs in the outer surface of the nanofibrous mat confer an important change in the wettability with an intrinsic hydrophobic character, showing a variation in the water contact angle value from 23.4° $\pm$ 1.3 (PEO/CS) up to 97.7° $\pm$ 5.3 (PEO/CS/AgNPs), respectively. The antifungal activity was evaluated against *P. ostreatus* (showing a clearly apical growth) and the experimental results indicate that both samples PEO/CS and PEO/CS/AgNPs showed an inhibition zone after 10 days of exposure to the fungus, indicating the antifungal effect of the chitosan in comparison with both PEO and reference glass slide samples. Finally, after 13 days of exposure to the fungus, only PEO/CS/AgNPs sample clearly showed an antifungal efficacy because a mean decrease in the growth rate of 39.7% was observed in comparison with the rest of the samples. Finally, the great versatility of both nanofabrication techniques with a high degree of compatibility between them enables the design of advanced functional coatings in terms of antimicrobial surfaces for a wide variety of biomedical or industrial sectors.

Author Contributions: Conceptualization, X.S.; Methodology, L.M., P.J.R., X.S., G.P. and J.F.P.; Validation, P.J.R. and R.J.R.; Formal analysis, P.J.R., J.F.P. and R.J.R.; Investigation, L.M., P.J.R., X.S., G.P. and J.F.P.; Writing—original draft, L.M., P.J.R. and G.P.; Writing—review & editing, R.J.R.; Visualization, R.J.R.; Supervision, P.J.R.; Funding acquisition, P.J.R. All authors have read and agreed to the published version of the manuscript.

Funding: This research received no external funding.

Institutional Review Board Statement: Not applicable.

Acknowledgments: The authors would express their gratitude to Pablo Pujol from UCTAI for his support in the TGA experiments.

Conflicts of Interest: The authors declare no conflict of interest.

References

1. Li, D.; Xia, Y. Electrospinning of Nanofibers: Reinventing the Wheel? *Adv. Mater.* **2004**, *16*, 1151–1170. [CrossRef]
2. Greiner, A.; Wendorff, J.H. Electrospinning: A Fascinating Method for the Preparation of Ultrathin Fibers. *Angew. Chem. Int. Ed.* **2007**, *46*, 5670–5703. [CrossRef] [PubMed]
3. Rivero, P.J.; Rosagaray, I.; Fuertes, J.P.; Palacio, J.F.; Rodríguez, R.J. Designing Multifunctional Protective PVC Electrospun Fibers with Tunable Properties. *Polymers* **2020**, *12*, 2086. [CrossRef]
4. Rivero, P.J.; Yurrita, D.; Berlanga, C.; Palacio, J.F.; Rodríguez, R. Functionalized Electrospun Fibers for the Design of Novel Hydrophobic and Anticorrosive Surfaces. *Coatings* **2018**, *8*, 300. [CrossRef]
5. Deitzel, J.M.; Kleinmeyer, J.; Harris, D.; Tan, N.C.B. The effect of processing variables on the morphology of electrospun nanofibers and textiles. *Polymer* **2001**, *42*, 261–272. [CrossRef]
6. Reneker, D.H.; Chun, I. Nanometre diameter fibres of polymer, produced by electrospinning. *Nanotechnology* **1996**, *7*, 216–223. [CrossRef]
7. Luo, C.J.; Stoyanov, S.D.; Stride, E.; Pelan, E.; Edirisinghe, M. Electrospinning versus fibre production methods: From specifics to technological convergence. *Chem. Soc. Rev.* **2012**, *41*, 4708–4735. [CrossRef]
8. Haider, A.; Haider, S.; Kang, I.-K. A comprehensive review summarizing the effect of electrospinning parameters and potential applications of nanofibers in biomedical and biotechnology. *Arab. J. Chem.* **2018**, *11*, 1165–1188. [CrossRef]
9. Iribarren, A.; Rivero, P.J.; Berlanga, C.; Larumbe, S.; Miguel, A.; Palacio, J.F.; Rodriguez, R. Multifunctional protective PVC-ZnO nanocomposite coatings deposited on aluminum alloys by electrospinning. *Coatings* **2019**, *9*, 264. [CrossRef]
10. Vicente, A.; Rivero, P.J.; García, P.; Mora, J.; Carreño, F.; Palacio, J.F.; Rodríguez, R. Icephobic and Anticorrosion Coatings Deposited by Electrospinning on Aluminum Alloys for Aerospace Applications. *Polymers* **2021**, *13*, 4164. [CrossRef]
11. Albistur, A.; Rivero, P.J.; Esparza, J.; Rodríguez, R. Evaluation of the Photocatalytic Activity and Anticorrosion Performance of Electrospun Fibers Doped with Metallic Oxides. *Polymers* **2021**, *13*, 2011. [CrossRef] [PubMed]
12. Sandua, X.; Rivero, P.J.; Esparza, J.; Fernández-Palacio, J.; Conde, A.; Rodríguez, R.J. Design of Photocatalytic Functional Coatings Based on the Immobilization of Metal Oxide Particles by the Combination of Electrospinning and Layer-by-Layer Deposition Techniques. *Coatings* **2022**, *12*, 862. [CrossRef]
13. Agarwal, S.; Wendorff, J.H.; Greiner, A. Use of electrospinning technique for biomedical applications. *Polymer* **2008**, *49*, 5603–5621. [CrossRef]
14. Zong, X.; Kim, K.; Fang, D.; Ran, S.; Hsiao, B.S.; Chu, B. Structure and process relationship of electrospun bioabsorbable nanofiber membranes. *Polymer* **2002**, *43*, 4403–4412. [CrossRef]
15. Goh, Y.-F.; Shakir, I.; Hussain, R. Electrospun fibers for tissue engineering, drug delivery, and wound dressing. *J. Mater. Sci.* **2013**, *48*, 3027–3054. [CrossRef]
16. Liu, M.; Duan, X.P.; Li, Y.M.; Yang, D.P.; Long, Y.Z. Electrospun nanofibers for wound healing. *Mater. Sci. Eng. C* **2017**, *76*, 1413–1423. [CrossRef]
17. Ding, F.; Deng, H.; Du, Y.; Shi, X.; Wang, Q. Emerging chitin and chitosan nanofibrous materials for biomedical applications. *Nanoscale* **2014**, *6*, 9477–9493. [CrossRef]
18. Qasim, S.B.; Zafar, M.S.; Najeeb, S.; Khurshid, Z.; Shah, A.H.; Husain, S.; Rehman, I.U. Electrospinning of Chitosan-Based Solutions for Tissue Engineering and Regenerative Medicine. *Int. J. Mol. Sci.* **2018**, *19*, 407. [CrossRef]
19. Kalantari, K.; Afifi, A.M.; Jahangirian, H.; Webster, T.J. Biomedical applications of chitosan electrospun nanofibers as a green polymer—Review. *Carbohydr. Polym.* **2019**, *207*, 588–600. [CrossRef]
20. Nada, A.A.; James, R.; Shelke, N.B.; Harmon, M.D.; Awad, H.M.; Nagarale, R.K.; Kumbar, S.G. A smart methodology to fabricate electrospun chitosan nanofiber matrices for regenerative engineering applications. *Polym. Adv. Technol.* **2014**, *25*, 507–515. [CrossRef]
21. Ma, G.; Liu, Y.; Fang, D.; Chen, J.; Peng, C.; Fei, X.; Nie, J. Hyaluronic acid/chitosan polyelectrolyte complexes nanofibers prepared by electrospinning. *Mater. Lett.* **2012**, *74*, 78–80. [CrossRef]
22. Ribeiro, A.S.; Costa, S.M.; Ferreira, D.P.; Abidi, H.; Fangueiro, R. Development of Chitosan-Gelatin Nanofibers with Cellulose Nanocrystals for Skin Protection Applications. *Key Eng. Mater.* **2021**, *893*, 45–55. [CrossRef]

23. Wongsasulak, S.; Puttipaiboon, N.; Yoovidhya, T. Fabrication, Gastromucoadhesivity, Swelling, and Degradation of Zein-Chitosan Composite Ultrafine Fibers. *J. Food Sci.* **2013**, *78*, N926–N935. [CrossRef]
24. He, J.; Cheng, Y.; Li, P.; Zhang, Y.; Zhang, H.; Cui, S. Preparation and characterization of biomimetic tussah silk fibroin/chitosan composite nanofibers. *Iran. Polym. J.* **2013**, *22*, 537–547. [CrossRef]
25. Zhang, Y.; Huang, X.; Duan, B.; Wu, L.; Li, S.; Yuan, X. Preparation of electrospun chitosan/poly(vinyl alcohol) membranes. *Colloid Polym. Sci.* **2007**, *285*, 855–863. [CrossRef]
26. Pakravan, M.; Heuzey, M.-C.; Ajji, A. A fundamental study of chitosan/PEO electrospinning. *Polymer* **2011**, *52*, 4813–4824. [CrossRef]
27. Panda, P.K.; Park, K.; Seo, J. Development of poly (vinyl alcohol)/regenerated chitosan blend film with superior barrier, antioxidant, and antibacterial properties. *Prog. Org. Coat.* **2023**, *183*, 107749. [CrossRef]
28. Desai, K.; Kit, K. Effect of spinning temperature and blend ratios on electrospun chitosan/poly(acrylamide) blends fibers. *Polymer* **2008**, *49*, 4046–4050. [CrossRef]
29. Hassiba, A.J.; El Zowalaty, M.E.; Webster, T.J.; Abdullah, A.M.; Nasrallah, G.K.; Khalil, K.A.; Luyt, A.S.; Elzatahry, A.A. Synthesis, characterization, and antimicrobial properties of novel double layer nanocomposite electrospun fibers for wound dressing applications. *Int. J. Nanomed.* **2017**, *12*, 2205–2213. [CrossRef]
30. An, J.; Zhang, H.; Zhang, J.; Zhao, Y.; Yuan, X. Preparation and antibacterial activity of electrospun chitosan/poly(ethylene oxide) membranes containing silver nanoparticles. *Colloid Polym. Sci.* **2009**, *287*, 1425–1434. [CrossRef]
31. Bhattarai, N.; Edmondson, D.; Veiseh, O.; Matsen, F.A.; Zhang, M. Electrospun chitosan-based nanofibers and their cellular compatibility. *Biomaterials* **2005**, *26*, 6176–6184. [CrossRef] [PubMed]
32. Kriegel, C.; Kit, K.M.; McClements, D.J.; Weiss, J. Electrospinning of chitosan–poly(ethylene oxide) blend nanofibers in the presence of micellar surfactant solutions. *Polymer* **2008**, *50*, 189–200. [CrossRef]
33. Duan, B.; Dong, C.; Yuan, X.; Yao, K. Electrospinning of chitosan solutions in acetic acid with poly(ethylene oxide). *J. Biomater. Sci. Polym. Ed.* **2004**, *15*, 797–811. [CrossRef] [PubMed]
34. Penchev, H.; Paneva, D.; Manolova, N.; Rashkov, I. Electrospun Hybrid Nanofibers Based on Chitosan or *N* -Carboxyethylchitosan and Silver Nanoparticles. *Macromol. Biosci.* **2009**, *9*, 884–894. [CrossRef]
35. Penchev, H.; Paneva, D.; Manolova, N.; Rashkov, I. Hybrid nanofibrous yarns based on N-carboxyethylchitosan and silver nanoparticles with antibacterial activity prepared by self-bundling electrospinning. *Carbohydr. Res.* **2010**, *345*, 2374–2380. [CrossRef]
36. Korniienko, V.; Husak, Y.; Yanovska, A.; Banasiuk, R.; Yusupova, A.; Savchenko, A.; Holubnycha, V.; Pogorielov, M. Functional and biological characterization of chitosan electrospun nanofibrous membrane nucleated with silver nanoparticles. *Appl. Nanosci.* **2021**, *12*, 1061–1070. [CrossRef]
37. Champer, J.; Patel, J.; Fernando, N.; Salehi, E.; Wong, V.; Kim, J. Chitosan against cutaneous pathogens. *AMB Express* **2013**, *3*, 37. [CrossRef]
38. Matica, M.A.; Aachmann, F.L.; Tøndervik, A.; Sletta, H.; Ostafe, V. Chitosan as a Wound Dressing Starting Material: Antimicrobial Properties and Mode of Action. *Int. J. Mol. Sci.* **2019**, *20*, 5889. [CrossRef]
39. Raafat, D.; von Bargen, K.; Haas, A.; Sahl, H.-G. Insights into the Mode of Action of Chitosan as an Antibacterial Compound. *Appl. Environ. Microbiol.* **2008**, *74*, 3764–3773. [CrossRef]
40. Rabea, E.I.; Badawy, M.E.-T.; Stevens, C.V.; Smagghe, G.; Steurbaut, W. Chitosan as Antimicrobial Agent: Applications and Mode of Action. *Biomacromolecules* **2003**, *4*, 1457–1465. [CrossRef]
41. Segala, K.; Nista, S.V.G.; Cordi, L.; Bizarria, M.T.M.; de Ávila, J.; Kleinubing, S.A.; Cruz, D.C.; Brocchi, M.; Lona, L.M.F.; Caballero, N.E.D.; et al. Silver nanoparticles incorporated into nanostructured biopolymer membranes produced by electrospinning: A study of antimicrobial activity. *Braz. J. Pharm. Sci.* **2015**, *51*, 911–921. [CrossRef]
42. AshaRani, P.V.; Mun, G.L.K.; Hande, M.P.; Valiyaveettil, S. Cytotoxicity and Genotoxicity of Silver Nanoparticles in Human Cells. *ACS Nano* **2009**, *3*, 279–290. [CrossRef] [PubMed]
43. Urrutia, A.; Rivero, P.J.; Ruete, L.; Goicoechea, J.; Matías, I.R.; Arregui, F.J. Single-stage in situ synthesis of silver nanoparticles in antibacterial self-assembled overlays. *Colloid Polym. Sci.* **2012**, *290*, 785–792. [CrossRef]
44. Rivero, P.J.; Urrutia, A.; Goicoechea, J.; Zamarreño, C.R.; Arregui, F.J.; Matías, I.R. An antibacterial coating based on a polymer/sol-gel hybrid matrix loaded with silver nanoparticles. *Nanoscale Res. Lett.* **2011**, *6*, 305. [CrossRef]
45. Liz-Marzán, L.M. Nanometals: Formation and color. *Mater. Today* **2004**, *7*, 26–31. [CrossRef]
46. Liz-Marzán, L.M. Tailoring Surface Plasmons through the Morphology and Assembly of Metal Nanoparticles. *Langmuir* **2005**, *22*, 32–41. [CrossRef]
47. Zhang, X.-F.; Liu, Z.-G.; Shen, W.; Gurunathan, S. Silver Nanoparticles: Synthesis, Characterization, Properties, Applications, and Therapeutic Approaches. *Int. J. Mol. Sci.* **2016**, *17*, 1534. [CrossRef] [PubMed]
48. Kim, J.S.; Kuk, E.; Yu, K.N.; Kim, J.-H.; Park, S.J.; Lee, H.J.; Kim, S.H.; Park, Y.K.; Park, Y.H.; Hwang, C.-Y.; et al. Antimicrobial effects of silver nanoparticles. *Nanomed. Nanotechnol. Boil. Med.* **2007**, *3*, 95–101, Erratum in *Nanomed. Nanotechnol. Biol. Med.* **2014**, *10*, e1119. [CrossRef]
49. Chernousova, S.; Epple, M. Silver as Antibacterial Agent: Ion, Nanoparticle, and Metal. *Angew. Chem. Int. Ed.* **2013**, *52*, 1636–1653. [CrossRef]

50. Benn, T.; Cavanagh, B.; Hristovski, K.; Posner, J.D.; Westerhoff, P. The Release of Nanosilver from Consumer Products Used in the Home. *J. Environ. Qual.* **2010**, *39*, 1875–1882. [CrossRef]
51. Abdelgawad, A.M.; Hudson, S.M.; Rojas, O.J. Antimicrobial wound dressing nanofiber mats from multicomponent (chitosan/silver-NPs/polyvinyl alcohol) systems. *Carbohydr. Polym.* **2014**, *100*, 166–178. [CrossRef] [PubMed]
52. Sánchez-López, E.; Gomes, D.; Esteruelas, G.; Bonilla, L.; Lopez-Machado, A.L.; Galindo, R.; Cano, A.; Espina, M.; Ettcheto, M.; Camins, A.; et al. Metal-Based Nanoparticles as Antimicrobial Agents: An Overview. *Nanomaterials* **2020**, *10*, 292. [CrossRef]
53. Sambhy, V.; MacBride, M.M.; Peterson, B.R.; Sen, A. Silver Bromide Nanoparticle/Polymer Composites: Dual Action Tunable Antimicrobial Materials. *J. Am. Chem. Soc.* **2006**, *128*, 9798–9808. [CrossRef] [PubMed]
54. Bagheri, M.; Validi, M.; Gholipour, A.; Makvandi, P.; Sharifi, E. Chitosan nanofiber biocomposites for potential wound healing applications: Antioxidant activity with synergic antibacterial effect. *Bioeng. Transl. Med.* **2021**, *7*, e10254. [CrossRef] [PubMed]
55. Kohsari, I.; Shariatinia, Z.; Pourmortazavi, S.M. Antibacterial electrospun chitosan–polyethylene oxide nanocomposite mats containing bioactive silver nanoparticles. *Carbohydr. Polym.* **2016**, *140*, 287–298. [CrossRef]
56. Aljohani, M.M.; Abu-Rayyan, A.; Elsayed, N.H.; Alatawi, F.A.; Al-Anazi, M.; Mustafa, S.K.; Albalawi, R.K.; Abdelmonem, R. One-pot microwave synthesis of chitosan-stabilized silver nanoparticles entrapped polyethylene oxide nanofibers, with their intrinsic antibacterial and antioxidant potency for wound healing. *Int. J. Biol. Macromol.* **2023**, *235*, 123704. [CrossRef]
57. Song, J.; Remmers, S.J.; Shao, J.; Kolwijck, E.; Walboomers, X.F.; Jansen, J.A.; Leeuwenburgh, S.C.; Yang, F. Antibacterial effects of electrospun chitosan/poly(ethylene oxide) nanofibrous membranes loaded with chlorhexidine and silver. *Nanomed. Nanotechnol. Biol. Med.* **2016**, *12*, 1357–1364. [CrossRef]
58. Fereydouni, N.; Zangouei, M.; Darroudi, M.; Hosseinpour, M.; Gholoobi, A. Antibacterial activity of chitosan-polyethylene oxide nanofibers containing silver nanoparticles against aerobic and anaerobic bacteria. *J. Mol. Struct.* **2023**, *1274*, 134304. [CrossRef]
59. Chatterjee, T.; Chatterjee, B.K.; Majumdar, D.; Chakrabarti, P. Antibacterial effect of silver nanoparticles and the modeling of bacterial growth kinetics using a modified Gompertz model. *Biochim. Biophys. Acta (BBA) Gen. Subj.* **2015**, *1850*, 299–306. [CrossRef]
60. Fouda, M.M.G.; El-Aassar, M.R.; Al-Deyab, S.S. Antimicrobial activity of carboxymethyl chitosan/polyethylene oxide nanofibers embedded silver nanoparticles. *Carbohydr. Polym.* **2013**, *92*, 1012–1017. [CrossRef]
61. Ezzat, M.; Ghanim, M.; Nageh, H.; Hassanin, A.H.; Abdel-Moneim, A. Antimicrobial Activity of O-Carboxymethyl Chitosan Nanofibers Containing Silver Nanoparticles Synthesized by Green Method. *J. Nano Res.* **2016**, *40*, 136–145. [CrossRef]
62. Borgognone, A.; Castanera, R.; Muguerza, E.; Pisabarro, A.G.; Ramírez, L. Somatic transposition and meiotically driven elimination of an active helitron family in *Pleurotus ostreatus*. *DNA Res.* **2017**, *24*, 103–115. [CrossRef] [PubMed]
63. Larray, L.M.; Pérez, G.; Peñas, M.M.; Baars, J.J.P.; Mikosch, T.S.P.; Pisabarro, A.G.; Ramírez, L. Molecular Karyotype of the White Rot Fungus *Pleurotus ostreatus*. *Appl. Environ. Microbiol.* **1999**, *65*, 3413–3417. [CrossRef]
64. Ing, L.Y.; Zin, N.M.; Sarwar, A.; Katas, H. Antifungal Activity of Chitosan Nanoparticles and Correlation with Their Physical Properties. *Int. J. Biomater.* **2012**, *2012*, 632698. [CrossRef] [PubMed]
65. Baker, L.G.; Specht, C.A.; Donlin, M.J.; Lodge, J.K. Chitosan, the Deacetylated Form of Chitin, Is Necessary for Cell Wall Integrity in *Cryptococcus neoformans*. *Eukaryot. Cell* **2007**, *6*, 855–867. [CrossRef]
66. Pérez, G.; Lopez-Moya, F.; Chuina, E.; Ibañez-Vea, M.; Garde, E.; López-Llorca, L.V.; Pisabarro, A.G.; Ramírez, L. Strain Degeneration in *Pleurotus ostreatus*: A Genotype Dependent Oxidative Stress Process Which Triggers Oxidative Stress, Cellular Detoxifying and Cell Wall Reshaping Genes. *J. Fungi* **2021**, *7*, 862. [CrossRef]
67. Rivero, P.J.; Urrutia, A.; Goicoechea, J.; Rodríguez, Y.; Corres, J.M.; Arregui, F.J.; Matías, I.R. An antibacterial submicron fiber mat with in situ synthesized silver nanoparticles. *J. Appl. Polym. Sci.* **2012**, *126*, 1228–1235. [CrossRef]
68. Rivero, P.J.; Goicoechea, J.; Matias, I.R.; Arregui, F.J. A comparative study of two different approaches for the incorporation of silver nanoparticles into layer-by-layer films. *Nanoscale Res. Lett.* **2014**, *9*, 301. [CrossRef]
69. Zeinali, T.; Alemzadeh, E.; Zarban, A.; Khorashadizadeh, M.; Ansarifar, E. Fabrication and characterization of jujube extract-loaded electrospun polyvinyl alcohol nanofiber for strawberry preservation. *Food Sci. Nutr.* **2021**, *9*, 6353–6361. [CrossRef]
70. Cruz, E.P.D.; Pires, J.B.; Santos, F.N.D.; Fonseca, L.M.; Radünz, M.; Dal Magro, J.; Gandra, E.A.; Zavareze, E.D.R.; Dias, A.R.G. Encapsulation of lemongrass essential oil into cassava starch fibers for application as antifungal agents in bread. *Food Hydrocoll.* **2023**, *145*, 109105. [CrossRef]
71. Han, B.; Chen, P.; Guo, J.; Yu, H.; Zhong, S.; Li, D.; Liu, C.; Feng, Z.; Jiang, B. A Novel Intelligent Indicator Film: Preparation, Characterization, and Application. *Molecules* **2023**, *28*, 3384. [CrossRef]
72. Wardhani, R.A.K.; Asri, L.; Nasir, M.; Purwasasmita, B.S. Preparation of Chitosan-Polyethylene Oxide-Colocasia esculenta Flour Nanofibers using Electrospinning Method. *J. Mech. Eng. Sci. Technol.* **2019**, *3*, 1–7. [CrossRef]
73. Nate, Z.; Moloto, M.J.; Mubiayi, P.K.; Sibiya, P.N. Green synthesis of chitosan capped silver nanoparticles and their antimicrobial activity. *MRS Adv.* **2018**, *3*, 2505–2517. [CrossRef]
74. Gao, S.; Zong, L.; Zhang, Y.; Guo, X.; Guo, G.; Zhao, L.; Ye, F.; Fu, Y. Antifungal pentachloronitrobenzene/hydroxypropyl-beta-cyclodextrin inclusion complex nanofibers by electrospun with no polymer: Fabrication and characterization. *J. Clean. Prod.* **2023**, *413*, 137499. [CrossRef]
75. Ogawa, T.; Ding, B.; Sone, Y.; Shiratori, S. Super-hydrophobic surfaces of layer-by-layer structured film-coated electrospun nanofibrous membranes. *Nanotechnology* **2007**, *18*, 165607. [CrossRef]

76. Ridolfi, D.M.; Lemes, A.P.; De Oliveira, S.; Justo, G.Z.; Palladino, M.V.; Durán, N. Electrospun poly(ethylene oxide)/chitosan nanofibers with cellulose nanocrystals as support for cell culture of 3T3 fibroblasts. *Cellulose* **2017**, *24*, 3353–3365. [CrossRef]
77. Pakravan, M.; Heuzey, M.-C.; Ajji, A. Core–Shell Structured PEO-Chitosan Nanofibers by Coaxial Electrospinning. *Biomacromolecules* **2012**, *13*, 412–421. [CrossRef] [PubMed]
78. Rivero, P.J.; Urrutia, A.; Goicoechea, J.; Matias, I.; Arregui, F. A Lossy Mode Resonance optical sensor using silver nanoparticles-loaded films for monitoring human breathing. *Sens. Actuators B Chem.* **2013**, *187*, 40–44. [CrossRef]
79. Abalde-Cela, S.; Ho, S.; Rodríguez-González, B.; Correa-Duarte, M.A.; Álvarez-Puebla, R.A.; Liz-Marzán, L.M.; Kotov, N.A. Loading of Exponentially Grown LBL Films with Silver Nanoparticles and Their Application to Generalized SERS Detection. *Angew. Chem. Int. Ed.* **2009**, *48*, 5326–5329. [CrossRef]
80. Malikova, N.; Pastoriza-Santos, I.; Schierhorn, M.; Kotov, N.A.; Liz-Marzán, L.M. Layer-by-Layer Assembled Mixed Spherical and Planar Gold Nanoparticles: Control of Interparticle Interactions. *Langmuir* **2002**, *18*, 3694–3697. [CrossRef]
81. Rivero, P.J.; Goicoechea, J.; Urrutia, A.; Arregui, F.J. Effect of both protective and reducing agents in the synthesis of multicolor silver nanoparticles. *Nanoscale Res. Lett.* **2013**, *8*, 101. [CrossRef] [PubMed]
82. Neukäufer, J.S.; Seyfang, B.; Gruetzner, T. Investigation of Contact Angles and Surface Morphology of 3D-Printed Materials. *Ind. Eng. Chem. Res.* **2020**, *59*, 6761–6766. [CrossRef]
83. Tuteja, A.; Choi, W.; Ma, M.; Mabry, J.M.; Mazzella, S.A.; Rutledge, G.C.; McKinley, G.H.; Cohen, R.E. Designing Superoleophobic Surfaces. *Science* **2007**, *318*, 1618–1622. [CrossRef]
84. Kharat, Z.; Goushki, M.A.; Sarvian, N.; Asad, S.; Dehghan, M.M.; Kabiri, M. Chitosan/PEO nanofibers containing Calendula officinalis extract: Preparation, characterization, in vitro and in vivo evaluation for wound healing applications. *Int. J. Pharm.* **2021**, *609*, 121132. [CrossRef] [PubMed]
85. Trinca, R.B.; Westin, C.B.; da Silva, J.A.F.; Moraes, M. Electrospun multilayer chitosan scaffolds as potential wound dressings for skin lesions. *Eur. Polym. J.* **2017**, *88*, 161–170. [CrossRef]
86. Fratoddi, I. Hydrophobic and Hydrophilic Au and Ag Nanoparticles. Breakthroughs and Perspectives. *Nanomaterials* **2017**, *8*, 11. [CrossRef]
87. Sandua, X.; Rivero, P.J.; Palacio, J.F.; Esparza, J.; Rodríguez, R. An Alternative Methodology for the Evaluation of Photocatalytic Activity of Polymeric Coatings by Monitoring Dye Degradation. *Coatings* **2022**, *12*, 1529. [CrossRef]
88. Bürgers, R.; Eidt, A.; Frankenberger, R.; Rosentritt, M.; Schweikl, H.; Handel, G.; Hahnel, S. The anti-adherence activity and bactericidal effect of microparticulate silver additives in composite resin materials. *Arch. Oral Biol.* **2009**, *54*, 595–601. [CrossRef]
89. Ashok, D.; Cheeseman, S.; Wang, Y.; Funnell, B.; Leung, S.; Tricoli, A.; Nisbet, D. Superhydrophobic Surfaces to Combat Bacterial Surface Colonization. *Adv. Mater. Interfaces* **2023**, *10*, 2300324. [CrossRef]
90. Mansoor, S.; Zahoor, I.; Baba, T.R.; Padder, S.A.; Bhat, Z.A.; Koul, A.M.; Jiang, L. Fabrication of Silver Nanoparticles Against Fungal Pathogens. *Front. Nanotechnol.* **2021**, *3*, 679358. [CrossRef]
91. El-Samad, L.M.; Hassan, M.A.; Bakr, N.R.; El-Ashram, S.; Radwan, E.H.; Aziz, K.K.A.; Hussein, H.K.; El Wakil, A. Insights into Ag-NPs-mediated pathophysiology and ultrastructural aberrations in ovarian tissues of darkling beetles. *Sci. Rep.* **2022**, *12*, 13899. [CrossRef] [PubMed]
92. Matras, E.; Gorczyca, A.; Przemieniecki, S.W.; Oćwieja, M. Surface properties-dependent antifungal activity of silver nanoparticles. *Sci. Rep.* **2022**, *12*, 18046. [CrossRef] [PubMed]

polymers

MDPI

Article

Electrospinning Membrane with Polyacrylate Mixed Beta-Cyclodextrin: An Efficient Adsorbent for Cationic Dyes

Chunling Zheng [1,2,*], Wei Zhao [1,2], Xiaoqian Tu [2,3] and Shaoqiang Zhou [2,4]

1 College of Food Science and Light Industry, Nanjing Tech University, 30 South Puzhu Road, Nanjing 211816, China
2 Institute of Textile Auxiliary and Ecological Dyeing Finishing, Nanjing Tech University, 30 South Puzhu Road, Nanjing 211816, China
3 Jiangsu Textile Product Quality Supervision and Inspection Institute, 3 East Guanghua Street, Nanjing 210007, China
4 Nanjing Customs Industrial Products Testing Center, 39 Chuangzhi Road, Nanjing 210019, China
* Correspondence: zhengchunling@njtech.edu.cn; Tel.: +86-25-58139426

Abstract: A simple and non-chemical binding nanofiber (β-CD/PA) adsorbent was obtained by electrospinning a mixture of β-cyclodextrin (β-CD) and polyacrylate (PA). The cationic dyes in wastewater were removed by the host–guest inclusion complex of the β-cyclodextrin and the electrostatic interaction between the polyacrylate and the dyes groups. The influence of the content of β-cyclodextrin on the surface morphology and adsorption capacity of the nanofiber membrane was discussed, and the optimized adsorption capacity of nanofiber adsorption material was determined. The adsorption capacity of nanofiber adsorbents for basic red 9, basic red 14, basic red 46, basic blue 9, basic yellow 19 and basic yellow 28 was 86.71 mg/g, 21.513 mg/g, 18.926 mg/g, 44.525 mg/g, 116.516 mg/g and 155.206 mg/g, respectively. The effects of different initial concentrations and pH values on the adsorption properties of adsorbent materials were studied. The kinetic analysis showed that the adsorption process of nanofibers for cationic dyes was more in line with the pseudo-second-order kinetic adsorption model. Moreover, nanofiber adsorbent could be easily separated from the dye solution and showed high recycling efficiency. These results indicated that the β-cyclodextrin/polyacrylate composite nanofibers are expected to be recyclable adsorbents in dye wastewater treatment.

Keywords: β-cyclodextrin; polyacrylate; electrospinning; electrospinning; adsorption; cationic dye

1. Introduction

Water pollution has always been a serious environmental problem. The printing industry and the textile industry produce a large amount of dye wastewater, which is one of the main sources of water pollution [1–3]. If not properly treated, direct discharge into the environment will bring many environmental problems. Various methods, including electrochemical, photocatalytic degradation, adsorption and ultra-filtration, have been widely investigated for treating dye polluting wastewater [4–8]. Some studies have shown that the adsorption method has advantages of high efficiency, economy, simplicity, and good selectivity in treating dye wastewater [9–11]. There are many ways to prepare adsorbent materials, and electrostatic spinning is one of the common methods for preparing high-efficiency adsorbent materials [12]. Multifunctional nanofibers can be electro-spun by using polymer composites or polymer blends that have many unique advantages, such as relatively high surface area and porosity, and fiber diameters at the nanometer scale [13,14].

https://doi.org/10.3390/polym17020243

Therefore, the electrospinning technology is used in many fields, including biological materials, filtration membranes, electroanalysis, etc. [15–17]. In addition, the adsorption effect of electro-spun nanofiber membranes is superior to conventional adsorption membranes because they can be easily designed to obtain functional surface adsorption sites or specific structures, which are beneficial for providing better adsorption capacity [18].

β-cyclodextrin (β-CD) is a cyclic oligosaccharide formed by the linkage of seven glucose monomers via α-1,4 glycosidic bonds. β-CD has a special molecular configuration: the whole molecule looks like a conical ring with some hydrophilic hydroxyl groups on the outside and a hydrophobic cavity inside. Due to its special molecular configuration, β-CD can form a non-covalent inclusion compound with some compounds [19,20]. Therefore, β-CD can also be used as an adsorbent for treating dye wastewater, because it can form inclusion complexes with dye molecules in contaminated wastewater [21,22].

For instance, Fan et al. [5] obtained a grafted product of cyclodextrin and chitosan (β-CD–chitosan) by connecting one end of maleic anhydride to the hydroxyl group at the sixth position of cyclodextrin and the other end to chitosan. Then, β-CD–chitosan and magnetic Fe_3O_4 were combined under the action of glutaraldehyde to form magnetic modified chitosan nanoparticles. It was found that this nanoparticle material had a good adsorption capacity for methyl blue. Ebadi et al. [19] self-assembled mesoporous silica particles using cetyltrimethylammonium bromide, ammonia water and ethyl silicate. Then, these particles were combined with 3-aminopropyltriethoxysilane in toluene to form amino-functionalized silica particles. Subsequently, the amino-functionalized particles were conjugated with β-CD modified by p-toluene-sulfonyl chloride to obtain a functionalized silica particle (β-CD-SNHS) adsorbent with multiple cyclodextrins, which was significantly superior to other existing adsorbents in removing methylene blue from aqueous solutions.

However, as a powder solid, β-CD is difficult to be used in water treatment. In recent years, some researchers have prepared a nanofiber membrane with water resistance by electrospinning a mixture of β-CD and some high molecular polymers (dissolving β-cyclodextrin and high molecular polymer in an organic solvent) [14,23], in order to achieve synergistic adsorption of the dye molecules by the β-CD and the polymer [2]. The copolymer of methyl methacrylate and butyl acrylate has characteristics of good water resistance, good toughness, strong compatibility and easy preparation, and the properties of the copolymer can be changed by adjusting the ratio of the monomer [24,25].

In this research, polyacrylate (PA) was synthesized by emulsion polymerization with methyl methacrylate, butyl acrylate, and acrylic acid as monomers, and the adsorption materials of cationic dyes were prepared by electrospinning a mixture (DMF as solvent) of cyclodextrin and polyacrylate. The effect of β-CD content (mass ratio of β-CD to PA) on the morphology and adsorption properties of electro-spun fibers was investigated, and the nanofiber materials with the best adsorption effect were determined. The distribution of cyclodextrin in fiber materials was analyzed by XRD. The adsorption properties of fiber materials at different initial solubility and pH were discussed experimentally, and the kinetics of the adsorption process was also studied. Finally, the desorption capacity and recycling efficiency of the fiber material were tested. The dynamic process of the possible adsorption and desorption of the nanofiber materials was shown.

2. Experimental

2.1. Materials

β-cyclodextrin (β-CD), potassium peroxydisulfate ($K_2S_2O_8$), OP-10, sodium dodecyl sulfate (SDS), methyl methacrylate (MMA), butyl acrylate (BA), acrylic acid (AA) and N,N-dimethylformamide (DMF)were purchased from LingFeng Chemical Inc., Shanghai, China. Cationic dyes were obtained from XingWu Chemical Industry Co., Ltd., Rugao,

Jiangsu, China. The structures of six cationic dyes, including basic red 14, basic red 46, basic red 9, basic yellow 19, basic yellow 28 and basic blue 9, are shown in Figure 1 [26]. All chemicals are analytical grade and used without further purification and all the solutions are prepared with homemade deionized water.

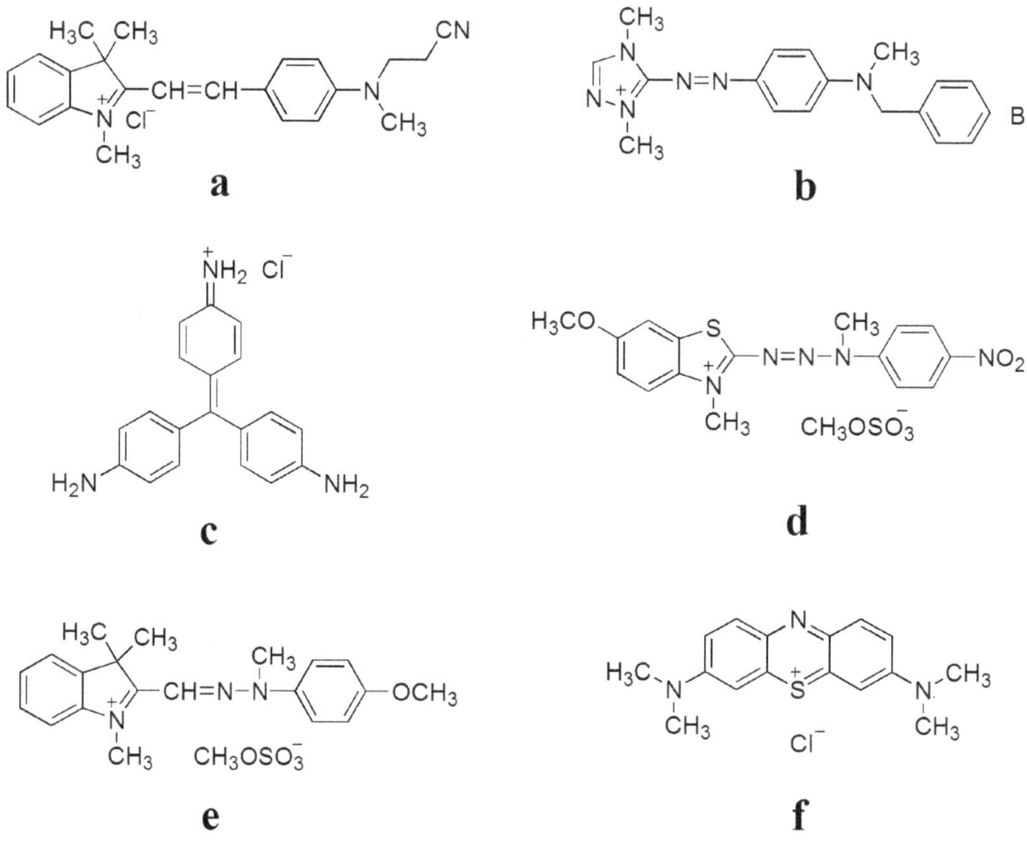

Figure 1. Molecular structures of basic red 14 (**a**), basic red 46 (**b**), basic red 9 (**c**), basic yellow 19 (**d**), basic yellow 28 (**e**) and basic blue 9 (**f**).

2.2. Synthesis of Polyacrylate (PA) Copolymer

90 mL deionized water, 0.9 g OP-10, and 0.45 g SDS were added into the beaker and stirred fully, and 21 g MMA, 19 g BA, and 5 g AA were added continuously and stirred well. The mixture in the beaker was placed under a high-speed rotating homogenizer at 14,000 r/min, and a milky emulsion was obtained after high-speed homogenization for 6 min.

A proportion of 2/3 emulsion was poured into a constant pressure drip hopper, and the remainder was heated in a four-mouth flask to 75 °C. $K_2S_2O_8$ was added to the reaction emulsion as an initiator and the timing began at this point as the reaction starting point. A period of uniform stirring made the reaction emulsion pale blue, then the reserved 2/3 emulsion kept dribbling to the flask slowly at a constant speed, for around 2 h. After the pre-emulsion was added, $K_2S_2O_8$ continued to initiate the subsequent polymerization at a strictly controlled 75–80 °C for 1 h until the system temperature reached 85 °C, and insulation needed another 1 h. Finally, the target emulsion was cooled to room temperature and filtered with a Buchner funnel.

2.3. Preparation of Electrospun β-CD/PA Nanofiber Membrane

The polyacrylate emulsion was taken into the evaporating dish and placed in the oven to obtain a colorless and transparent copolymer solid. An electro-spun solution consisting of 2 g of the copolymer solid and a different experimental quantity of β-CD dissolved in 18 mL DMF was mixed mechanically with stirring for 10 h. A 10 mL plastic syringe with a metal needle was filled with the resulting mixture and carried forward at an optimized speed of 1.5 mL/h during the electrospinning. A charged voltage of 13 kV was loaded between the needle and the grounded metal collector covered with a sheet of aluminum foil, which was served as a counter electrode. The distance between the needle and the collector was 12 cm. The electrospinning process was performed at relatively low humidity (40–60%) at room temperature.

2.4. Characterization

The scanning electron microscope (S-3400N II, Hitachi Production Co., Ltd., Shanghai, China) was used to observe the morphology of the β-CD/PA fiber membrane, and the samples were gold sputter coated to guarantee the stability of SEM imaging. XRD diffraction of the samples was obtained by an X-ray diffraction spectrometer (D8 Advance, Bruker Corporation, Beijing, China) using Cu-Kα radiation. The physical properties of the electrospinning solution were measured by conductivity meter (DDS-307A, INESA Instruments Co., Ltd., Shanghai, China) and viscosity meter respectively (SNB-3, Shanghai Precision Instrument Co., Ltd., Shanghai, China). The adsorption number of nanofibers could be obtained via a difference method by measuring the concentrations of dyes solution before and after adsorption with ultraviolet spectrophotometer (Cary 60, Agilent Technologies (China) Co., Ltd., Beijing, China).

2.5. Batch Adsorption Experiments

All experiments were carried out at room temperature and all adsorption processes were static adsorption. The absorbability of nanofibers was studied first. A total of 30 mg β-CD/PA nanofiber membrane with different β-CD contents was suspended in 20 mL dye solution (100 mg/L) of basic red 14 or basic yellow 28 to compare their respective adsorption effects. The maximum adsorption capacity of nanofibers to various cationic dyes was obtained when adsorption reached equilibrium. Next, the pH of the dye's solution was adjusted with 1.0 M sulfuric acid or sodium hydroxide to witness the effect of pH on the adsorption of nanofiber membrane. Finally, the adsorption kinetics were researched. The typical batch adsorption experiment was carried out as follows: 50 mg nanofiber membrane was placed in 20 mL basic yellow 28 (50 mg/L), basic yellow 19 (30 mg/L), basic red 46 (15 mg/L), basic red 14 (15 mg/L), basic blue 9 (10 mg/L) and basic red 9 (10 mg/L) solutions, respectively, and the adsorption capacity was measured periodically.

The adsorption capacity of the nanofiber membrane is calculated according to the concentration difference of dye solution before and after adsorption, and the calculation formula is:

$$q_e = (C_i - C_e)V/M_S \quad (1)$$

where q_e (mg/g) is the adsorption capacity, C_i (mg/L) is the initial concentration of dye solution, C_e (mg/L) is the equilibrium concentration of dye solution after adsorption, V (L) is the volume of dye solution and M_s (g) is the mass of solid adsorbent.

3. Results and Discussion

3.1. Determination of Optimum Adsorption Capacity of Nanofibers

During the electrospinning process, the concentration of PA was kept at 12% (Mass of PA/DMF volume, g/mL), and the mass of β-CD added to the solution was changed to obtain five different spinning fluids. The conductivity and viscosity of the five spinning fluids and the diameter of five nanofibers are shown in Table 1. Five different nanofiber membranes were obtained by these spinning fluids through electrostatic spinning, as shown in Figure 2.

Table 1. Details of 5 kinds of electrospinning solutions and fibers.

Sample	PA/g	β-CD/g	DMF/mL	$(m_{\beta\text{-CD}}/m_{PA})$/%	Conductivity/μS·cm^{-1}	Viscosity/mPa·s	Fiber's Diameter/μm
a	2	0	16	0	138.9	143.3	25
b	2	0.5	16	25	125.1	154.1	12.2
c	2	1	16	50	106.3	168.5	11.3
d	2	1.5	16	75	94.3	187.5	10.8
e	2	2	16	100	72.7	235.0	11.8

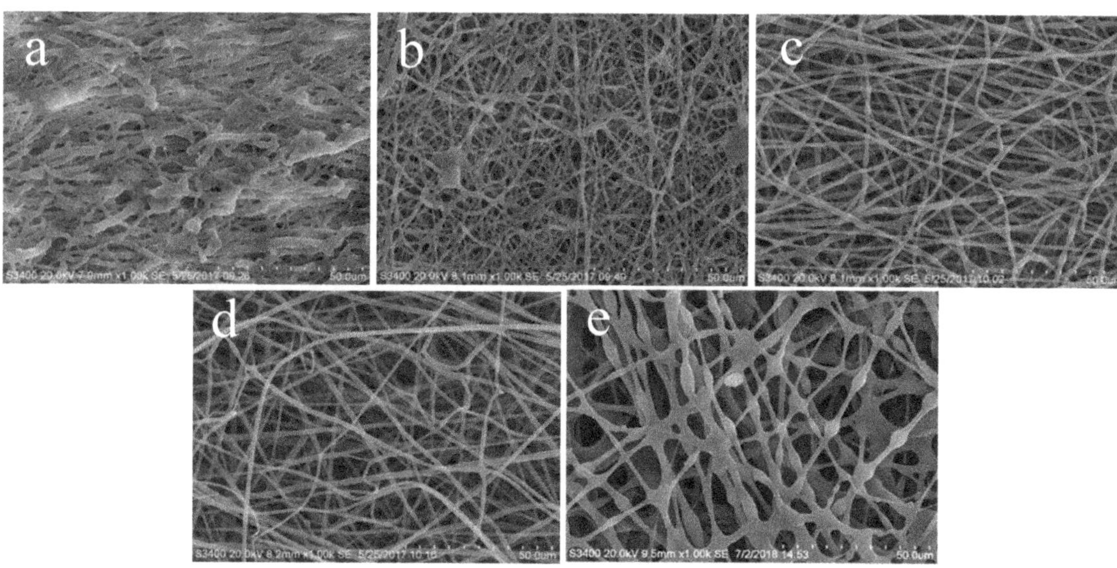

Figure 2. SEM images of PA electro-spun nanofibers without β-CD (**a**); β-CD/PA composite electro-spun nanofibers with β-CD mass accounting for 25% (**b**), 50% (**c**), 75% (**d**) and 100% (**e**) of PA quality.

The morphology of the electro-spun fiber can be observed in Figure 2. When the content of β-CD increased from 0 to 25% (Figure 2a,b), the filamentous shape of the spun fiber went from uneven to uniform and the diameter of the fiber decreased considerably. This indicated that the introduction of β-CD improved the viscosity of the system and significantly enhanced the filament formation of the nanofibers. When the content of β-CD increased to 50% (Figure 2c), the beads on the surface of the nanofibers decreased significantly, the individual fibers became more uniform and the gap between the fibers was more obvious. As the content of β-CD approached 75% (Figure 2d), the morphology of the nanofiber did not change significantly, but the diameter of some nanofibers continued to decrease and the gap between the fibers kept widening. Therefore, keeping the content of β-CD in the spinning liquid system between 50% and 75%, the silky uniform nanofibers could be favorably obtained. When the β-CD content reached 100% (Figure 2e), the viscosity of the spinning solution increased but the conductivity decreased, which reduced the charge density during the electrostatic spinning and caused the electrostatic repulsion on the surface of droplet. As a result, the draft ratio of jet flow decreased to make the diameter of the electro-spun fiber thicker and beads appear.

The adsorption for two cationic dyes were tested by using nanofiber membranes with different β-CD contents of the same quality under the same conditions, and the results are shown in Figure 3. With an increasing amount of β-CD, the adsorption of the nanofiber membranes performed increasingly well, and reached a peak when the content of β-CD was 75%. This showed that the participation of β-CD not only improved the morphology of the nanofiber but also increased its adsorptive property for cationic dyes. However, the excessive β-CD did not further perpetuate this effect, which might make spinning difficult and the diameter of nanofibers large and uneven. As a result, the contact sites among the dye molecules and the nanofibers reduced. Taken together, when the content of cyclodextrin reached 75%, nanofibers with uniform morphology and maximum adsorption capacity could be obtained.

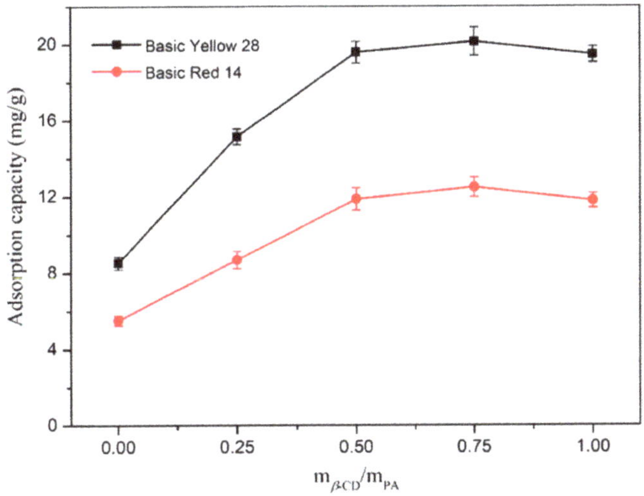

Figure 3. Adsorption of β-CD/PA nanofiber membrane with different β-CD content for basic yellow 28 (black line) and basic red 14 (red line).

Figure 4 shows the SEM images of nanofiber membrane before and after adsorption for basic yellow 28. It is obvious in Figure 4b that the adsorbed nanofibers have a large amount of cationic dye enrichment, indicating that the prepared adsorbent had good adsorption for cationic dyes. At the same time, because of the spatial size and steric resistance of the long side chain of the basic red 14 molecule, the difficulty of β-CD inclusion is obviously greater than that of the basic yellow 28 molecule. Therefore, during the adsorption process, the adsorption capacity of basic yellow 28 is significantly higher than that of basic red 14.

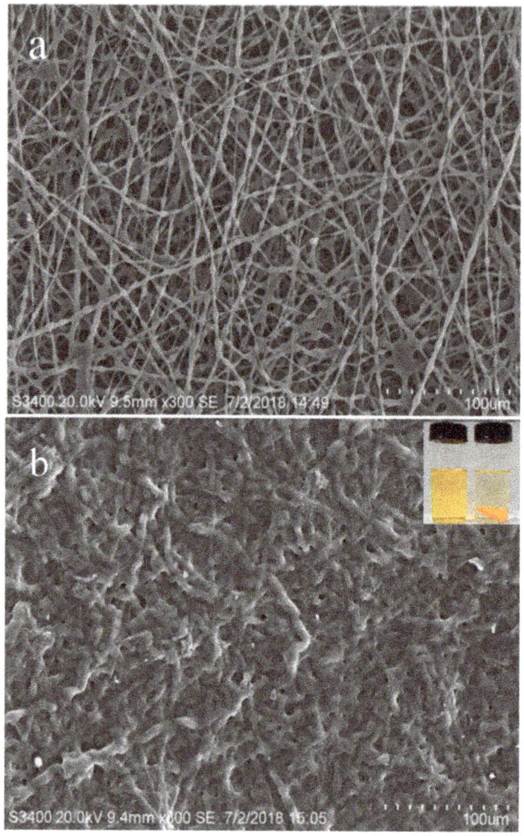

Figure 4. SEM images of nanofibers before (**a**) and after (**b**) adsorbing basic yellow 28.

3.2. XRD Characterization of the Obtained Nanofibers

In order to better explore the crystallization and distribution of cyclodextrin in nanofiber, the XRD diffraction analysis was performed on β-CD powder, PA nanofiber, and PA/β-CD composite nanofiber, as shown in Figure 5.

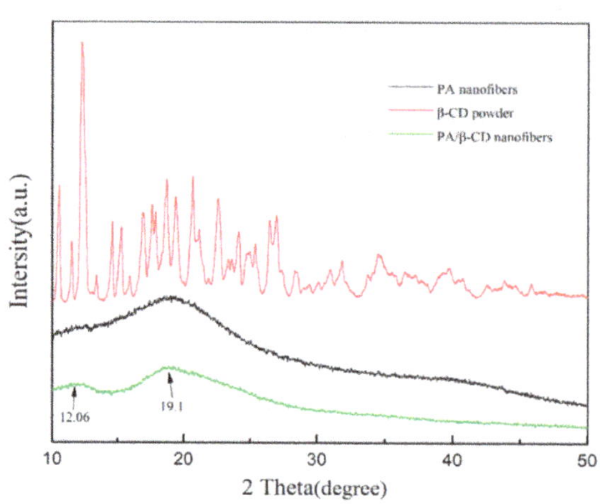

Figure 5. XRD of β-CD powder, PA nanofiber, and PA/β-CD nanofiber.

It can be seen from the figure that, in the range of 2θ from 10° to 50°, the β-CD powder had many strong diffraction peaks, but the PA nanofiber had a broad diffraction peak only at 19.1°. The XRD of the PA/β-CD nanofiber membrane is similar to that of the polyacrylate film, which indicated that the β-CD molecules should be uniformly distributed in the composite nanofibers in an amorphous form. The difference between the XRD of the PA/β-CD nanofiber and the PA nanofiber was that the PA/β-CD nanofiber membrane exhibited a new diffraction peak at 2θ of 12.06° with the addition of β-CD. This might be due to the nonuniform distribution of excess β-CD, which finally existed in the form of a small number of crystals [24]. This further indicated that the mixing of PA and β-CD was only physical, and the cavity structure of β-CD was retained on the nanofiber, which was beneficial to the adsorption of this composite material.

3.3. *Effect of Initial Concentration of Cationic Dye Solution on the Adsorption of Nanofibers*

The initial concentration of cationic dye solution was an important factor in influencing the adsorption rate and the maximum equilibrium adsorption amount. The relationship between the adsorption amount of six cationic dyes and their initial concentrations are shown in Figure 6. It can be seen from the diagram that the unit adsorption capacity increases with the initial concentration until reaching equilibrium. Therefore, the maximum adsorption capacity of the optimized electro-spun fibers for basic red 9, basic red 14, basic red 46, basic blue 9, basic yellow 19 and basic yellow 28 is 86.71 mg/g, 21.513 mg/g, 18.926 mg/g, 44.525 mg/g, 116.516 mg/g and 155.206 mg/g, respectively.

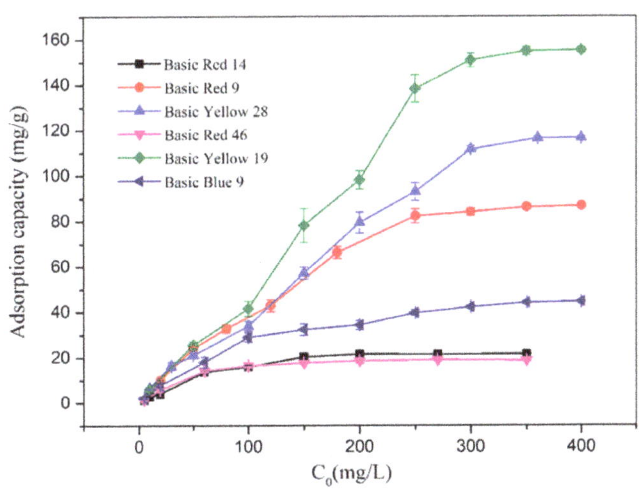

Figure 6. The maximum adsorption capacity of various dyes under different initial concentrations.

From the molecular structure of dyes, the effect of electrostatic attraction due to the charge carried by the molecules being adsorbed by the β-CD/PA nanofiber membrane was similar. A more significant influence should stem from the molecular size and the length of the flexible side chains, which resulted in steric hindrance. The top three dyes in terms of adsorption capacity, i.e., basic yellow 19, basic yellow 28, and basic red 9, all possessed relatively compact molecular structures, moderate sizes, and relatively short flexible side chains with low steric hindrance. Basic blue 9 with a moderately rigid main chain structure was slightly affected by the side chains, thereby demonstrating a medium adsorption capacity. However, basic red 14 and basic red 46, which had stronger molecular flexibility and more side chains, brought about the lowest adsorption capacity. All these factors suggest that, in the process of adsorbing cationic dyes by the β-CD/PA nanofiber membrane, the inclusion of dye molecules by β-CD should play a principal role.

3.4. Effect of pH on the Adsorption of Nanofibers and the Adsorption Mechanism

The pH value was another important reason to affect the adsorption process because the concentration of hydrogen ion (H^+) would influence the surface charge of an adsorbent and the ionization behaviors of both the adsorbent and dyes [27]. In the range of pH from 3 to 9, the effect of pH value on the adsorption process is shown in Figure 7. It can be seen that the adsorption amount of the nanofiber membrane for cationic dyes increased with the increase of pH, and the most attractive adsorption appeared when the dye solution was neutral. Lower pH values provided more positive charges, which would generate stronger repulsive interaction with the positive groups in a dye molecule and the hydroxyl groups with positive charge on cyclodextrin. Protonated dye molecules were not conducive to forming the host–guest inclusion complexes with protonated β-cyclodextrin [28,29]. In addition, because of the high concentration of H^+, the competitive adsorption with carboxyl groups contained in polyacrylate between the available H^+ and dye molecules would also decrease the adsorption capacity. As the pH increased, the positively charged groups in the dye molecules gradually became neutral, which enhanced the interaction among the dye molecules with β-cyclodextrin, and more activated deprotonation carboxyl groups could form an electrostatic interaction with the dye molecules [2]. When the solution was neutral or alkaline, the carboxylic acid of polyacrylate would interact with the hydroxyl ion (OH^-), and the adsorption performance would decrease because of the electrostatic interaction between the dye molecules and the absorbent material. As the alkalinity of the dye solution increased further, part of the β-cyclodextrin dissolved in the aqueous solution and the adsorption material was damaged, resulting in a significant decrease in adsorption performance.

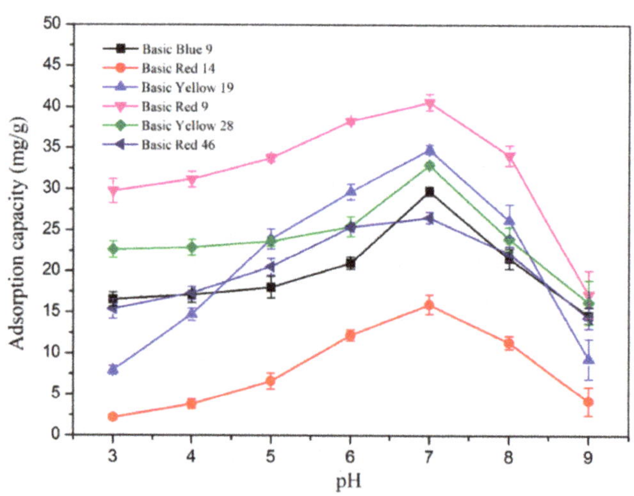

Figure 7. Effect of initial pH value on the adsorption capacity of β-CD/PA nanofiber membrane for six cationic dyes.

As the pH value decreased, the adsorption capacity of the fiber gradually decreased, which provided an opportunity for the desorption of the adsorbent. The possible dynamic adsorption process of the nanofiber is shown in Figure 8. In short, the adsorption performance of nanofiber membrane for dye molecules can be attributed to two processes: electrostatic interactions and the host–guest inclusion complexes. For electrostatic interactions, positively charged dye groups could be attached to β-CD/PA nanofibers through positive and negative electrostatic interactions with carboxyl groups in polyacrylate. For inclusion cooperation, the cavity of β-cyclodextrin in β-CD/PA fibers could form a host–object inclusion compound with dye molecules. Based on the above adsorption mechanism, β-CD/PA fibers possessed favorable adsorption capacity for cationic dyes.

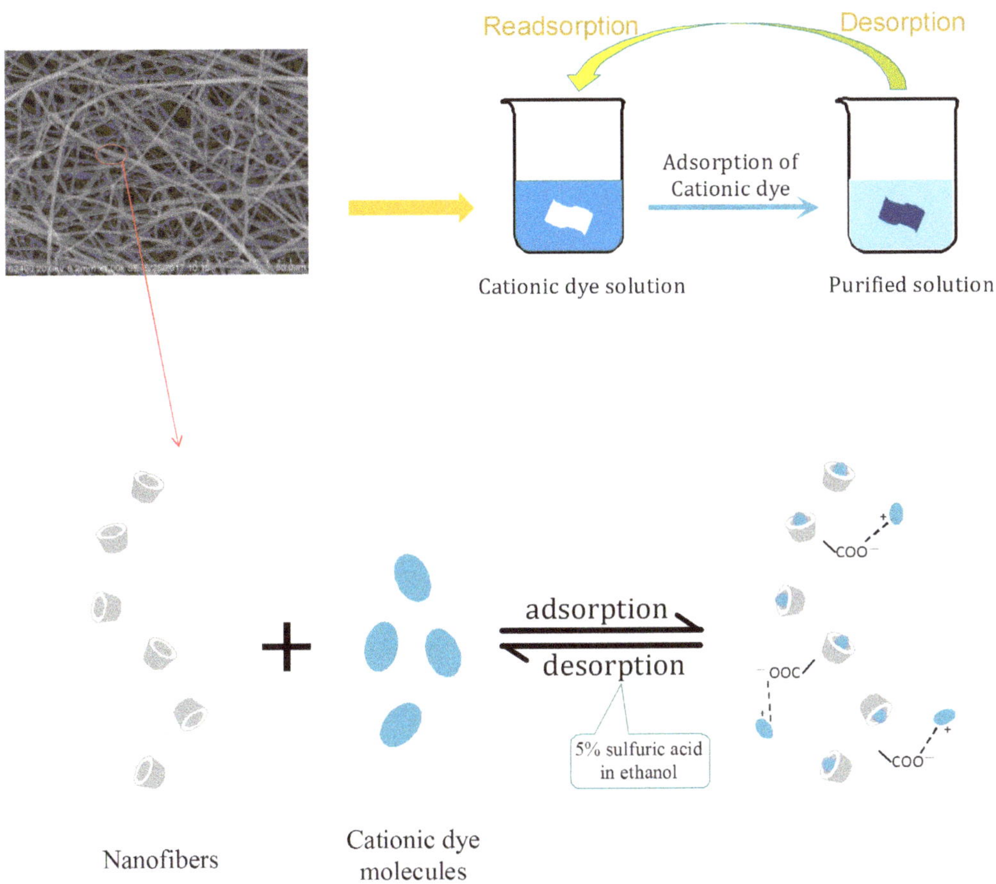

Figure 8. The structure of β-CD/PA composite fiber and its adsorption mechanism for cationic dyes.

3.5. Adsorption Kinetics of β-CD/PA Nanofiber for Cationic Dyes

In order to understand the dynamics of the adsorption process, the adsorption kinetics of β-CD/PA fibers for six cationic dyes were studied using two kinetic models, the pseudo-first-order model and the pseudo-second-order model. The common equation for the pseudo-first-order kinetic model is [18]:

$$Ln(q_e - q_t) = Ln(q_e) - k_1 t \quad (2)$$

where q_e and q_t are the amounts of dye adsorbed (mg/g) at equilibrium and at any instant of time t (h), respectively, and k_1 is the rate constant of pseudo-first-order adsorption (h^{-1}). The plot of $Ln(q_e - q_t)$ versus t gave a straight line for the pseudo-first-order adsorption kinetics, as shown in Figure 9a, and the values of the pseudo-first-order rate constants k_1 and q_e are given in Table 2.

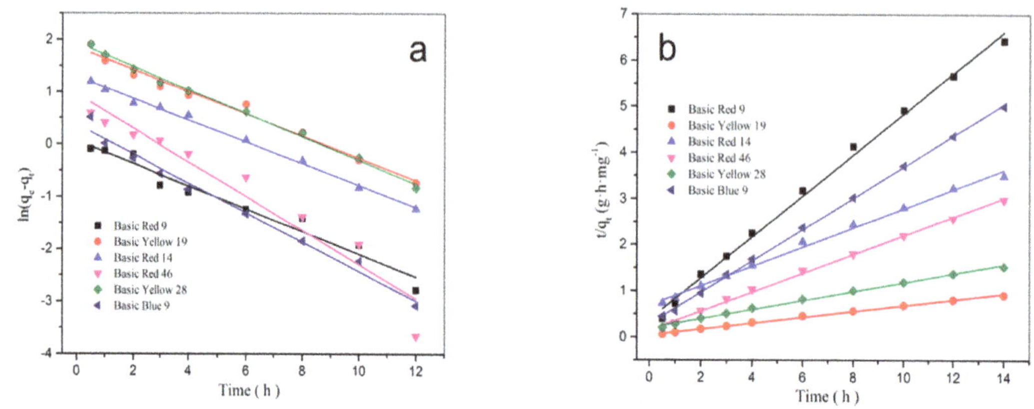

Figure 9. Adsorption kinetic curves of β-CD/PA nanofiber for six cationic dyes. (**a**) pseudo-second-order kinetics; (**b**) pseudo-first-order kinetics.

Table 2. Kinetics parameters for the adsorption of β-CD/PA fibers for six cationic dyes.

Cationic Dyes	Pseudo-First-Order Kinetic			Pseudo-Second-Order Kinetic		
	q_e	k_1	R^2	q_e	k_2	R^2
Basic Red 9	2.11	0.2165	0.958	2.18	0.4452	0.995
Basic Yellow 19	14.78	0.2116	0.983	15.26	0.0639	0.997
Basic Red 14	3.71	0.2073	0.995	4.01	0.2095	0.996
Basic Red 46	4.66	0.3256	0.925	4.68	0.2050	0.995
Basic Yellow 28	8.73	0.2245	0.995	9.16	0.0978	0.997
Basic Blue 9	2.75	0.2807	0.983	2.80	0.3402	0.999

The equation for the pseudo-second-order kinetic model is as follows [18]:

$$t/q_t = 1/\left(k_2 {q_e}^2\right) - 1/q_e \tag{3}$$

where q_e is the equilibrium adsorption capacity (mg/g), q_t is the amount of dye adsorbed at time t (mg/g), and k_2 is the rate constant of pseudo-second-order adsorption ($g{\cdot}mg^{-1}{\cdot}h^{-1}$). Figure 9b is a pseudo-second-order kinetic model line of the adsorption rate for six cationic dyes, and its kinetic parameters are shown in Table 2. Comparing the pseudo-first-order and pseudo-second-order kinetic parameters, the correlation coefficient (R^2) of the pseudo-second-order model is larger, which indicates that the adsorption process of the cationic dyes in the β-CD/PA fibers was closer to the pseudo-second-order kinetic model.

3.6. Desorption and Recycling Ability

From the perspective of economic efficiency, the desorption and recycling of adsorbents is vital for the adsorption process. Therefore, two kinds of desorption liquids, 5% sulfuric acid aqueous solution and 5% sulphate alcohol solution (ethanol volume accounted for 60% of the total volume) were used to understand the desorption capacity of the adsorbent materials. The latter obtained better desorption performance, which reached desorption equilibrium after desorption for 10 h, and the desorption rate was 99.13% (Figure 10a). This may be because the solubility of β-CD and cationic dyes in ethanol was smaller than their solubility in water, so the cationic dyes adsorbed on the β-CD/PA nanofiber membrane by coating were more difficult to get rid of the cavity binding of β-CD during the analysis process but would be resolved together with β-CD under the action of sulfuric acid. The desorption in this case would be higher.

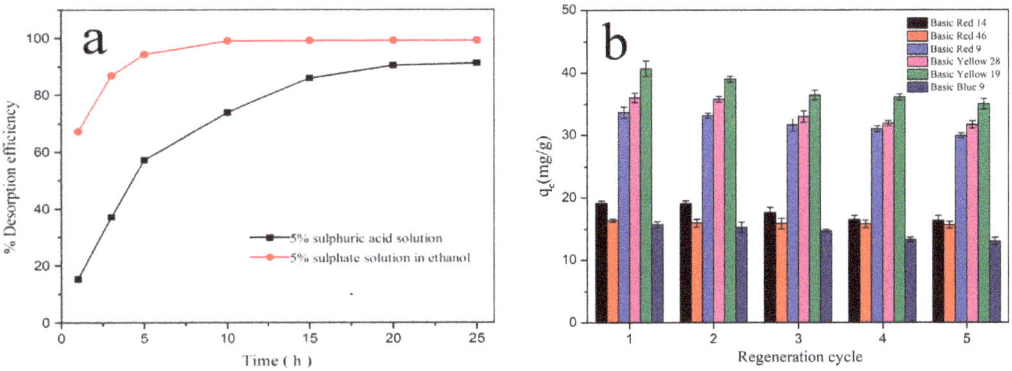

Figure 10. (**a**) Desorption of different desorption fluids for basic yellow 28; (**b**) cyclic adsorption capacity of β-CD/PA nanofiber membrane adsorbent.

Using 5% sulphate alcohol solution as the desorption solution, the adsorption and desorption cycles of six cationic dyes were studied, and the adsorption amount of the adsorbent is shown in Figure 10b. The results show that the adsorption capacity for six cationic dyes after five cycles was higher than 85% of the original adsorption effect. These showed that the β-CD/PA nanofiber adsorbent had good recycling performance

4. Conclusions

Our research has demonstrated a convenient and feasible way to prepare a β-CD/PA composite nanofiber membrane adsorbent by electrostatic spinning. Cationic dyes pollutants in water were obviously removed by the synergistic adsorption of β-cyclodextrin and polyacrylate in this nanofiber membrane. The adsorption properties of the adsorbent materials under different initial concentrations and pH conditions were studied, and it was found that the adsorbent performed better under high initial concentration and a neutral condition. The adsorption kinetics for cationic dyes showed that the adsorption process could be consistent with the pseudo-second-order kinetic model. In addition, nanofiber membrane adsorbent had good desorption after adsorbing cationic dyes and a high recycling efficiency. These results demonstrated that β-CD/PA nanofiber membrane could be used as a promising adsorbent for dye wastewater treatment.

Author Contributions: Methodology, C.Z.; Validation, C.Z.; Formal analysis, C.Z.; Investigation, C.Z.; Data curation, C.Z. and X.T.; Writing—original draft, W.Z.; Writing—review & editing, C.Z.; Supervision, C.Z.; Project administration, S.Z.; Funding acquisition, X.T. and S.Z. All authors have read and agreed to the published version of the manuscript.

Funding: This research was funded by General Program of National Natural Science Foundation of China (Grant No. 51003047), Natural Science Foundation of Jiangsu Province (Grant No. SBK201121603), Higher Education Institutions Natural Science Foundation of Jiangsu Educational Commission (Grant No. 09KJB540001) and Open Foundation of Hubei Key Laboratory of biomass fiber and Ecological Dyeing and Finishing (Grant No. 201603).

Institutional Review Board Statement: This research does not involve human or animal studies.

Data Availability Statement: Data are contained within the article.

Conflicts of Interest: The authors declare that they have no known competing financial interests or personal relationships that could have appeared to influence the work reported in this paper.

References

1. Li, J.; Feng, J.; Yan, W. Synthesis of polypyrrole-modified TiO_2 composite adsorbent and its adsorption performance on acid Red G. *J. Appl. Polym. Sci.* **2013**, *128*, 3231–3239. [CrossRef]
2. Zhao, R.; Wang, Y.; Li, X.; Sun, B.; Jiang, Z.; Wang, C. Water-insoluble sericin/β-cyclodextrin/PVA composite electrospun nanofibers as effective adsorbents towards methylene blue. *Colloids Surf. B Biointerfaces* **2015**, *136*, 375–382. [CrossRef]
3. McMullan, G.; Meehan, C.; Conneely, A.; Kirby, N.; Robinson, T.; Nigam, P.; Banat, I.M.; Marchant, R.; Smyth, W.F. Microbial decolourisation and degradation of textile dyes. *Appl. Microbiol. Biot.* **2001**, *56*, 81–87. [CrossRef] [PubMed]
4. Ghaedi, M.; Jah, A.H.; Khodadoust, S.; Sahraei, R.; Daneshfar, A.; Mihandoost, A.; Purkait, M.K. Cadmium telluride nanoparticles loaded on activated carbon as adsorbent for removal of sunset yellow. *Spectrochim. Acta Part A Mol. Biomol. Spectrosc.* **2012**, *90*, 22–27. [CrossRef] [PubMed]
5. Fan, L.; Zhang, Y.; Luo, C.; Lu, F.; Qiu, H.; Sun, M. Synthesis and characterization of magnetic β-cyclodextrin–chitosan nanoparticles as nano-adsorbents for removal of methyl blue. *Int. J. Biol. Macromol.* **2012**, *50*, 444–450. [CrossRef] [PubMed]
6. Feng, Y.; Feng, N.; Wei, Y.; Zhang, G. An in situ gelatin-assisted hydrothermal synthesis of ZnO-reduced graphene oxide composites with enhanced photocatalytic performance under ultraviolet and visible light. *RSC Adv.* **2014**, *4*, 7933. [CrossRef]
7. Ayekoe, C.Y.P.; Robert, D.; Lanciné, D.G. Combination of coagulation-flocculation and heterogeneous photocatalysis for improving the removal of humic substances in real treated water from Agbô River (Ivory-Coast). *Catal. Today* **2017**, *281*, 2–13. [CrossRef]
8. Liu, Y.; Zhu, W.; Guan, K.; Peng, C.; Wu, J. Freeze-casting of alumina ultra-filtration membranes with good performance for anionic dye separation. *Ceram. Int.* **2018**, *44*, 11901–11904. [CrossRef]
9. Hua, Y.; Xiao, J.; Zhang, Q.; Cui, C.; Wang, C. Facile synthesis of surface-functionalized magnetic nanocomposites for effectively selective adsorption of cationic dyes. *Nanoscale Res. Lett.* **2018**, *13*, 99. [CrossRef]
10. Vakili, M.; Rafatullah, M.; Salamatinia, B.; Abdullah, A.Z.; Ibrahim, M.H.; Tan, K.B.; Gholami, Z.; Amouzgar, P. Application of chitosan and its derivatives as adsorbents for dye removal from water and wastewater: A review. *Carbohyd. Polym.* **2014**, *113*, 115–130. [CrossRef] [PubMed]
11. Li, T.; Liu, Y.; Wang, T.; Wu, Y.; He, Y.; Yang, R.; Zheng, S. Regulation of the surface area and surface charge property of MOFs by multivariate strategy: Synthesis, characterization, selective dye adsorption and separation. *Micropor. Mesopor. Mat.* **2018**, *272*, 101–108. [CrossRef]
12. Martín, D.M.; Faccini, M.; García, M.A.; Amantia, D. Highly efficient removal of heavy metal ions from polluted water using ion-selective polyacrylonitrile nanofibers. *J. Environ. Chem. Eng.* **2018**, *6*, 236–245. [CrossRef]
13. Wu, J.; Wang, N.; Zhao, Y.; Jiang, L. Electrospinning of multilevel structured functional micro-/nanofibers and their applications. *J. Mater. Chem. A* **2013**, *1*, 7290. [CrossRef]
14. Celebioglu, A.; Uyar, T. Electrospinning of nanofibers from non-polymeric systems: Polymer-free nanofibers from cyclodextrin derivatives. *Nanoscale* **2012**, *4*, 621–631. [CrossRef] [PubMed]
15. Fang, Q.; Zhu, M.; Yu, S.; Sui, G.; Yang, X. Studies on soy protein isolate/polyvinyl alcohol hybrid nanofiber membranes as multi-functional eco-friendly filtration materials. *Mater. Sci. Eng. B* **2016**, *214*, 1–10. [CrossRef]
16. Zhang, W.; Chen, M.; Diao, G. Electrospinning β-cyclodextrin/poly(vinyl alcohol) nanofibrous membrane for molecular capture. *Carbohyd. Polym.* **2011**, *86*, 1410–1416. [CrossRef]
17. Uyar, T.; Nur, Y.; Hacaloglu, J.; Besenbacher, F. Electrospinning of functional poly(methyl methacrylate) nanofibers containing cyclodextrin-menthol inclusion complexes. *Nanotechnology* **2009**, *20*, 125703. [CrossRef] [PubMed]
18. Chen, M.; Wang, C.; Fang, W.; Wang, J.; Zhang, W.; Jin, G.; Diao, G. Electrospinning of Calixarene-Functionalized Polyacrylonitrile Nanofiber Membranes and Application as an Adsorbent and Catalyst Support. *Langmuir* **2013**, *29*, 11858–11867. [CrossRef] [PubMed]
19. Ebadi, A.; Rafati, A.A. Preparation of silica mesoporous nanoparticles functionalized with β-cyclodextrin and its application for methylene blue removal. *J. Mol. Liq.* **2015**, *209*, 239–245. [CrossRef]
20. Mallard, I.; Städe, L.W.; Ruellan, S.; Jacobsen, P.A.L.; Larsen, K.L.; Fourmentin, S. Synthesis, characterization and sorption capacities toward organic pollutants of new β-cyclodextrin modified zeolite derivatives. *Colloids Surf. A Physicochem. Eng. Asp.* **2015**, *482*, 50–57. [CrossRef]
21. Badruddoza, A.Z.M.; Hazel, G.S.S.; Hidajat, K.; Uddin, M.S. Synthesis of carboxymethyl-β-cyclodextrin conjugated magnetic nano-adsorbent for removal of methylene blue. *Colloids Surf. A Physicochem. Eng. Asp.* **2010**, *367*, 85–95. [CrossRef]
22. Liu, X.; Yan, L.; Yin, W.; Zhou, L.; Tian, G.; Shi, J.; Yang, Z.; Xiao, D.; Gu, Z.; Zhao, Y. A magnetic graphene hybrid functionalized with beta-cyclodextrins for fast and efficient removal of A magnetic graphene hybrid functionalized with beta-cyclodextrins for fast and efficient removal of organic dyes. *J. Mater. Chem. A* **2014**, 12296–12303. [CrossRef]
23. Liu, M.; Xue, Y.; Cai, Y.L.; Zhu, Y.Q.; Li, L.F.; Zheng, C.L. Water resistance of β-cyclodextrin/polyacrylate nanofiber films and their supramolecular capturing performance for heavy metal ions. *J. Text. Res.* **2016**, *37*, 1–5.
24. Wang, X.; Shen, Y.; Lai, X. Micromorphology and mechanism of polyurethane/polyacrylate membranes modified with epoxide group. *Prog. Org. Coat.* **2014**, *77*, 268–276. [CrossRef]

25. Guo, Y.; Li, S.; Wang, G.; Ma, W.; Huang, Z. Waterborne polyurethane/poly(n-butyl acrylate-styrene) hybrid emulsions: Particle formation, film properties, and application. *Prog. Org. Coat.* **2012**, *74*, 248–256. [CrossRef]
26. Sun, J.; Yuan, A.L.; Zheng, C.L. *Dyes Chemistry*; Chemical Industry Press: Beijing, China, 2022.
27. Liu, L.; Xu, S.Q. The adsorption performance of β-cyclodextrin functionalized polyacrylonitrile (PAN) electrospun nanofibers. *Shanghai Text. Sci. Technol.* **2014**, *42*, 5–8.
28. Zhao, L.; Yu, B.; Xue, F.; Xie, J.; Zhang, X.; Wu, R.; Wang, R.; Hu, Z.; Yang, S.; Luo, J. Facile hydrothermal preparation of recyclable S-doped graphene sponge for Cu^{2+} adsorption. *J. Hazard. Mater.* **2015**, *286*, 449–456. [CrossRef] [PubMed]
29. Zhang, X.; Shi, L.; Xu, G.; Chen, C. Synthesis of β-cyclodextrin-calix[4]arene coupling product and its adsorption of basic fuchsin and methylene blue from water. *Jincl. Phenom. Macro* **2013**, *75*, 147–153. [CrossRef]

polymers

MDPI

Article

Theory-Driven Tailoring of the Microenvironment of Quaternary Ammonium Binding Sites on Electrospun Nanofibers for Efficient Bilirubin Removal in Hemoperfusion

Xingyu Fu [1,†], Minsi Shi [1,†], Dingyang Chen [1], Xinyue Zhao [1], Tingting Jiang [2,*] and Rui Zhao [1,*]

1 Key Laboratory of Polyoxometalate and Reticular Material Chemistry of Ministry of Education, Faculty of Chemistry, Northeast Normal University, Changchun 130024, China; fuxingyu@nenu.edu.cn (X.F.); shiminsi@126.com (M.S.); chendy799@nenu.com (D.C.); zxy2022aaa@163.com (X.Z.)
2 School of Chemical Engineering, Northeast Electric Power University, Jilin 132012, China
* Correspondence: jiangting_chem@neepu.edu.cn (T.J.); zhaor814@nenu.edu.cn (R.Z.)
† These authors contributed equally to this work.

Abstract: Efficient adsorbents for excess bilirubin removal are extremely important for the treatment of hyperbilirubinemia. However, traditional adsorbents, such as activated carbons and ion-exchange resins, still suffer from dissatisfactory adsorption performance and poor blood compatibility. Herein, we adopted a rational design strategy guided by density functional theory (DFT) calculations to prepare blood-compatible quaternary ammonium group grafted electrospun polyacrylonitrile nanofiber adsorbents. The calculation analysis and adsorption experiments were used to investigate the structure–function relationship between group types and bilirubin adsorption, both indicating that quaternary ammonium groups with suitable configurations played a crucial role in bilirubin binding. The obtained nanofiber adsorbents showed the bilirubin removal efficiency above 90% even at a coexisting BSA concentration of 50 g L^{-1}. The maximum adsorption capacities were 818.9 mg g^{-1} in free bilirubin solution and 163.7 mg g^{-1} in albumin bound bilirubin solution. The nanofiber adsorbents also showed considerable bilirubin removal in dynamic adsorption to reduce the bilirubin concentration to a normal level, which was better than commercial activated carbons. Our study demonstrates the high feasibility of a theory-driven design method for the development of grafted electrospun nanofibers, which have good potential as bilirubin adsorbents in hemoperfusion applications.

Keywords: electrospinning; grafting; adsorption; bilirubin; hemoperfusion

1. Introduction

Bilirubin is a metabolite of aging red blood cells, and is transported by the blood to the liver for metabolism and excreted through the biliary tract [1]. If the body experiences symptoms such as liver damage or blockage of the biliary tract, excessive production of bilirubin can occur in the blood, resulting in so-called hyperbilirubinemia. Hyperbilirubinemia can lead to jaundice, hepatitis, brain damage, and in severe cases, death. Meanwhile, other complications in patients have been clinically found to be accompanied by a massive accumulation of bilirubin in the human body [2,3]. How to quickly and effectively remove excess bilirubin from blood is an urgent issue that needs to be solved. At present, blood purification therapy has the most therapeutic effect on patients with hyperbilirubinemia in clinical practice [4]. Blood purification therapy mainly includes blood replacement, hemodialysis, and hemoperfusion adsorption [5]. Blood replacement requires a large amount of fresh plasma, which is expensive and difficult in terms of clinical application. Hemodialysis has a good removal effect on small blood poisons; however, it has little effect on the removal of albumin-bound bilirubin. Among them, hemoperfusion adsorption is regarded as a simple, efficient, and practical bilirubin treatment method, which can remove albumin-bound bilirubin to the maximum extent. The core component of the hemoperfusion method is the specific bilirubin adsorbents [6].

In the past few decades, various adsorbents, including activated carbons [7], polymer resins [8], natural polymers [2], inorganic mesoporous materials [5], etc., have been widely studied in hemoperfusion therapy to remove bilirubin. However, most of them have problems such as low adsorption performance, complex separation, high cost, and poor blood compatibility. Therefore, it is necessary to develop novel adsorption materials with high adsorption efficiency and blood safety for bilirubin adsorption, but this remains challenging. Electrospinning technology is a feasible method to produce continuous one-dimensional nanofibers with the action of a high-voltage electrostatic field [9]. This method exhibits simple equipment, controllable preparation process, abundant raw materials, and a relatively low cost. The obtained nanofibers have high porosity, high pore connectivity, easy separation, and facile modification properties [10]. As a result, electrospun fibers are good candidates for adsorbents in aqueous solution systems [11]. Recently, electrospun fibers have been used as bilirubin adsorbents during hemoperfusion to purify blood. Natural polymers and synthetic polymers have been electrospun for bilirubin adsorption. Ma and co-workers applied electrospun cellulose nanofibers as an affinity membrane, and this nanofiber membrane showed a bilirubin adsorption capacity of 4.0 mg g^{-1} [12]. In Jiao's group, nanoporous polyethersulfone (PES) fiber mats were prepared via electrospinning and this adsorbent displayed an adsorption capacity of 176.0 mg g^{-1} for bilirubin [13]. They also immobilized bovine serum albumin on electrospun PES fibers, which showed an excellent bilirubin adsorption capacity of 192.4 mg g^{-1} [14]. Moreover, grafted electrospun polyacrylonitrile nanofibers also showed promising bilirubin removal. In our previous work, branched polyethylenimine was grafted on electrospun polyacrylonitrile fibers and they exhibited a maximum adsorption capacity toward albumin-bonded bilirubin of 112.9 mg g^{-1} [15]. Liu and co-workers grafted lysine on polyacrylonitrile electrospun nanofibers. The maximum adsorption capacity toward bilirubin was 64.0 mg g^{-1} [16]. Studies have found that necessary modifications need to be carried out on electrospun fibers before their usage as bilirubin adsorbents. Nevertheless, there has been a lack of systematic research into the interaction strength between functional groups and bilirubin molecules. Investigation of this structure–function relationship is needed. Based on the chemical structure (Figure 1a), bilirubin has anionic groups and conjugated five-membered rings. Moreover, bilirubin is also a hydrophobic molecule [17]. Thus, modifying the electrospun fiber adsorbents with positively charged functional groups with different microenvironments is an effective approach to study the interaction between cationic adsorption sites and bilirubin, which is beneficial to the design desirable adsorbents.

Based on the above reasons, in this work, our proposed method was firstly to utilize quantitative calculations for functional group design, avoiding the consumption of significant human and financial resources [18]. Three quaternary ammonium groups with different microenvironments of ethyl, tolyl, and octyl structures, which considered electrostatic interaction, $\pi-\pi$ interaction, and hydrophobic interaction, were designed. The interaction energy of three quaternary ammonium groups on bilirubin was calculated and compared to find the suitable functional groups. To verify the calculations, the designed three quaternary ammonium groups were grafted on the electrospun nanofibers for bilirubin adsorption. Both calculation and experiment results suggested that the quaternary ammonium group with low steric resistance showed high bilirubin affinity. The adsorption rate of the obtained nanofiber adsorbent was faster than that of commercial activated carbon and resin. The obtained maximum adsorption capacities for free bilirubin solution and albumin-bound bilirubin solution were 818.9 mg g^{-1} and 163.7 mg g^{-1}, respectively. The nanofiber adsorbents could also remove more than 90% of bilirubin from albumin-bound bilirubin solution. In the simulated hemoperfusion experiments, the designed nanofibers could effectively decrease the bilirubin concentration to a normal level, which was much better than commercial activated carbon. The overall results demonstrate that our designed electrospun nanofiber adsorbent is expected to be used as a high-affinity bilirubin adsorbent in hemoperfusion.

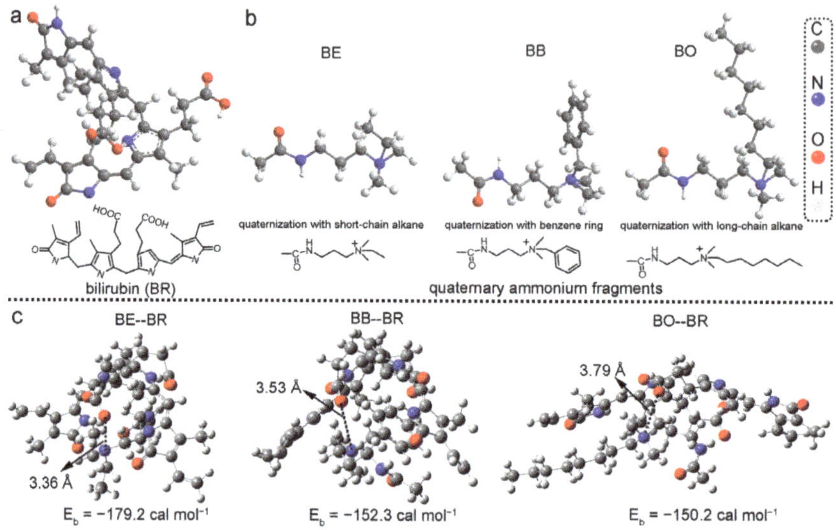

Figure 1. (**a**) Chemical structure of bilirubin. (**b**) Chemical structures of the designed quaternary ammonium groups. (**c**) The optimized adsorption complexes of bilirubin with the three quaternary ammonium groups via density functional theory.

2. Materials and Methods

2.1. Materials and Instruments

Polyacrylonitrile (PAN, Mw = 150,000 g mol^{-1}), bilirubin (98%), and multi-walled carbon nanotubes (CNTs), were purchased from Aladdin Biochemical Technology Co., Ltd., Shanghai, China. *N*,*N*-dimethylformamide (DMF) was purchased from Tiantai Chemical Corporation, Guangzhou, China. *N*,*N*-dimethyl-1,3-propanediamine (98%), bromoethane (99%), benzyl bromide (98%), and 1-bromooctane (98%) were obtained from Energy Chemical, Shanghai, China. All reagents were used without further purification. FT-IR spectra were recorded on a Nicolet iS50 Fourier transform infrared spectrometer from 4000 to 400 cm^{-1}. Field emission scanning electron microscopy (SEM, Shimadzu SSX-550, Kyoto, Japan) was used to observe the morphology.

2.2. Synthesis of Grafted Electrospun Nanofibers

PAN nanofibers were obtained from the electrospinning process. Briefly, PAN (0.8 g), and CNTs (0.8 g) powders were dissolved in 10 mL of DMF for 8 h with magnetic stirring at 90 °C to form a homogeneous solution. Then, the obtained solution was loaded into a 20 mL plastic syringe with a stainless-steel needle (internal diameter: 0.5 mm) with a distance between the needle tip and the collector of 20 cm and an electrospinning voltage of 16 kV. The flow rate of the solution was controlled at 0.5 mL h^{-1} by a syringe pump. PAN nanofibers were collected from the collector.

The grafting processes were conducted as follows: 500 mg of obtained PAN nanofibers, 14 mL of *N*,*N*-dimethyl-1,3-propanediamine, and 12 mL of water were added into a three-neck flask for reflux process for 3 h. After the reaction mixture was cooled, the aminated PAN nanofibers were washed by water and ethanol in turn, then dried in vacuum oven at 60 °C for 12 h.

The aminated PAN nanofibers were added into ethanol solution containing the quaternary reagent (bromoethane, benzyl bromide, and 1-bromooctane; the weight ratio of quaternary reagent and aminated PAN nanofibers was 10:1). The reaction mixture was stirred for 24 h under reflux. Then, the grafted electrospun nanofibers were filtered and washed repeatedly with ethanol and water. After washing, the products were dried in a vacuum oven at 60 °C for 12 h.

2.3. Bilirubin Adsorption Experiments

Due to the high sensitivity of bilirubin to light, all solution preparation and adsorption experiments were carried out in the dark. To obtain a homogeneous bilirubin solution, a certain amount of bilirubin was initially dissolved in a small volume of dimethyl sulfoxide and 0.1 M Na_2CO_3 solution, and then diluted with phosphate buffer saline (PBS, pH = 7.4) to obtain the desired concentration. The concentration of bilirubin was determined using a UV–visible spectrophotometer. Each adsorption experiment was conducted in triplicate to obtain reproductive results. The amount of adsorbed bilirubin was analyzed according to the following equation:

$$q\ (\text{mg/g}) = \frac{(C_0 - C_e)V}{W} \quad (1)$$

where C_0 and C_e are the initial concentration and equilibrium concentration of bilirubin in the test solution (mg L^{-1}), V is the volume of the test solution (L), and W is the mass of the adsorbent (g).

For the dynamic adsorption experiment, an appropriate amount of nanofiber adsorbent was placed into the adsorption column, and the dynamic adsorption was performed by the power pump at an appropriate speed for 4 h. At a certain time, the bilirubin concentration was determined.

2.4. Blood Compatibility Experiments

Hemolysis Assay: Hemolytic ability was determined by incubating the nanofibers with red blood cells at 37 °C for 1 h. Then, 1 mL rat blood was placed in the centrifuge tube and centrifuged at 3000 rpm for 10 min. The serum (upper layer) was discarded, and the blood cells (under layer) were washed three times with PBS solution and then diluted to 10 mL. Fiber materials were washed with PBS and then fibers with different amounts (10, 20, 60, and 100 mg, respectively) were added into the as-mentioned suspension. Deionized water and PBS were used as positive and negative controls, respectively. After incubation at 37 °C for 1 h, the mixture was centrifuged at 3000 rpm 10 min and the absorbance of the supernatant solution at 541 nm was recorded by UV–vis spectroscopy. The hemolysis rate was calculated using the following equation:

$$\text{Hemolysis rate}\ (\%) = \frac{OD_s - OD_{nc}}{OD_{pc} - OD_{nc}} \times 100 \quad (2)$$

in which OD_s, OD_{pc}, and OD_{nc} are the absorbances of the sample, positive control, and negative control, respectively.

Coagulation time tests: The activated partial time (APTT), thrombin time (TT), and prothrombin time (PT) were used to evaluate the anticoagulation property. Firstly, different amounts of fiber materials were put into 24-well plate, and the physiological saline was added into the plates for immersion overnight and then incubated at 37 °C for 1 h. After removing the physiological saline, 100 μL fresh platelet-poor plasma (PPP) was introduced and incubated with the nanofibers at 37 °C for 20 min. Platelet-poor plasma (PPP) could be obtained by centrifuging the whole anticoagulant blood at 4000 rpm for 15 min. For APTT measurement, 100 μL of APTT agent was added to the wells at 37 °C and incubated at 37 °C for another 5 min. Then, 100 μL of 0.025 M $CaCl_2$ was added. Then, the coagulation time was measured as an APTT value. For the TT test, 100 μL of thrombin agent (incubated 10 min at 37 °C before use) was added to the wells, and then TT values were measured. To test PT, 100 μL Thromborel S (incubated 10 min before use) was added to the wells at 37 °C and further incubated at 37 °C for 2 min, and then the PT values were measured. Pure PPP without samples was the control group. All measurements were carried out six times.

Platelet adhesion test: To study platelet adhesion, the fiber membranes (1 cm × 1 cm) were placed in a 24-well cell culture plate. One milliliter of fresh platelet-rich plasma (PRP) was dropped in each well and then incubated at 37 °C for 2 h. The PRP was removed with an aspirator, and the membrane was rinsed three times with normal saline, and then the adhered platelets were fixed with 2.5 wt% glutaraldehyde in normal saline at 4 °C for 12 h. Finally, the samples were washed with normal saline, and then dehydrated with a series of ethanol/normal saline mixtures of increasing ethanol concentration (0, 25, 50, 75, and 100 wt% ethanol, 15 min in each mixture) and then dried at room temperature. The platelet adhesion was observed using scanning electron microscopy.

2.5. Computational Method

The interaction energies between the quaternary ammonium fragments and bilirubin were evaluated using quantum chemical calculations. The 3D structures of the fragments were built with GaussView5.0.8, and bilirubin 3D structure was downloaded from the pubchem website. All calculations were carried out using the density functional theory (DFT) with the B3LYP function [19,20], as implemented in the Gaussian 09 program [21]. All geometry structures were optimized with the 6-31G(d,p) basis set. On the basis of optimized geometries, the best binding structures of bilirubin with the quaternary ammonium fragments were searched with the AutoDock Vina program [22], in which all parameters are default. Following this, all binding structures were further optimized using B3LYP with 6-31G(d,p) basis set. To obtain more accurate energies, single-point calculations were performed based on these optimized geometries with the 6-311++G(2d,2p) basis set on all atoms. All binding energies (BEs) were calculated to evaluate the adsorption strengths of bilirubin in the different quaternary ammonium fragments. Basis set superposition error (BSSE) was considered in the calculations of binding energies.

3. Results

3.1. Design and Theoretical Calculations

Cationic quaternary ammonium groups are the common functional groups used for bilirubin adsorption in the previous literature [23] and commercially available adsorbents. Moreover, the chemical environments surrounding the quaternary ammonium sites are also important for bilirubin (BR) binding [17]. In this work, three quaternary ammonium groups with different quaternary structures were selected as the theoretical models to assess their binding ability toward bilirubin molecules through performing density functional theory (DFT) calculations. The different quaternary structures were ethyl, tolyl, and octyl building units, respectively (Figure 1b). The ethyl quaternary structure (BE) could only provide electrostatic interaction. The octyl quaternary structure (BO) could provide electrostatic interaction and hydrophobic interaction. The tolyl quaternary structure (BB) could provide electrostatic interaction, $\pi-\pi$ interaction, and hydrophobic interaction. As a result, the influence of chemical structures of quaternary ammonium groups involving the multiple interactions in the bilirubin adsorption was comprehensively investigated. The optimized adsorption geometries of these functional groups toward bilirubin are displayed in Figure 1c. Interestingly, we found that the calculated binding energy (E_b) of BE--BR (-179.2 kcal mol^{-1}), which only involved electrostatic interaction, was higher than that of the other two systems (BB--BR (-152.3 kcal mol^{-1}) and BO--BR (-150.2 kcal mol^{-1})). The results indicated that electrostatic interaction was the main driving force for bilirubin binding. Though other van der Waals forces ($\pi-\pi$ interaction and hydrophobic interaction) were also attributed to bilirubin adsorption, the geometry of the quaternary ammonium group was more important. It was observed that the bond distance belonging to the electrostatic interaction of BE--BR was lower than that of BB--BR and BO--BR. Introducing a large quaternary structure, such as tolyl and octyl units, significantly increased the steric hindrance. The calculated data suggested that quaternary ammonium group with low steric hindrance showed the strongest binding affinity toward bilirubin, which guided the design of more efficient adsorbents for bilirubin.

3.2. Preparation and Characterization of Three Electrospun Nanofiber Adsorbents

Based on the theoretical analysis and preparation feasibility, the above three quaternary ammonium groups were grafted onto electrospun nanofibers. Electrospun polyacrylonitrile (PAN) nanofibers (PAN-NFs) were selected as the substrates because cyano groups of PAN-NFs have high chemical reactivity for grafting. To improve the mechanical strength of the PAN-NFs, carbon nanotubes served as a reinforcement component in the electrospinning solution (Figure 2a). A simple three-step approach was applied to prepare the three quaternary ammonium grafted electrospun nanofibers (Figure 2b). First, PAN-NF was reacted with *N,N*-dimethyl-1,3-propanediamine to introduce tertiary amine on the nanofibers, yielding PAN-DA-NF. Then, PAN-DA-NF was reacted with different quaternary reagents (bromoethane, benzyl bromide, and 1-bromooctane). To reduce the toxicity, counter-balancing Br^- ions were ion-exchanged by Cl^- ions, finally yielding the targeted PAN-BE-NF, PAN-BB-NF, and PAN-BO-NF, respectively. These grafted quaternary ammonium groups were consistent with the fragments in theoretical calculations.

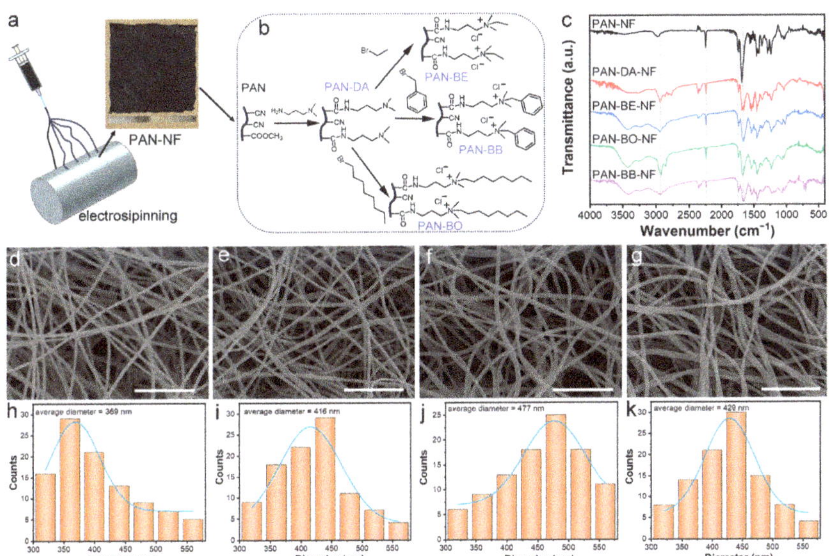

Figure 2. (**a**) Schematic illustration of the electrospinning process. (**b**) The grafting routes of the quaternary ammonium group modified electrospun nanofibers. (**c**) FT-IR spectra of the obtained fibers. SEM images and diameter distributions of PAN-NF (**d**,**h**), PAN-BE-NF (**e**,**i**), PAN-BO-NF (**f**,**j**), and PAN-BB-NF (**g**,**k**) (the scale bar is 10 μm).

The chemical structures of the prepared nanofibers were characterized by FT-IR spectra. As shown in Figure 2c, PAN-NF displayed typical characteristic peaks of polyacrylonitrile. Peaks at 2247 cm^{-1} and 1738 cm^{-1} were assigned to the vibrations of cyano groups and carbonyl groups of methacrylate copolymers [9]. The peaks at 2930 cm^{-1} and 2859 cm^{-1} belonged to the C-H stretching vibration of alkane groups [24]. After the amination reaction, the intensities of peaks at 2247 cm^{-1} and 1738 cm^{-1} were reduced in the spectrum of PAN-DA-NF. Moreover, PAN-DA-NF showed a new strong absorption peak at 1659 cm^{-1}, which was due to the stretching of the C=O amide group [3]. These results demonstrated the hydrolysis of cyano and methacrylate groups, and the amidation reaction occurred between the amino group of *N,N*-dimethyl-1,3-propanediamine and the hydrolyzed carboxyl group. After the quaternization reaction, the spectrum of PAN-BE-NF showed few changes compared with that of PAN-DA-NF. However, the peak intensities at 2930 cm^{-1} and 2859 cm^{-1} were evidently enhanced for PAN-BO-NF, owing to the grafted long alkyl chain. The spectrum of PAN-BB NF exhibited a bending vibration of the benzene ring at 698 cm^{-1} and 733 cm^{-1} [25]. The degree of modification of grafted nanofibers can be expressed by weight gain (%) and functionality (mmol g^{-1}). Their expressions are showed as follows:

$$\text{Weight gain } (\%) = \frac{W_m - W_0}{W_0} \times 100 \tag{3}$$

$$\text{Functionality} = \frac{W_m - W_0}{W_m M} \tag{4}$$

where W_0 and W_m represent the weight of the nanofibers before and after grafting, respectively, and M is the molecular weight caused by functional organic molecules. Correspondingly, the weight gain rate and functional group grafting rate were calculated, and each grafting rate of the quaternary ammonium groups was stable at about 1.8 mmol g^{-1} (Table 1), suggesting that the content of quaternary ammonium groups in the three nanofibers was nearly the same. Combined with FT-IR results, it was proved that different quaternary ammonium groups were grafted onto the electrospun PAN nanofibers successfully. The morphology of the nanofibers was observed using a scanning electron microscope (SEM). All the grafted nanofibers maintained a good fibrous shape, and a large number of pores existed among the fibers (Figure 2d–g), which made them suitable for adsorption applications. As shown in Figure 2h–k, the average diameter of PAN-NF was 369 nm. After the grafting reaction, due to the grafted groups and swelling, the fibers became more disordered and the average diameter increased.

Table 1. Modification degree of different nanofibers.

Sample	Weight Gain (%)	Functionality (mmol g^{-1})
PAN-NF	--	--
PAN-DA-NF	25.30%	1.96
PAN-BE-NF	5.50%	1.79
PAN-BB-NF	19.80%	1.82
PAN-BO-NF	25.70%	1.81

3.3. Bilirubin Adsorption Experiments

We conducted a comprehensive examination of the adsorption performance of the obtained grafted electrospun nanofibers toward bilirubin. In order to understand the adsorption rate and saturation adsorption time, the adsorption kinetics were studied. As can be seen from the kinetic curve shown in Figure 3a, the results showed that the PAN-BE-NF displayed the fastest adsorption rate and most efficient adsorption ability for free bilirubin. In the initial concentration of 150 mg L^{-1}, the saturation equilibrium time was reached at 40 min, and the residual bilirubin concentration was lower than the normal bilirubin level of the human body (10 mg L^{-1}). The equilibrium time for PAN-BB-NF and PAN-BO-NF was 60 min and 120 min, respectively, and their residual concentrations all exceeded 10 mg L^{-1}. However, based on the curves, the adsorption performance of the three electrospun nanofibers were all better than the commercial activated carbon and ion-exchange resin. This was because the abundant pores among the fibers were beneficial to the bilirubin molecule diffusion, which accelerated the adsorption rate.

In order to further directly compare the adsorption capacity of the nanofibers, isothermal experiments of free bilirubin adsorption were carried out at the different initial bilirubin concentrations (Figure 3b). Two classical isotherm models, Langmuir and Freundlich equations, were used to describe the equilibrium properties of adsorption. The fitting curves and R^2 results indicated that these three equilibrium processes were more suitable to be described by the Langmuir model, suggesting that the adsorption sites on the fiber surfaces were uniformly distributed, and the adsorption process could be considered as monolayer adsorption [26]. As summarized in Table 2, PAN-BE-NF had the largest adsorption capacity for free bilirubin uptake, with a maximum adsorption capacity up to 818.9 mg g^{-1}. This value was much higher than PAN-BB-NF (623.8 mg g^{-1}) and PAN-BO-NF (536.8 mg g^{-1}). PAN-BE-NF's quaternary ammonium groups showed low steric hindrance and high binding energy, resulting in high bilirubin uptake. The adsorption kinetics and isotherms results agreed with the theory calculation analysis, suggesting that the theory-driven method was an effective way to rationally design the adsorbents and avoid the trial-and-error process.

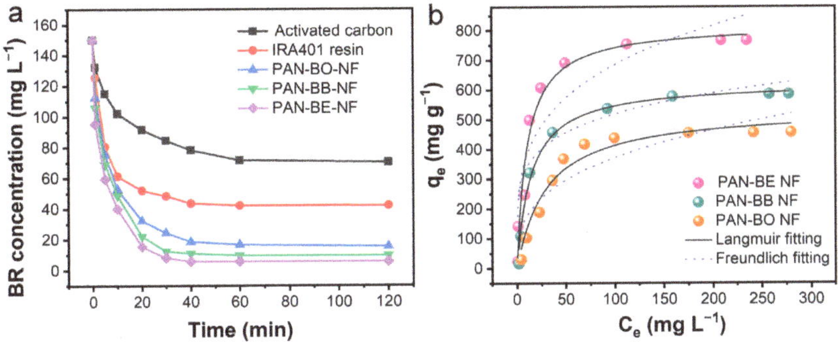

Figure 3. (**a**) Bilirubin (BR) removal versus contact time by different adsorbents (C_{BR} = 150 mg L^{-1}, adsorbent dosage = 1 g L^{-1}, pH = 7.4, temperature = 37 °C, shaking speed = 120 rpm) and (**b**) bilirubin removal versus contact time by different nanofiber adsorbents (C_{BR} = 20~350 mg L^{-1}, adsorbent dosage = 0.1 g L^{-1}, pH = 7.4, temperature = 37 °C, shaking speed = 120 rpm).

Table 2. Langmuir and Freundlich constants of different nanofibers for free bilirubin.

Isotherm Model	PAN-BE-NF	PAN-BB-NF	PAN-BO-NF
Langmuir isotherm			
q_{max} (mg g^{-1})	818.9	623.8	536.8
b (L mg^{-1})	0.096	0.07	0.033
R^2	0.9662	0.9886	0.9561
Freundlich isotherm			
K_F	210.9	222.9	83.2
n	3.91	5.44	3.07
R^2	0.8452	0.8782	0.7982

Because PAN-BE-NF had the best bilirubin adsorption performance among the three nanofibers, the following experiments about the practical applications focused on PAN-BE-NF. In the blood, the albumin–bilirubin complex is the existing form that interferes with the bilirubin removal by the adsorbents [1]. To simulate the blood environment, bovine serum albumin (BSA) was added into the above bilirubin solution to investigate the effect of albumin on bilirubin adsorption. In the presence of BSA at different concentrations (0–50 g L^{-1}), the adsorption efficiency of PAN-BE-NF for bilirubin was above 90% even at a BSA concentration of 50 g L^{-1}, and the residual concentration of bilirubin was lower than 10 mg L^{-1} (Figure 4a). The isothermal experiment of the adsorption of albumin-bound bilirubin was also studied (Figure 4b). The concentration of BSA in all albumin-bound bilirubin solutions was fixed at 50 g L^{-1}. The fitting curves and R^2 results (Figure 4b and Table 3) show that the equilibrium process is more suitable to be described by the Langmuir model, and the maximum adsorption capacity of PAN-BE-NF fibers for albumin-bound bilirubin could reach up to 163.7 mg g^{-1}. Although the adsorption capacity showed a dramatic decrease owing to competition from BSA, this adsorption capacity for albumin-bound bilirubin still surpassed many reported bilirubin adsorbents (Table S1), suggesting that PAN-BE-NF exhibited a good ability to resist albumin interference.

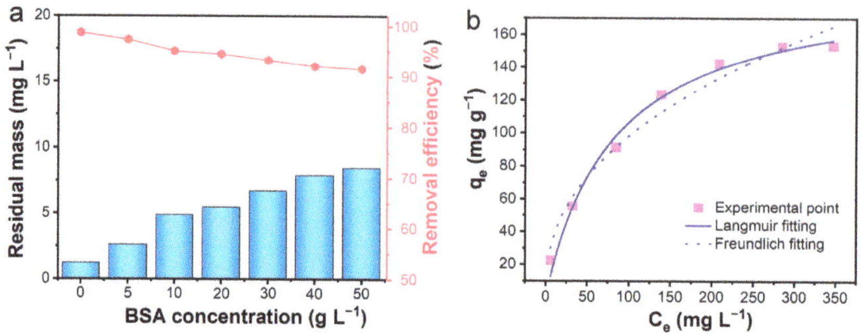

Figure 4. (**a**) Experimental results of adsorption of bilirubin under the coexistence of BSA (C_{BR} = 100 mg L^{-1}, adsorbent dosage = 3 g L^{-1}, pH = 7.4, temperature = 37 °C, shaking speed = 120 rpm). (**b**) Adsorption isotherm of PAN-BE-NF toward albumin-bound bilirubin (C_{BR} = 25~500 mg L^{-1}, C_{BSA} = 50 g L^{-1}, adsorbent dosage = 1 g L^{-1}, pH = 7.4, temperature = 37 °C, shaking speed = 120 rpm).

Table 3. Langmuir and Freundlich constants of different nanofibers from albumin-bound bilirubin.

Isotherm Model	PAN-BE-NF
Langmuir isotherm	
q_{max} (mg g^{-1})	163.7
b (L mg^{-1})	0.012
R^2	0.9868
Freundlich isotherm	
K_F	14.4
n	2.4
R^2	0.9281

3.4. Blood Compatibility Test

For blood purification materials, due to the direct contact with blood, blood compatibility is very important. In this study, the blood compatibility of PAN-BE-NF was characterized by a hemolysis test, platelet adhesion, activated partial thromboplastin time (APTT), plasma thrombin time (TT), and plasma prothrombin time (PT). The hemolysis test is a visual indicator used to evaluate blood compatibility, and materials with a hemolytic value of <5% are considered to be safe as blood-contact materials [15]. The results in Figure 5a indicate that PAN-BE-NF had almost no hemolytic activity, and the hemolysis rate of PAN-BE-NF, even with a concentration of 10 mg mL^{-1}, did not exceed 5%. SEM imaging was used to observe the nanofibers after the platelet adhesion experiment, and it was found that fewer platelets adhered to the nanofiber membrane (Figure 5b), indicating its low blood coagulation effect. At the same time, the coagulability of the fiber material upon contact with blood was measured (Figure 5c–e). APTT is the activated partial thromboplastin time used to assess the activity of inherent and common clotting pathways. TT is the plasma thrombin time, which refers to the time required for plasma fibrinogen to be converted into fibrin after adding "standardized" thrombin to the tested plasma. PT is the plasma prothrombin time, which can reflect the function of the exogenous coagulation system, and its adsorption level is used to evaluate the exogenous and common coagulation pathways. Compared with the control group, the APTT activity of PAN-BE-NF showed only a slight increase, and the TT and PT activities did not significantly decrease. These results confirmed that PAN-BE-NF was safe in the application of blood purification.

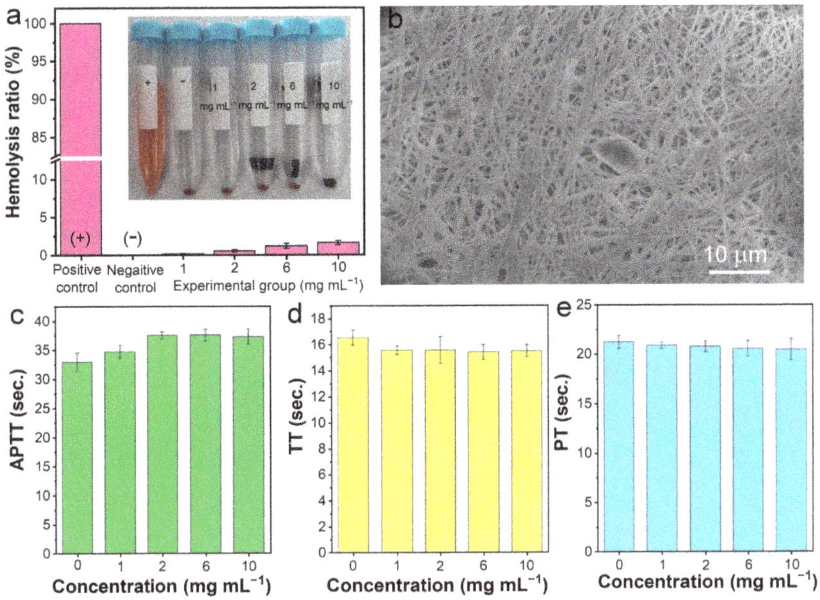

Figure 5. (**a**) Hemolysis test for PAN-BE-NF, in which deionized water and phosphate buffer saline were used as positive and negative controls, respectively. (**b**) SEM images of platelet adhesion on PAN-BE-NF. (**c**) APTT experiment, (**d**) TT experiment, and (**e**) PT experiment results.

3.5. Dynamic Bilirubin Adsorption

To simulate the real hemoperfusion process, a dynamic bilirubin adsorption experiment was conducted. PAN-BE-NF was placed into the adsorption column (Figure 6a), and bilirubin solution with 50 g L^{-1} of BSA was forced through the column. The adsorption effect of PAN-BE-NF was much more efficient than that of commercial activated carbon in the circulation system (Figure 6b). PAN-BE-NF could reduce the bilirubin concentration to a normal level of 10 mg L^{-1} within 4 h. However, at this time point, the residual concentration by activated carbon was 70.8 mg L^{-1}. The obtained results indicated that the adsorption effect of PAN-BE-NF was still considerable in the state of dynamic adsorption, and PAN-BE-NF possessed potential in practical hemoperfusion application.

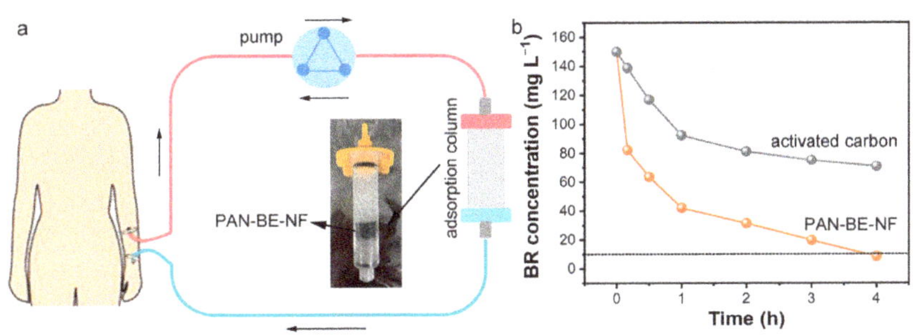

Figure 6. (**a**) Schematic diagram of dynamic cyclic adsorption experiment. (**b**) Experimental results of dynamic adsorption of PAN-BE NF and activated carbon toward albumin-bound bilirubin (C_{BR} = 150 mg L^{-1}, mass of adsorption column = 1.0 g, volume of bilirubin solution = 500 mL, flow rate = 100 mL min^{-1}).

4. Conclusions

In conclusion, we reasonably designed and synthesized quaternary ammonium group grafted electrospun nanofibers for the efficient adsorption of bilirubin. With the aid of theoretical calculations, we forecasted the effective functional group as the grafting component. Both theoretical analysis and experiment data indicated that the quaternary ammonium group with suitable structural configuration was attributed high bilirubin binding. As a result, the designed nanofibers showed faster bilirubin uptake than commercial activated carbon and ion-exchange resin. The obtained maximum adsorption capacity from free bilirubin was up to 818.9 mg g^{-1}. In further experiments, in the presence of different concentrations of BSA, the removal rate of bilirubin was still above 90%, and the concentration of residual bilirubin was lower than 10 mg L^{-1}. The maximum adsorption capacity toward albumin-bound bilirubin was 163.7 mg g^{-1}, which was higher than most of the reported bilirubin adsorbents. In the dynamic adsorption test, obtained nanofibers also showed better performance than activated carbon, which could reduce the bilirubin concentration to below 10 mg L^{-1}. In addition, the blood compatibility indicated the safety of as-prepared nanofibers as hemoperfusion adsorbents. Based on the obtained results, the calculation-guided strategy adopted could offer a universally applicable paradigm for the construction of bilirubin adsorbent in hemoperfusion.

Supplementary Materials: The following supporting information can be downloaded at: https://www.mdpi.com/article/10.3390/polym16111599/s1, Additional experimental section; Table S1: Comparison of bilirubin adsorption results in the presence of BSA. References [15,27–31] are cited in the supplementary materials.

Author Contributions: Conceptualization, X.F. and R.Z.; methodology, M.S. and D.C.; formal analysis, X.Z.; data curation, M.S.; funding acquisition, R.Z.; supervision, T.J. and R.Z.; writing—review and editing, R.Z. All authors have read and agreed to the published version of the manuscript.

Funding: This research was funded by the National Natural Science Foundation of China (22375032), Project of Education Department of Jilin Province (JJKH20241416KJ), Natural Science Foundation of Department of Science and Technology of Jilin Province (YDZJ202101ZYTS060), and Fundamental Research Funds for the Central Universities (2412021D008).

Institutional Review Board Statement: Not applicable.

Data Availability Statement: Data are contained within the article.

Conflicts of Interest: The authors declare no conflicts of interest.

References

1. Du, Y.; Chen, M.; Wang, B.; Chai, Y.; Wang, L.; Li, N.; Zhang, Y.; Zhuang, L.; Guo, C.; Jiang, X.; et al. TiO_2/polystyrene nanocomposite antibacterial material as a hemoperfusion adsorbent for efficient bilirubin removal and prevention of bacterial infection. *ACS Biomater. Sci. Eng.* **2024**, *10*, 1494–1506. [CrossRef] [PubMed]
2. Zhou, W.; Zhang, M.; Liu, X.; Hussain, W.; Zhu, B.; Yu, H.; Wang, S.; Zhou, L. Collagen-modified chitosan microsphere as a novel hemoperfusion adsorbent for efficient removal of bilirubin from human plasma. *Sep. Purif. Technol.* **2024**, *337*, 126303. [CrossRef]
3. Li, X.; Wei, F.; Li, Y.; Qi, F.; Wang, J.; Yao, Z.; Zhang, L.; Zhang, B. Fabricating polyacrylonitrile-polyethyleneimine nanofiber membranes with exceptional hemocompatibility and efficient bilirubin removal for hemoperfusion. *React. Funct. Polym.* **2024**, *194*, 105793. [CrossRef]
4. Wang, Y.; Wei, R.; Zhao, W.; Zhao, C. Bilirubin removal by polymeric adsorbents for hyperbilirubinemia therapy. *Macromol. Biosci.* **2023**, *23*, 2200567. [CrossRef]
5. Zhang, M.; Liu, X.; Zhou, W.; Zheng, X.; Wang, S.; Zhou, L. Ordered porous materials for blood purification. *Sep. Purif. Technol.* **2023**, *327*, 124844. [CrossRef]
6. Li, C.; Zhang, X.; Bao, C.; Zhang, J.; Tian, Y.; Shen, J.; Feng, X. Freezing-induced chemical crosslinking to fabricate nanocellulose-based cryogels for efficient bilirubin removal. *Sep. Purif. Technol.* **2022**, *300*, 121865. [CrossRef]
7. Zhang, J.; Zhong, Y.; Zhou, Y.; Zhang, Q.; Wang, S. Influence of the acid groups on activated carbon on the adsorption of bilirubin. *Mater. Chem. Phys.* **2021**, *260*, 124133.
8. Gao, X.; Liu, K.; Liu, P.; Bai, X.; Li, A.; Lyu, Z.; Li, Q. Preparation and properties of cellulose acetate graft copolymer-coated adsorbent resin for hemoperfusion device. *J. Appl. Polym. Sci.* **2023**, *140*, 53895. [CrossRef]

9. Zhao, R.; Chen, D.; Gao, N.; Yuan, L.; Hu, W.; Cui, F.; Tian, Y.; Shi, W.; Ma, S.; Zhu, G. Porous cationic electrospun fibers with sufficient adsorption sites for effective and continuous $^{99}TcO_4^-$ uptake. *Adv. Funct. Mater.* **2022**, *32*, 2200618. [CrossRef]
10. Liu, H.; Lai, W.; Shi, Y.; Tian, L.; Li, K.; Bian, L.; Xi, Z.; Lin, B. One-step fast fabrication of electrospun fiber membranes for efficient particulate matter removal. *Polymers* **2024**, *16*, 209. [CrossRef] [PubMed]
11. Huang, C.; Fu, M.; Ma, L.; Mao, L.; Yao, Y.; Xu, X.; Xu, L.; Han, J.; Xue, X.; Xu, G.; et al. Ultrafast strategy to self-assemble polyamidoxime PVA nanofibers with solar-induced synergy for enhancing uranium extraction from seawater. *Chem. Eng. J.* **2023**, *474*, 145718. [CrossRef]
12. Ma, Z.; Kotaki, M.; Ramakrishna, S. Electrospun cellulose nanofiber as affinity membrane. *J. Membr. Sci.* **2005**, *265*, 115–123. [CrossRef]
13. Wu, K.; Yang, W.; Liu, X.; Jiao, Y.; Zhou, C. Electrospun porous polyethersulfone (PES) fiber mats with high bilirubin adsorption capacity. *Mater. Lett.* **2016**, *185*, 252–255. [CrossRef]
14. Wu, K.; Song, X.; Cui, S.; Li, Z.; Jiao, Y.; Zhou, C. Immobilization of bovine serum albumin via mussel-inspired polydopamine coating on electrospun polyethersulfone (PES) fiber mat for effective bilirubin adsorption. *Appl. Surf. Sci.* **2018**, *451*, 45–55. [CrossRef]
15. Zhao, R.; Li, Y.; Li, X.; Li, Y.; Sun, B.; Chao, S.; Wang, C. Facile hydrothermal synthesis of branched polyethylenimine grafted electrospun polyacrylonitrile fiber membrane as a highly efficient and reusable bilirubin adsorbent in hemoperfusion. *J. Colloid Interface Sci.* **2018**, *514*, 675–685. [CrossRef] [PubMed]
16. Liu, M.; Zhang, W.; Zhang, P.; Yuan, L.; Li, H.; Zhan, J. Enhanced hemocompatibility of lysine grafted polyacrylonitrile electrospun nanofiber membranes as a potential bilirubin adsorption in hemoperfusion. *J. Polym. Environ.* **2023**, *31*, 1553–1567. [CrossRef]
17. Zhao, R.; Ma, T.; Cui, F.; Tian, Y.; Zhu, G. Porous aromatic framework with tailored binding sites and pore sizes as a High-Performance hemoperfusion adsorbent for bilirubin removal. *Adv. Sci.* **2020**, *7*, 2001899. [CrossRef] [PubMed]
18. Ma, Z.; He, L.; Zhao, F.; Tao, Y.; Sun, W.; Tai, B.; Guo, Q.; Zhang, W.; Peng, F.; Chen, J.; et al. Theory-driven design of phosphazene-based porous polymer beads for enhanced iodine adsorption. *Sep. Purif. Technol.* **2024**, *345*, 127321. [CrossRef]
19. Becke, A.D. Density-functional exchange-energy approximation with correct asymptotic behavior. *Phys. Rev. A* **1988**, *38*, 3098. [CrossRef] [PubMed]
20. Lee, C.; Yang, W.; Parr, R.G. Development of the colle-salvetti correlation-energy formula into a functional of the electron density. *Phys. Rev. B* **1988**, *37*, 785–789. [CrossRef] [PubMed]
21. Frisch, M.J.; Trucks, G.W.; Schlegel, H.B.; Scuseria, G.E.; Robb, M.A.; Cheeseman, J.R.; Scalmani, G.; Barone, V.; Mennucci, B.; Petersson, G.A.; et al. *Gaussian 09*; Gaussian, Inc.: Wallingford, CT, USA, 2009.
22. Trott, O.; Olson, A.J. Autodock vina: Improving the speed and accuracy of docking with a new scoring function, efficient optimization, and multithreading. *J. Comput. Chem.* **2010**, *31*, 455–461. [CrossRef] [PubMed]
23. Li, X.; Zhang, H.; Dong, J.; Ma, S.; Ou, J. One-pot synthesis of glucose-derived carbonaceous material with high hydrophilicity and adsorption capacity as bilirubin adsorbent. *J. Mater. Sci.* **2021**, *56*, 18006–18018. [CrossRef]
24. Lei, H.; Huang, S.; Pan, N.; Zou, H.; Wang, X.; Tuo, X. Novel Phytic Acid-Based Supramolecular Material for Uranium Removal from Highly Acidic Media: Synthesis, Performance, and Mechanistic Insights. *ACS EST Water* **2023**, *3*, 2522–2534. [CrossRef]
25. Li, L.; Ren, H.; Yuan, Y.; Yu, G.; Zhu, G. Construction and adsorption properties of porous aromatic frameworks via $AlCl_3$-triggered coupling polymerization. *J. Mater. Chem. A* **2014**, *2*, 11091. [CrossRef]
26. Wu, H.; Zhou, J.; Zhang, S.; Niu, P.; Li, H.; Liu, Z.; Zhang, N.; Li, C.; Wang, L.; Wang, Y. A comparative study of removal of acid red 27 by adsorption on four different chitosan morphologies. *Polymers* **2024**, *16*, 1019. [CrossRef] [PubMed]
27. Zhou, W.; Hu, W.; Zhan, Q.; Zhang, M.; Liu, X.; Hussain, W.; Yu, H.; Wang, S.; Zhou, L. Novel hemoperfusion adsorbents based on collagen for efficient bilirubin removal–A thought from yellow skin of patients with hyperbilirubinemia. *Int. J. Biol. Macromol.* **2023**, *253*, 127321. [CrossRef] [PubMed]
28. Wang, W.; Zhang, H.; Zhang, Z.; Luo, M.; Wang, Y.; Liu, Q.; Chen, Y.; Li, M.; Wang, D. Amine-functionalized PVA-co-PE nanofibrous membrane as affinity membrane with high adsorption capacity for bilirubin. *Colloids. Surf. B.* **2017**, *150*, 271–278. [CrossRef] [PubMed]
29. Yuan, Z.; Li, Y.; Zhao, D.; Zhang, K.; Wang, F.; Wang, C.; Wen, Y. High efficiency 3D nanofiber sponge for bilirubin removal used in hemoperfusion. *Colloids. Surf. B.* **2018**, *172*, 161–169. [CrossRef] [PubMed]
30. Ma, C.; Gao, Q.; Zhou, J.; Chen, Q.; Han, B.; Xia, K.; Zhou, C. Facile one-pot synthesis of magnetic nitrogen-doped porous carbon for high-performance bilirubin removal from BSA-rich solution. *RSC Adv.* **2017**, *7*, 2081. [CrossRef]
31. Ma, C.F.; Gao, Q.; Xia, K.S.; Huang, Z.Y.; Han, B.; Zhou, C.G. Three-dimensionally porous graphene: A high-performance adsorbent for removal of albumin-bonded bilirubin. *Colloids. Surf. B.* **2017**, *149*, 146–153. [CrossRef] [PubMed]

polymers

MDPI

Article

One-Step Fast Fabrication of Electrospun Fiber Membranes for Efficient Particulate Matter Removal

Huanliang Liu [1,2,†], Wenqing Lai [1,2,†], Yue Shi [1,2], Lei Tian [1,2], Kang Li [1,2], Liping Bian [1,2], Zhuge Xi [1,2,*] and Bencheng Lin [1,2,*]

1 Tianjin Institute of Environmental and Operational Medicine, Tianjin 300050, China; tjliuhuanliang@126.com (H.L.); laiwenqing0316@126.com (W.L.); october2144@163.com (Y.S.); tjtianlei@126.com (L.T.); tjlikang@126.com (K.L.); bmoglp@126.com (L.B.)

2 Tianjin Key Laboratory of Risk Assessment and Control Technology for Environment & Food Safety, Tianjin 300050, China

* Correspondence: zhugexi2003@sina.com (Z.X.); linbencheng123@126.com (B.L.); Tel.: +86-022-84655324 (Z.X.); +86-022-84655336 (B.L.)

† These authors contributed equally to this work.

Abstract: Rapid social and industrial development has resulted in an increasing demand for fossil fuel energy, which increases particulate matter (PM) pollution. In this study, we employed a simple one-step electrospinning technique to fabricate polysulfone (PSF) fiber membranes for PM filtration. A 0.3 g/mL polymer solution with an *N,N*-dimethylformamide:tetrahydrofuran volume ratio of 3:1 yielded uniform and bead-free PSF fibers with a diameter of approximately 1.17 μm. The PSF fiber membrane exhibited excellent hydrophobicity and mechanical properties, including a tensile strength of 1.14 MPa and an elongation at break of 116.6%. Finally, the PM filtration performance of the PSF fiber membrane was evaluated. The filtration efficiencies of the membrane for $PM_{2.5}$ and $PM_{1.0}$ were approximately 99.6% and 99.2%, respectively. The pressure drops were 65.0 and 65.2 Pa, which were significantly lower than those of commercial air filters. Using this technique, PSF fiber membrane filters can be easily fabricated over a large area, which is promising for numerous air filtration systems.

Keywords: electrospinning; air filtration; polysulfone; $PM_{2.5}$; $PM_{1.0}$

1. Introduction

With the development of human society, air pollution, particularly particulate matter (PM), has significantly increased, causing serious health problems [1]. PM includes solid and liquid particles in air. It can be categorized as PM_{10}, $PM_{2.5}$, and $PM_{1.0}$ based on the particle size [2–4]. Among them, $PM_{2.5}$ and $PM_{1.0}$ are the main constituents of airborne PM. Because of their diminutive sizes, they can easily permeate into human lung tissue and circulatory systems upon inhalation, causing heart and respiratory tract disorders [5,6]. To date, filtration is the most popular and effective method of removing PM from the air. Thus, there is an urgent need for simple methods for fabricating high-performance air filters.

Fiber filters have attracted significant attention for addressing PM pollution because of their high interception efficiency, favorable air permeability, and scalability [7,8]. Fiber filters employ four mechanisms to filter PM: interception, inertial impact, Brownian diffusion, and electrostatic effect [9]. Currently, meltblown polypropylene (PP) nonwoven fabric is commonly used in air filters [10]. Commercial filters are fabricated via meltblowing. To produce meltblown polypropylene (PP) fabrics, PP granules and electret masterbatches are blended and meltblown into PP fibers, which are then electrotreated via corona charging. According to Patra et al. [11], the filtration efficiencies of surgical masks and N95 masks, commercially available filters, for PM with particle sizes of 0.3–10 μm are 77.8% and 91.8%, respectively. Stahl et al. [12] also reported that the filtration efficiencies of N95 and N99 masks for PM with a particle size of approximately 0.3 μm are 54.1% and 90.5%, respectively. Because of their relatively large fiber diameter, these commercial filters cannot efficiently filter small particles in the air [13]. In addition, their fabrication processes are complex and expensive. Moreover, PP filters are associated with the rapid depletion of electrostatic charges, which reduces their PM filtration efficiency [14]. Recently, various polymeric fibers, including polyacrylonitrile (PAN) [15], polyvinyl alcohol (PVA) [16], polyvinylidene chloride (PVC) [17], and polyvinylidene fluoride (PVDF) [18], have been explored for PM filtration. Zhang et al. [19] synthesized PVA-based nanofiber membranes using a spinning method and fabricated a novel fiber filter using a thermal crosslinking technique. The $PM_{2.5}$ removal efficiency of the fiber membrane was above 95%, and the pressure drop was less than 100 Pa. Liu et al. [20] combined spinning and spraying techniques to fabricate a PVA/ethyl cellulose composite membrane, which recorded filtration efficiencies of 99.74% and 99.77% for $PM_{1.0}$ and $PM_{2.5}$, respectively. Yao et al. [21] prepared monodispersed ZnO particles using template- and surfactant-free hydrothermal methods and fabricated PVA-co-PE fibers using a melt extrusion phase-separation technique. The air filter was obtained by coating a mixture of ZnO particles and a PVA-co-PE fiber suspension onto a PP meltblown nonwoven fabric, and the filtration efficiency for sodium chloride aerosol with a particle size of 1.0 μm was more than 99.9%. Zheng et al. [22] synthesized self-supporting three-dimensional (3D) PVDF nanofiber membranes with crimped patterns and an optimum filtration efficiency of 93.6% for PM. Zhang et al. [23] fabricated a novel polyvinyl alcohol (PVA)/cellulose nanocrystal (CNC) composite nanofiber filter via electrowinning and reported a $PM_{2.5}$ filtration efficiency of 99.1%. Despite their excellent filtration performance, these filters are limited by their complex preparation processes, the high cost of raw materials, and poor scalability. Therefore, there is an urgent need to develop new filter materials that can be fabricated via simple and low-cost process, which will promote their large-scale production and diverse applications.

Polysulfone (PSF) is a thermoplastic polymer with remarkable chemical stability and excellent mechanical and thermal properties, which are attributed to its unique molecular structure [24]. A PSF molecular unit consists of a sulfonyl group, an ether group, and isopropylidene groups in the main chain [25]. Because of its special molecular structure and excellent physicochemical properties, PSF is widely used in water and blood filtration. Li et al. [26] synthesized a PSF/rGO/ZnO composite ultrafiltration membrane using a simple and green method, and the membrane showed good performance in dye wastewater treatment. Zhong et al. [27] fabricated a PSF-based hemodialysis membrane with excellent hemocompatibility and biocompatibility, and the membrane exhibited excellent hemodialysis performance superior to that of conventional membranes. However, to date, PSF has not been explored for air filtration.

Compared with other fiber fabrication techniques, including self-assembly [28], the template method [29], and phase separation [30], electrospinning is simple and versatile [31]. In principle, electrospinning involves the use of high-voltage electrostatic force to generate a polymer melt or solution jet [32]. The jet initially moves in a straight path and gradually transitions into a spiral trajectory as it travels. Concurrently, the solvent volatilizes, and a fiber membrane is formed on the collector. Although considerable progress has been made in single-nozzle electrospinning, it can only fabricate small-fiber membranes. To fabricate large-scale fiber membranes, needleless electrospinning has emerged as a promising technique. Electrospun fiber membranes offer several advantages, including small and uniform diameters, high specific surface areas, interconnected pore structures, and good mechanical properties, making them promising for air filtration [33]. Moreover, in situ charge injection into fiber membranes can be achieved during electrospinning. The electrospinning process enables bulk charges to be captured in the membrane and generates dipole charges. Charges in the membrane improve the attraction of PM to the fibers, thus increasing the PM filtration efficiency [34,35].

In this study, we employed one-step electrospinning to fabricate PSF-based fiber filters and investigated their PM filtration performance. Figure 1 shows the synthesis process and filtration mechanisms of the filter membranes. We investigated the effect of PSF solution concentration and solvents on the PSF fiber synthesis. Then, we fabricated PSF fiber membranes with different thicknesses by adjusting the spinning time. Furthermore, the physical and chemical properties of PSF fiber membranes were comprehensively examined using scanning electron microscopy (SEM), Fourier-transform infrared spectroscopy (FTIR), and water contact angle (WCA) analysis. Finally, the PM filtration performance of the PSF fiber membranes and the underlying mechanism were investigated.

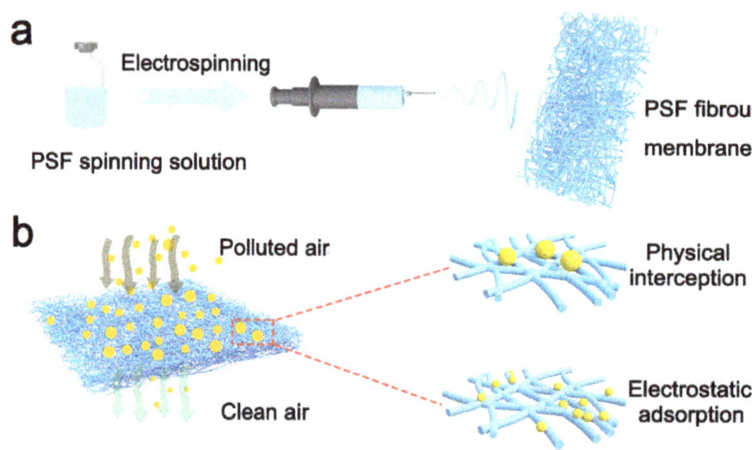

Figure 1. (**a**) Preparation process and (**b**) filtration mechanism of the polysulfone (PSF) fiber membranes.

2. Materials and Methods

2.1. Materials

PSF (P-1800; Mw: 68,102 g/mol; ρ: 1.24 g/cm^3) was purchased from Solvary Company (Long Beach, CA, USA). *N,N*-dimethylformamide (DMF) and tetrahydrofuran (THF) were purchased from Aladdin. All materials and solvents were used as received without further purification.

2.2. Fabrication of the PSF Fiber Membranes

PSF fiber membranes were fabricated using a homemade electrospinning system. Various masses of PSF (25, 30, and 35 wt%) were dissolved in 4 mL of solvent (DMF:THF volume ratio = 3:1) for 12 h. Subsequently, the solutions were placed on an electrospinning device, and fibers were formed on aluminum foil under a high-voltage power supply of 8 kV. The distance between the nozzle and the collector was 20 cm, and the flow rate was maintained at 0.8 mL/h. Ultrathin PSF fiber membranes with different thicknesses were obtained by varying the electrospinning time (60, 90, 120, and 150 min), and the corresponding samples were denoted as PSF-1, PSF-2, PSF-3, and PSF-4, respectively. The fabricated PSF fiber membranes were vacuum-dried at 50 °C for 24 h to eliminate the residual solvent.

2.3. Characterization

The morphology of the prepared PSF fibers was characterized using SEM (SU8010, Hitachi, Tokyo, Japan). To measure the average diameter of the PSF fibers, 100 fibers were randomly selected from the SEM images. Each fiber diameter was measured using ImageJ software (v1.8.0), and the average diameter was calculated. The average pore size of the fiber membrane was determined by randomly selecting 100 voids in the fibers from the SEM images. The diameters of the inner circles in the voids were obtained using ImageJ, and the average value was recorded as the pore size of the fiber membrane. The fiber membranes were further characterized using FTIR (FTIR, Nicolet iS10, Billerica, MA, USA). The surface wettability of the membranes was determined by WCA measurements (DSA25, Kruss, Hamburg, Germany). A 2 μL of water droplet was dropped at three locations on the sample surface, and the mean WCA was determined. An electrostatic tester (JH-TEST, Shenzhen, China) was used to measure the surface potential of the fiber membranes in situ. The probe was slowly moved close to the targeted measurement point until two LED light spots from the electrostatic tester completely overlapped, and the value was displayed. The mechanical properties of the fiber membranes, including tensile strength, elongation at break, and Young's modulus, were measured using a tensile testing machine (Sunszy-2000, utm2103, Suns, Shenzhen, China). Furthermore, the porosity of the fiber membranes was determined using the volume replacement method. The samples were immersed in anhydrous ethanol for 5 min until saturation, and their masses after ethanol absorption were recorded. The porosity of the samples was then calculated using Equation (1).

$$P = \frac{(M_1 - M_0)}{(\rho \times V)} \times 100\%, \tag{1}$$

where M_0 and M_1 are the masses of the sample before and after ethanol absorption, respectively, V the volume of the sample, and ρ the density of ethanol. All samples were measured three times.

2.4. Filtration Performance of the PSF Fiber Membranes

The TSI 8532 automated filtration tester was used to evaluate the filtration performance of the PSF fiber membranes. The unit was connected to a particle generator that produced particles with diameters of 0.3–10.0 μm. A fan was connected to the particle generator to propel the generated aerosol particles through the filters. The aerosol wind speed was measured using an anemometer (FMA1002R-V1, OMEGA, New York, NY, USA). Two solid-state laser photometers were employed to measure the aerosol concentrations and particle size distributions before and after passing through the filters. The pressure drop across the filter was measured using a pressure transducer (DP-CLCTM5825, TSI, Missoula, MT, USA). Particulate aerosol tests were conducted at room temperature with a fixed aerosol wind speed of 0.25 km/h, and the filtration efficiency (η) of the filters was calculated using Equation (2):

$$\eta = \frac{\xi 1 - \xi 2}{\xi 1} \times 100\%, \tag{2}$$

where $\xi 1$ and $\xi 2$ are the particle concentrations upstream and downstream of the filter, respectively. The pressure drop ΔP (the difference in pressure before and after filtration) was calculated as follows:

$$\Delta P = P_2 - P_1, \tag{3}$$

where P_1 and P_2 are the pressures before and after filtration, respectively.

The quality factor (QF) is a crucial parameter for evaluating the filtration performance of filters and can be calculated as follows:

$$QF = \frac{-\ln(1-\eta)}{\Delta P}, \tag{4}$$

To evaluate the durability of PSF-2, the surface electrostatic potential, filtration efficiency, pressure drop, and QF were evaluated for 28 days. The fiber membranes were stored at room temperature and 50% relative humidity, and the indices, including surface electrostatic potential, filtration efficiency, pressure drop, and QF, were tested every 7 days.

2.5. Statistical Analysis

We used Microsoft Excel 2021 to calculate the average values and standard deviations (SDs) of the indices. The obtained SD was used as the error bar, and the final results were expressed as mean ± SD.

3. Results and Discussion

3.1. Effects of Solution Concentration and Solvent Composition on the Spinnability of the PSF Solutions

We first investigated the effects of the polymer solution concentration and solvent composition on the morphology of the resulting fibers. Poor spinnability is indicated by the presence of beads in the spun fibers, and good spinnability is indicated by homogeneous but discontinuous fibers (some charged droplets deposit on the fiber membrane). Excellent spinnability is indicated by homogeneous and continuous fibers. The PSF solution with a concentration of 25 wt% showed poor spinnability (Table 1). Regardless of the solvent (pure DMF, pure THF, or a mixture of DMF and THF), PSF fibers with a beaded structure were obtained, as shown in Figure S1. Polymer solutions play a key role in electrostatic spinning. When the concentration of the polymer solution is low, the entanglement between the polymer chains is weak; thus, fibers with many beads are obtained [36,37]. Therefore, we increased the concentration of the polymer solution to 30 wt%. Pure THF was used as the solvent, and good spinnability was achieved, as indicated by homogeneous but discontinuous fibers. This is attributed to occasional nozzle clogging and the deposition of charged droplets on the fiber membranes, as shown in Figure S2. When a mixture of 3 mL of DMF and 1 mL of THF was used as the solvent, the PSF solution exhibited excellent spinnability, as indicated by the homogeneous and continuous fibers (Figure S3). Furthermore, we increased the polymer concentration to 35 wt%, but the spinning needles were completely blocked, preventing the formation of fibers. The optimal spinning conditions included a solution of 30 wt% and a mixed solvent containing 3 mL of DMF and 1 mL of THF.

Table 1. The spinnability of various PSF solutions.

PSF Solution Concentration (wt%)	Solvent Composition (4 mL)	Spinnability
25 wt%	Pure THF	Poor
	Pure DMF	Poor
	DMF:THF = 3:1	Poor
30 wt%	Pure THF	Good
	Pure DMF	Poor
	DMF:THF = 3:1	Excellent
35 wt%	Pure THF	No
	Pure DMF	No
	DMF:THF = 3:1	No

3.2. Morphology of the PSF Fiber Membranes

Based on the optimal conditions, we further synthesized PSF fiber membranes with different thicknesses by varying the fiber deposition time from 60 to 150 min. The thicknesses of the four PSF fiber membranes obtained were 32, 60, 100, and 120 μm. We further characterized the PSF fiber membranes using FTIR spectroscopy (Figure S4). All PSF fiber membranes exhibited the same characteristic bands. The peaks at 1588 and 1242 cm^{-1} were attributed to the C=C and C–O–C stretching vibrations, respectively, and the peak at 1151 cm^{-1} was ascribed to the symmetrical O=S=O telescoping vibration.

Figure 2 shows the microscopic morphology of the PSF fiber membranes. All fibers were uniform and bead-free (Figure 2a). The average fiber diameters in the four PSF fiber membranes were approximately 1.17 μm (Figure 2c). This is because the PSF fibers were fabricated using the same concentration and solution composition. The pore-size distribution was evaluated, and the average pore size of the PSF fiber membranes was calculated, as shown in Figure 2b,d. For the four samples, the pore size slightly decreased with increasing thickness (from 1.34 μm for PSF-1 to 1.17 μm for PSF-4). PSF-1, PSF-2, PSF-3, and PSF-4 showed porosities of 87.5%, 84.9%, 80.5%, and 75.1%, respectively (Figure 2e). These results indicate that the pore structures (pore size and porosity) of the fiber membranes can be tuned by varying their thickness. During the electrospinning process, fibers are deposited on the collector layer by layer, and as the thickness of the membrane increases, the fibers stack tightly, resulting in a gradual decrease in both the pore size and porosity [38]. The obtained porosity and pore size were comparable to those reported for electrospun fiber membranes. For example, Zhu and Cai reported a relatively small pore size of approximately 1 μm and high porosity (77.0–84.9%) for electrospun fiber membranes [39,40].

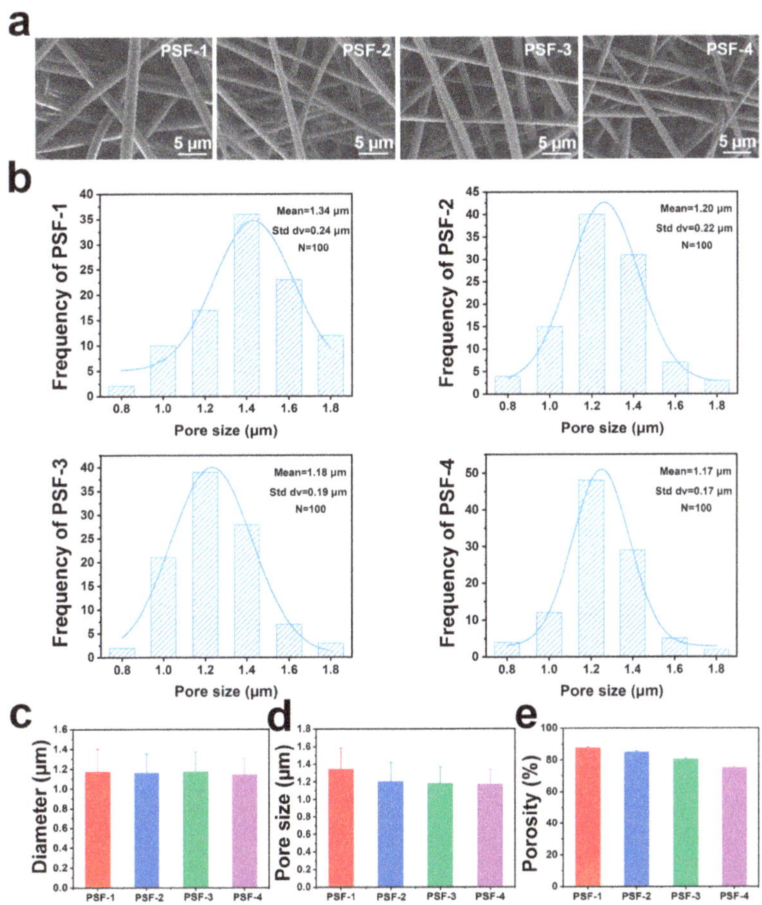

Figure 2. (**a**) SEM image, (**b**) pore-size distribution, (**c**) average fiber diameters, (**d**) average pore sizes, and (**e**) porosity of the PSF fiber membranes.

3.3. Mechanical Properties of the PSF Fiber Membranes

Figure 3 shows the mechanical properties of the synthesized PSF fiber membranes. Figure 3a shows the stress–strain curves. The tensile strength of PSF-1, PSF-2, PSF-3, and PSF-4 was 0.92, 1.14, 1.21, and 1.29 MPa (Figure 3b), and their elongation values at break were approximately 106.2%, 116.6%, 141.3%, and 229.0%, respectively (Figure 3c). As shown in Figure 3d, the Young's modulus of the membranes slightly decreased with increasing thickness (0.86, 0.85, 0.76, and 0.56 MPa for PSF-1, PSF-2, PSF-3, and PSF-4, respectively). These results show that the mechanical properties of PSF membranes can be improved by tuning the thickness of the membrane. During the electrospinning process, the fibers stack more tightly as the deposition time increases. Friction between fibers increases with the thickness of the membrane, which makes thicker membranes more prone to fracture [41]. Zhu et al. [42] reported an electrospun fiber membrane with good mechanical properties, including a tensile strength of 0.8 MPa and an elongation at break of 16%. Zhang et al. [43] also reported an electrospun fiber membrane with a tensile strength of 0.89 MPa and an elongation at break of 8.99%. According to Zhang et al. [44], the tensile strength of disposable masks is 2.5 MPa. The mechanical properties of our PSF fiber membranes were comparable to those reported for fiber filters.

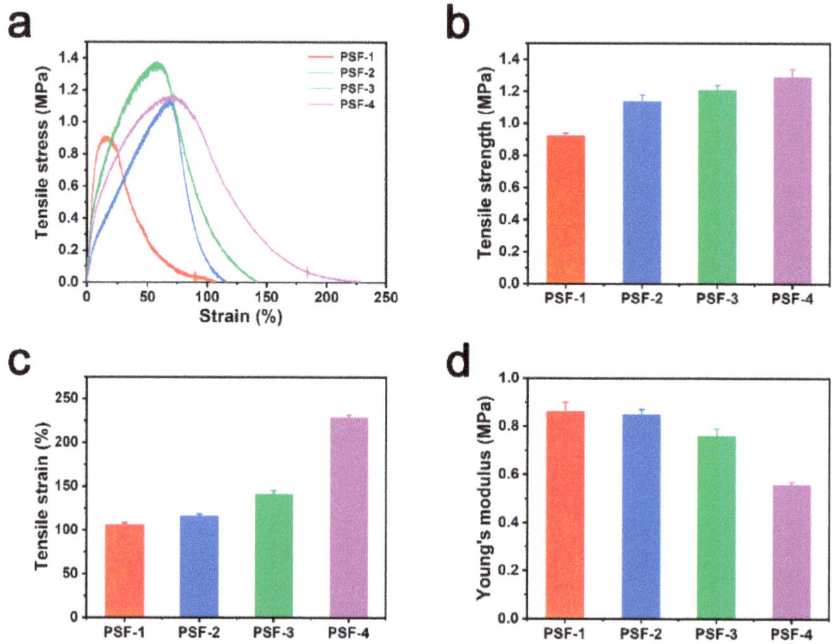

Figure 3. (**a**) Stress–strain curve, (**b**) tensile strength, (**c**) elongation at break, and (**d**) Young's modulus of the PSF fiber membranes.

Electrospun PSF fibers have been reported in the literature [45,46]. These studies focused on some spinning parameters that influenced the fiber morphology; however, in this study, we investigated in detail the effects of spinning parameters, including polymer solution concentration, solvent composition, spinning voltage, and tip-to-collector distance, on the morphology of the fiber. In addition, previous studies have considered the morphology of single fibers, especially the roughness and diameter, whereas we considered the surface morphology of single fibers (fiber diameter and roughness) and the fiber aggregate structure, including pore size and porosity. Furthermore, we investigated the chemical composition, wetting behavior, and mechanical properties of the PSF fibers.

3.4. Filtration Performance of the PSF Fiber Membranes and the Underlying Mechanism

Figure 4 shows the filtration performance of the PSF fiber membranes with different thicknesses and the corresponding filtration mechanism. The filtration efficiency of the membranes for $PM_{2.5}$ and $PM_{1.0}$ increased with the thickness of the membrane. For $PM_{2.5}$,

PSF-1, PSF-2, PSF-3, and PSF-4 exhibited filtration efficiencies of 84.7%, 96.7%, 97.5%, and 99.6%, respectively, and for $PM_{1.0}$, their efficiencies were 82.4%, 94.2%, 97.3%, and 99.2%, respectively. Previous studies have shown that electrospun fiber membranes exhibit good particle-trapping effects, making them promising for air filtration. Shao et al. [47] reported above 98.5% PM filtration efficiency for modified PLA nanofiber membranes. Zhang et al. [48] fabricated a novel dendritic cellulose nanofiber membrane with a PM filtration efficiency of 98.37%. Peng et al. [49] prepared a self-charging air filtration mask using the friction electric effect between electrospun poly(vinylidene fluoride) nanofiber film and nylon fabric. The membrane showed a filtration efficiency of 95.8% for 0.3 μm particles.

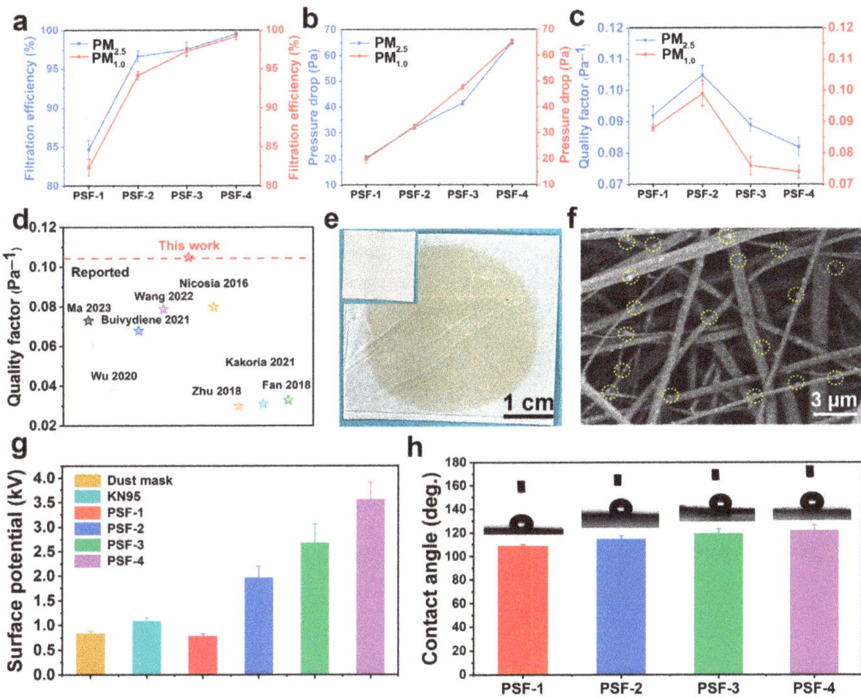

Figure 4. (**a**) Filtration efficiency, (**b**) pressure drop, and (**c**) *QF* of the PSF fiber membranes. (**d**) Comparison of the *QF*s of the PSF fiber membranes with those of the reported filters [16,38,50–55]. (**e**) Appearance and (**f**) SEM image of the PSF fiber membranes after $PM_{2.5}$ filtration (the inset shows the appearance of the membranes before filtration). (**g**) Surface potential of the PSF fiber membranes. (**h**) WCAs of the PSF fiber membranes.

A drop in pressure is the change in static pressure as air moves through a filter. It reflects the respiratory resistance and breathing comfort of the filter [56,57]. Therefore, we studied the pressure drop in all PSF fiber membranes. For $PM_{2.5}$, the pressure drop in the fiber membranes increased from 20.4 Pa for PSF-1 to 65.0 Pa for PSF-4 (Figure 4b). A similar trend was obtained for $PM_{1.0}$ filtration. This is because, compared with a thin membrane, a thick membrane can easily intercept PM, thereby increasing the respiratory resistance of the membrane. Segovia et al. [58] demonstrated that the pressure drop of an ordinary mask for particles was approximately 200 Pa, whereas that of an N95 mask was as high as 600 Pa. These results show that PSF fiber filters exhibit lower respiratory resistance than both ordinary and N95 masks. To achieve a better balance between filtration efficiency and pressure drop, the *QF* values of the fiber membranes were calculated, as shown in Figure 4c. For $PM_{2.5}$ filtration, the *QF* of the PSF-1 fiber membrane was 0.092 Pa^{-1}, which was lower than that of PSF-2 (0.105 Pa^{-1}). The QF values of the PSF-3 and PSF-4 fiber membranes (0.089 and 0.082 Pa^{-1}, respectively) were lower than those of PSF-1 and PSF-2. A similar trend was observed for $PM_{1.0}$ filtration (0.088, 0.099, 0.076, and 0.074 Pa^{-1} for PSF-1, PSF-2, PSF-3, and PSF-4, respectively). Owing to its high QF for $PM_{2.5}$ and $PM_{1.0}$, PSF-2 is considered the optimal membrane and was chosen for further study. We also compared the QF of the fiber membranes with that of the reported filters. The QF of the PSF fiber membranes was higher than that of the reported filters, indicating the good filtration performance and low respiratory resistance of our fiber membranes (Figure 4d) [16,38,50–55].

To understand the filtration mechanism of the PSF membranes, the macroscopic appearance and microscopic morphology of the PSF fiber membranes after $PM_{2.5}$ filtration were observed. The PSF fiber membranes after filtration showed a distinct yellow area in the filtration region, indicating that a large amount of PM was intercepted by the PSF fiber membranes (Figure 4e). The SEM results further revealed that many small particulate matter particles were tightly absorbed on the outer surfaces of the PSF fibers (Figure 4f). Electrostatic charges involved in electrostatic adsorption by the PSF fiber membranes may play a vital role in PM interception; therefore, we investigated the variation of the surface electrostatic potential of the PSF fiber membranes via membrane thickness [59]. The surface electrostatic potential of two commercially available masks was also evaluated as a control (Figure 4g). The results showed that the surface electrostatic potential of a common dust mask and a KN95 filter was 0.84 and 1.09 kV, respectively, and that of PSF-1, PSF-2, PSF-3, and PSF-4 was 0.78, 1.97, 2.68, and 3.56 kV, respectively. This shows that the surface electrostatic potential of PSF-2, PSF-3, and PSF-4 was larger than that of the commercial filters, and thicker PSF membranes exhibited higher surface electrostatic potential.

PSF fiber formation during electrospinning is driven by electric field forces. A high voltage is generally exerted on the polymer solution. When the fibers form, electrical charges are simultaneously integrated into the fibers. When the spinning voltage is up to 8 kV, a high electrostatic charge remains in the fibers. Because of the large surface electrostatic potential, PSF fibers can highly absorb PM, resulting in excellent PM filtration efficiency of the membrane [60].

Previous studies have shown that the surface static charges of filters may quickly be lost in high-humidity environments, which affects the PM filtration of the filter [61]. In this study, the WCA values of the PSF-1, PSF-2, PSF-3, and PSF-4 fiber membranes were 108°, 114°, 119°, and 122°, respectively, indicating strong hydrophobic properties of the PSF membranes (Figure 4h). The hydrophobic properties of the PSF fiber membranes can be attributed to the chemical composition and rough surface structure of the fibers. PSF contains many hydrophobic groups, including phenyl and isopropyl [62]. After the electrospinning process, a rough fiber surface can be obtained, and air can be captured in the voids. Owing to the repellency effect of the trapped air, the PSF fiber membranes showed strong hydrophobic properties. The water resistance of the membrane prevents not only the dissipation of electrostatic charges from the PSF membrane but also the fouling of externally contaminated liquids [63,64].

Furthermore, the effects of high voltage and tip-to-collector distance on the porosity and PM filtration efficiency of the PSF fiber membranes were investigated (Figure S5). The results indicate that both high voltage and tip-to-collector distance influence the porosity and PM filtration efficiency of the PSF fiber membranes. As the voltage increased from 6 to 10 kV, the porosity of the PSF fiber membranes decreased from 86.0% to 83.8%, and the corresponding filtration efficiency increased from 95.2% to 96.9%. In addition, the PM filtration efficiency of the PSF fiber membranes decreased with increasing tip-to-collector distance. As the tip-to-collector distance increased from 15 to 25 cm, the porosity of the fiber membranes increased from 83.2% to 85.0%, and the PM filtration efficiency decreased from 97.2% to 95.6%. We further investigated the effects of the spinning voltage and tip-to-collector distance on the surface electrostatic potential (Figure S6). There was no significant change in the surface electrostatic potential of the fiber membranes as voltages and tip-to-collector distances varied. This further indicates that both high voltage and tip-to-collector distance affect the porosity and filtration efficiency of fiber membranes.

3.5. Comparison of the Filtration Performance of PSF-2 and Commercial Filters

Figure 5 compares the thickness, filtration efficiency, pressure drop, and *QF* of PSF-2, a common dust mask, and KN95 filters. As shown in Figure 5a, the filter layers of the common dust mask and KN95 filter had thicknesses of 156.2 and 485.4 μm, respectively, which were much higher than that of the PSF-2 fiber membrane (~60 μm). The filtration efficiencies of the dust mask and KN95 filter for $PM_{2.5}$ were 88.8% and 98.7% (Figure 5b). The efficiency of the PSF-2 fiber membrane was higher than that of the dust mask and comparable to that of KN95. A similar trend was observed for $PM_{1.0}$ filtration. Furthermore, the pressure drop of the dust mask and KN95 filter for $PM_{2.5}$ filtration was 49.6 and 149.3 Pa, respectively, which was significantly higher than that of the PSF-2 fiber membrane (32.3 Pa) (Figure 5c). Alternatively, as shown in Figure 5d, the *QF* of the PSF-2 fiber membrane was much larger than that of the two commercial filters (0.105 and 0.099 Pa^{-1} for $PM_{2.5}$ and $PM_{1.0}$ filtration, respectively). According to the National Standard GB 2626-2019 of the People's Republic of China [65], the filtration efficiency of a filter should be higher than 95% and the pressure drop should be less than 170 Pa, both of which were satisfied by the PSF-2 fiber membrane.

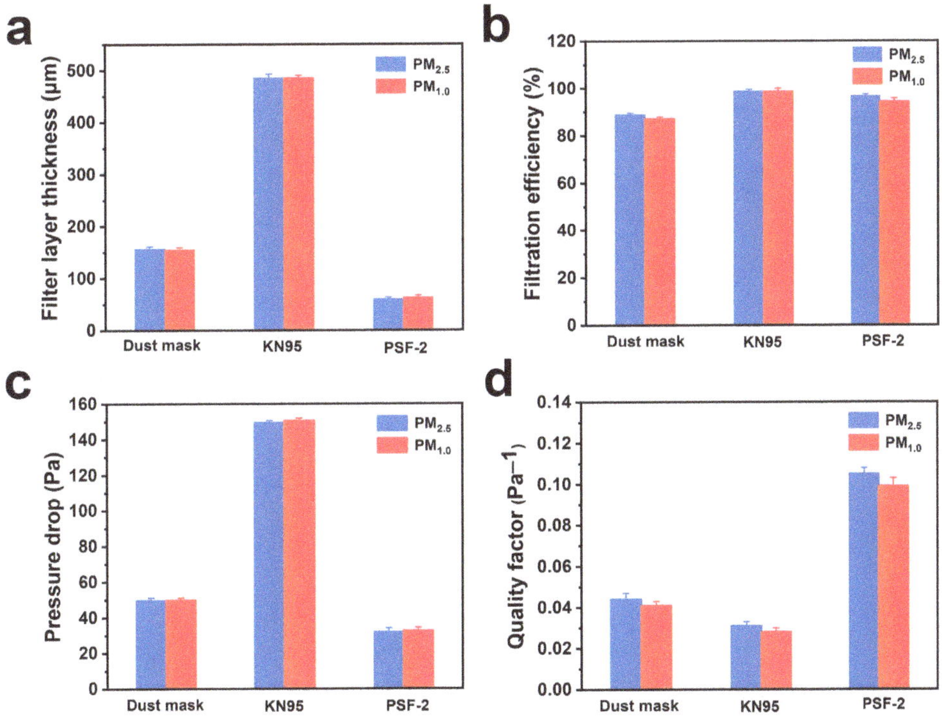

Figure 5. Comparison of the (**a**) thickness, (**b**) filtration efficiency, (**c**) pressure drop, and (**d**) *QF* of the PSF-2 fiber membrane with those of commercial filters.

3.6. Filtration Durability of the PSF-2 Fiber Membrane

To evaluate the durability of the membrane, the surface electrostatic potential, filtration efficiency, pressure drop, and *QF* of the PSF-2 fiber membrane were assessed for 28 days. Figure 6a shows that the electrostatic potential on the PSF fiber membrane surface remained stable over 28 days. In addition, no significant changes were observed in the filtration efficiency (Figure 6b), pressure drop (Figure 6c), and *QF* (Figure 6d) of the fiber membrane after $PM_{2.5}$ and $PM_{1.0}$ filtration. These results indicate the long-term durability of the PSF fiber membrane for PM filtration.

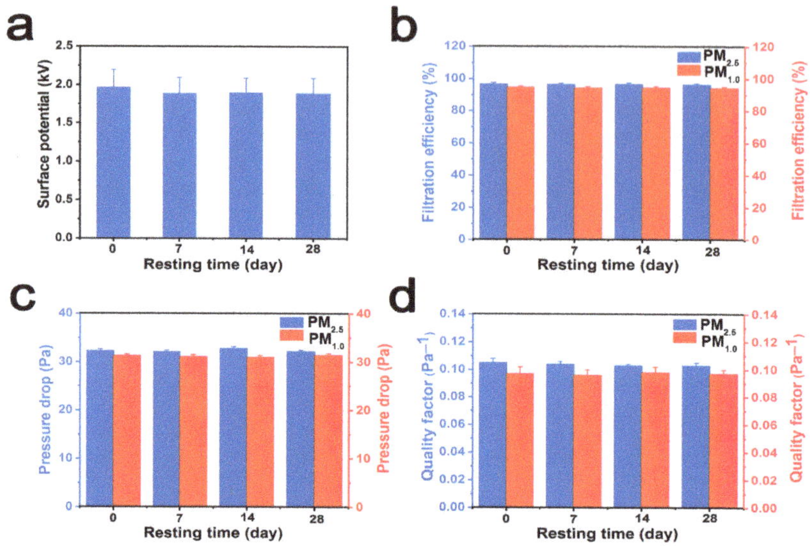

Figure 6. (**a**) Electrostatic dissipation, (**b**) filtration efficiency, (**c**) pressure drop, and (**d**) *QF* of the PSF-2 fiber membrane.

Recent studies have shown that the electrostatic trapping of surface charges plays an important role in PM filtration. Qiu et al. [66] prepared a self-powered air filter based on a breath-driven friction nanogenerator with filtration efficiencies of 95% for $PM_{0.3}$ and 86.3% after 2 h of use. Peng et al. [49] produced a respirator-inspired, self-charging air-filtering mask with a reduction filtration efficiency of 95.8% after 60 h of effective use. These show that the electrostatic charge generated by friction dissipates after use, thereby affecting the filtration efficiency. Zhang et al. [67] reported that corona charges are much more stable than frictional charges and exhibit long-term stability. This is because corona charges can transfer from a "shallow trap" to a "deep trap". In this study, fibers were formed by applying high voltage to a polymer solution. During this process, charges were simultaneously injected into the fibers because of the high spinning voltage of 8 kV, resulting in a large amount of electrostatic charge remaining in the fibers. In addition, high electrospinning voltage promotes corona charge formation; thus, numerous stable corona charges can be injected into the fiber. Because the WCA of a fiber membrane is approximately 120°, it is not easily wetted, which promotes the storage of electrostatic charge. Therefore, the surface electrostatic potential of the PSF fiber filters did not change significantly after 28 days of use. Finally, the PSF fiber membrane showed long-term stability because of its large and continuous surface electrostatic potential.

4. Conclusions

In summary, PSF fiber membranes were fabricated using a simple one-step electrospinning method. The fiber membranes exhibited highly porous structures and good mechanical and hydrophobic properties. The porosity of the fiber membranes could be tuned by varying the spinning parameters (high voltage and tip-to-collector distance). Because of its low porosity and abundant injected charges, the PSF-2 fiber membrane exhibited exceptional filtration efficiency (99.6% and 99.2% for $PM_{2.5}$ and $PM_{1.0}$, respectively) and minimal pressure drop (65.0 and 65.2 Pa for $PM_{2.5}$ and $PM_{1.0}$, respectively), which were much higher than those of commercial dusk masks and KN95 respirators. Furthermore, no obvious electrostatic dissipation was observed over 28 days. The filtration efficiency, pressure drop, and *QF* of the PSF fiber membrane remained unchanged after 28 days, indicating the long-term durability of the fiber membrane during PM filtration. Because of the simplicity of the manufacturing procedures and the low cost of the raw materials, the PSF fiber membrane can be fabricated into various shapes and sizes and assembled into high-performance masks and air conditioning filters. The simple preparation process, low cost, and high PM removal efficiency make PSF fiber membranes promising for various industrial filtration applications.

Supplementary Materials: The following supporting information can be downloaded at: https://www.mdpi.com/article/10.3390/polym16020209/s1.

Author Contributions: H.L.: conceptualization, methodology, validation, project administration, writing, original draft, funding acquisition. W.L.: writing: original draft, validation, data curation, methodology, formal analysis. Y.S.: methodology, validation, data curation, formal analysis. L.T.: methodology, validation, formal analysis K.L.: methodology, validation, visualization L.B.: methodology, validation, and data curation. Z.X.: conceptualization, supervision, writing, review, and editing. B.L.: conceptualization, writing, review, and editing, project administration. All authors have read and agreed to the published version of the manuscript.

Funding: This research was funded by the National Key Research and Development Program of China (20202A050462).

Data Availability Statement: All required data are in the article.

Conflicts of Interest: The authors declare that they have no known competing financial interests or personal relationships that could have influenced the work reported in this paper.

References

1. Liu, H.; Cao, C.; Huang, J.; Chen, Z.; Chen, G.; Lai, Y. Progress on particulate matter filtration technology: Basic concepts, advanced materials, and performances. *Nanoscale* **2020**, *12*, 437–453. [CrossRef]
2. Losacco, C.; Perillo, A. Particulate matter air pollution and respiratory impact on humans and animals. *Environ. Sci. Pollut. Res.* **2018**, *25*, 33901–33910. [CrossRef]
3. Arias-Pérez, R.D.; Taborda, N.A.; Gómez, D.M.; Narvaez, J.F.; Porras, J.; Hernandez, J.C. Inflammatory effects of particulate matter air pollution. *Environ. Sci. Pollut. Res.* **2020**, *27*, 42390–42404. [CrossRef]
4. Ji, X.; Huang, J.; Teng, L.; Li, S.; Li, X.; Cai, W.; Chen, Z.; Lai, Y. Advances in particulate matter filtration: Materials, performance, and application. *Green Energy Environ.* **2023**, *8*, 673–697. [CrossRef]
5. Newman, J.D.; Bhatt, D.L.; Rajagopalan, S.; Balmes, J.R.; Brauer, M.; Breysse, P.N.; Brown, A.G.M.; Carnethon, M.R.; Cascio, W.E.; Collman, G.W.; et al. Cardiopulmonary Impact of Particulate Air Pollution in High-Risk Populations: JACC State-of-the-Art Review. *J. Am. Coll. Cardiol.* **2020**, *76*, 2878–2894. [CrossRef]
6. Combes, A.; Franchineau, G. Fine particle environmental pollution and cardiovascular diseases. *Metabolism* **2019**, *100*, 153944. [CrossRef]
7. Chen, H.-W.; Kuo, Y.-L.; Chen, C.-H.; Chiou, C.-S.; Chen, W.-T.; Lai, Y.-H. Biocompatibile nanofiber based membranes for high-efficiency filtration of nano-aerosols with low air resistance. *Process Saf. Environ. Prot.* **2022**, *167*, 695–707. [CrossRef]
8. Ju, Y.; Han, T.; Yin, J.; Li, Q.; Chen, Z.; Wei, Z.; Zhang, Y.; Dong, L. Bumpy structured nanofibrous membrane as a highly efficient air filter with antibacterial and antiviral property. *Sci. Total Environ.* **2021**, *777*, 145768. [CrossRef]
9. Li, P.; Wang, C.; Zhang, Y.; Wei, F.J.S. Air filtration in the free molecular flow regime: A review of high-efficiency particulate air filters based on carbon nanotubes. *Small* **2014**, *10*, 4543–4561. [CrossRef]
10. Zhou, Z.J.; Zhou, B.; Tseng, C.H.; Hu, S.C.; Shiue, A.; Leggett, G. Evaluation of characterization and filtration performance of air cleaner materials. *Int. J. Environ. Sci. Technol.* **2021**, *18*, 2209–2220. [CrossRef]
11. Patra, S.S.; Nath, J.; Panda, S.; Das, T.; Ramasamy, B. Evaluating the filtration efficiency of commercial facemasks' materials against respiratory aerosol droplets. *J. Air Waste Manag. Assoc.* **2022**, *72*, 3–9. [CrossRef]
12. Stahl, C.; Frederick, K.; Chaudhary, S.; Morton, C.J.; Loy, D.; Muralidharan, K.; Sorooshian, A.; Parthasarathy, S. Comparison of the Filtration Efficiency of Different Face Masks Against Aerosols. *Front. Med.* **2021**, *8*, 23577. [CrossRef]
13. Yang, C.; Jiang, X.; Gao, X.; Wang, H.; Li, L.; Hussain, N.; Xie, J.; Cheng, Z.; Li, Z.; Yan, J.; et al. Saving 80% Polypropylene in Facemasks by Laser-Assisted Melt-Blown Nanofibers. *Nano Lett.* **2022**, *22*, 7212–7219. [CrossRef]
14. Xu, Y.; Zhang, X.; Teng, D.; Zhao, T.; Li, Y.; Zeng, Y. Multi-layered micro/nanofibrous nonwovens for functional face mask filter. *Nano Res.* **2022**, *15*, 7549–7558. [CrossRef]
15. Cheng, X.; Zhao, L.; Zhang, Z.; Deng, C.; Li, C.; Du, Y.; Shi, J.; Zhu, M. Highly efficient, low-resistant, well-ordered PAN nanofiber membranes for air filtration. *Colloids Surf. A Physicochem. Eng. Asp.* **2022**, *655*, 130302. [CrossRef]

16. Zhu, M.; Hua, D.; Pan, H.; Wang, F.; Manshian, B.; Soenen, S.J.; Xiong, R.; Huang, C. Green electrospun and crosslinked poly(vinyl alcohol)/poly(acrylic acid) composite membranes for antibacterial effective air filtration. *J. Colloid Interface Sci.* **2018**, *511*, 411–423. [CrossRef]
17. Wang, N.; Raza, A.; Si, Y.; Yu, J.; Sun, G.; Ding, B. Tortuously structured polyvinyl chloride/polyurethane fibrous membranes for high-efficiency fine particulate filtration. *J. Colloid Interface Sci.* **2013**, *398*, 240–246. [CrossRef]
18. Wang, L.-Y.; Yu, L.E.; Lai, J.-Y.; Chung, T.-S. Developing ultra-high gas permeance PVDF hollow fibers for air filtration applications. *Sep. Purif. Technol.* **2018**, *205*, 184–195. [CrossRef]
19. Zhang, Q.; Li, Q.; Zhang, L.; Wang, S.; Harper, D.P.; Wu, Q.; Young, T.M. Preparation of electrospun nanofibrous poly(vinyl alcohol)/cellulose nanocrystals air filter for efficient particulate matter removal with repetitive usage capability via facile heat treatment. *Chem. Eng. J.* **2020**, *399*, 125768. [CrossRef]
20. Liu, Z.; Qin, L.; Liu, S.; Zhang, J.; Wu, J.; Liang, X. Superhydrophobic and highly moisture-resistant PVA@EC composite membrane for air purification. *RSC Adv.* **2022**, *12*, 34921–34930. [CrossRef] [PubMed]
21. Yao, Z.; Xia, M.; Xiong, Z.; Wu, Y.; Cheng, P.; Cheng, Q.; Xu, J.; Wang, D.; Liu, K. A Hierarchical Structure of Flower-Like Zinc Oxide and Poly(Vinyl Alcohol-co-Ethylene) Nanofiber Hybrid Membranes for High-Performance Air Filters. *ACS Omega* **2022**, *7*, 3030–3036. [CrossRef]
22. Zheng, G.; Shao, Z.; Chen, J.; Jiang, J.; Zhu, P.; Wang, X.; Li, W.; Liu, Y. Self-Supporting Three-Dimensional Electrospun Nanofibrous Membrane for Highly Efficient Air Filtration. *Nanomaterials* **2021**, *11*, 2567. [CrossRef]
23. Zhang, Q.; Li, Q.; Young, T.M.; Harper, D.P.; Wang, S. A Novel Method for Fabricating an Electrospun Poly(Vinyl Alcohol)/Cellulose Nanocrystals Composite Nanofibrous Filter with Low Air Resistance for High-Efficiency Filtration of Particulate Matter. *ACS Sustain. Chem. Eng.* **2019**, *7*, 8706–8714. [CrossRef]
24. Serbanescu, O.S.; Voicu, S.I.; Thakur, V.K. Polysulfone functionalized membranes: Properties and challenges. *Mater. Today Chem.* **2020**, *17*, 100302. [CrossRef]
25. Kheirieh, S.; Asghari, M.; Afsari, M. Application and modification of polysulfone membranes. *Rev. Chem. Eng.* **2018**, *34*, 657–693. [CrossRef]
26. Li, N.; Wang, W.; Ma, C.; Zhu, L.; Chen, X.; Zhang, B.; Zhong, C. A novel conductive rGO/ZnO/PSF membrane with superior water flux for electrocatalytic degradation of organic pollutants. *J. Membr. Sci.* **2022**, *641*, 119901. [CrossRef]
27. Zhong, D.; Wang, Z.; Zhou, J.; Wang, Y. Additive-free preparation of hemodialysis membranes from block copolymers of polysulfone and polyethylene glycol. *J. Membr. Sci.* **2021**, *618*, 118690. [CrossRef]
28. Mendes, A.C.; Strohmenger, T.; Goycoolea, F.; Chronakis, I.S. Electrostatic self-assembly of polysaccharides into nanofibers. *Colloids Surf. A Physicochem. Eng. Asp.* **2017**, *531*, 182–188. [CrossRef]
29. Gan, Y.; Xu, F.; Luo, J.; Yuan, H.; Jin, C.; Zhang, L.; Fang, C.; Sheng, O.; Huang, H.; Xia, Y.; et al. One-pot Biotemplate Synthesis of FeS2 Decorated Sulfur-doped Carbon Fiber as High Capacity Anode for Lithium-ion Batteries. *Electrochim. Acta* **2016**, *209*, 201–209. [CrossRef]
30. Ghodsi, A.; Fashandi, H.; Zarrebini, M.; Mirzaei, M. Controlling the Morphology of PVDF Hollow Fiber Membranes by Promotion of Liquid–Liquid Phase Separation. *Adv. Eng. Mater.* **2018**, *20*, 1701169. [CrossRef]
31. Russo, F.; Castro-Muñoz, R.; Santoro, S.; Galiano, F.; Figoli, A. A review on electrospun membranes for potential air filtration application. *J. Environ. Chem. Eng.* **2022**, *10*, 108452. [CrossRef]
32. Mirjalili, M.; Zohoori, S. Review for application of electrospinning and electrospun nanofibers technology in textile industry. *J. Nanostruct. Chem.* **2016**, *6*, 207–213. [CrossRef]
33. Shao, Z.; Chen, H.; Wang, Q.; Kang, G.; Wang, X.; Li, W.; Liu, Y.; Zheng, G. High-performance multifunctional electrospun fibrous air filter for personal protection: A review. *Sep. Purif. Technol.* **2022**, *302*, 122175. [CrossRef]
34. Zhang, J.; Gong, S.; Wang, C.; Jeong, D.-Y.; Wang, Z.L.; Ren, K. Biodegradable Electrospun Poly(lactic acid) Nanofibers for Effective PM 2.5 Removal. *Macromol. Mater. Eng.* **2019**, *304*, 1900259. [CrossRef]
35. Al Rai, A.; Stojanovska, E.; Fidan, G.; Yetgin, E.; Polat, Y.; Kilic, A.; Demir, A.; Yilmaz, S. Structure and performance of electroblown PVDF-based nanofibrous electret filters. *Polym. Eng. Sci.* **2020**, *60*, 1186–1193. [CrossRef]
36. Shao, Z.; Chen, Y.; Jiang, J.; Xiao, Y.; Kang, G.; Wang, X.; Li, W.; Zheng, G. Multistage-Split Ultrafine Fluffy Nanofibrous Membrane for High-Efficiency Antibacterial Air Filtration. *ACS Appl. Mater. Interfaces* **2022**, *14*, 18989–19001. [CrossRef] [PubMed]
37. Deng, Y.; Lu, T.; Cui, J.; Keshari Samal, S.; Xiong, R.; Huang, C. Bio-based electrospun nanofiber as building blocks for a novel eco-friendly air filtration membrane: A review. *Sep. Purif. Technol.* **2021**, *277*, 119623. [CrossRef]
38. Nicosia, A.; Keppler, T.; Müller, F.A.; Vazquez, B.; Ravegnani, F.; Monticelli, P.; Belosi, F. Cellulose acetate nanofiber electrospun on nylon substrate as novel composite matrix for efficient, heat-resistant, air filters. *Chem. Eng. Sci.* **2016**, *153*, 284–294. [CrossRef]
39. Zhu, X.; Kim, K. Electrospun polyacrylonitrile fibrous membrane for dust removal. *Front. Mater.* **2022**, *9*, 973660. [CrossRef]
40. Cai, R.-R.; Li, S.-Z.; Zhang, L.-Z.; Lei, Y. Fabrication and performance of a stable micro/nano composite electret filter for effective $PM_{2.5}$ capture. *Sci. Total Environ.* **2020**, *725*, 138297. [CrossRef]
41. Al-Attabi, R.; She, F.; Zhao, S.; Dumée, L.F.; Schütz, J.A.; Xing, W.; Zhong, Z.; Kong, L. Durable and comfortable electrospun nanofiber membranes for face mask applications. *Sep. Purif. Technol.* **2023**, *322*, 124370. [CrossRef]
42. Zhu, Z.; Yu, M.; Ren, R.; Wang, H.; Kong, B. Thymol incorporation within chitosan/polyethylene oxide nanofibers by concurrent coaxial electrospinning and in-situ crosslinking from core-out for active antibacterial packaging. *Carbohydr. Polym.* **2024**, *323*, 121381. [CrossRef]

43. Zhang, Y.; Yang, M.; Zhou, Y.; Yao, A.; Han, Y.; Shi, Y.; Cheng, F.; Zhou, M.; Zhu, P.; Tan, L. Transition sandwich Janus membrane of cellulose acetate and polyurethane nanofibers for oil–water separation. *Cellulose* **2022**, *29*, 1841–1853. [CrossRef]
44. Zhang, T.; Zhao, C.; Chen, X.; Jiang, A.; You, Z.; Shah, K.J. Different weathering conditions affect the release of microplastics by masks. *Environ. Sci. Pollut. Res.* **2023**, *30*, 66102–66112. [CrossRef]
45. Mazoochi, T.; Hamadanian, M.; Ahmadi, M.; Jabbari, V. Investigation on the morphological characteristics of nanofiberous membrane as electrospun in the different processing parameters. *Int. J. Ind. Chem.* **2012**, *3*, 2. [CrossRef]
46. Chee, T.Y.; Mohd, Y.A.R.; Abdullah, F.; Wan Mahmood, W.M.A.; Jasni, M.J.F.; Malek, N.A.N.N.; Buang, N.A.; Govarthanan, M. Fabrication, characterization and application of electrospun polysulfone membrane for phosphate ion removal in real samples. *Chemosphere* **2022**, *303*, 135228. [CrossRef]
47. Shao, W.; Niu, J.; Han, R.; Liu, S.; Wang, K.; Cao, Y.; Han, P.; Li, X.; Zhang, H.; Yu, H.; et al. Electrospun Multiscale Poly(lactic acid) Nanofiber Membranes with a Synergistic Antibacterial Effect for Air-Filtration Applications. *ACS Appl. Polym. Mater.* **2023**, *5*, 9632–9641. [CrossRef]
48. Zhang, K.; Li, Z.; Kang, W.; Deng, N.; Yan, J.; Ju, J.; Liu, Y.; Cheng, B. Preparation and characterization of tree-like cellulose nanofiber membranes via the electrospinning method. *Carbohydr. Polym.* **2018**, *183*, 62–69. [CrossRef]
49. Peng, Z.; Shi, J.; Xiao, X.; Hong, Y.; Li, X.; Zhang, W.; Cheng, Y.; Wang, Z.; Li, W.J.; Chen, J.; et al. Self-charging electrostatic face masks leveraging triboelectrification for prolonged air filtration. *Nat. Commun.* **2022**, *13*, 7835. [CrossRef]
50. Ma, W.; Qi, H.; Zhang, Y.; Lin, M.; Qiu, Y.; Zhang, C. Fabrication of Laminated Micro/Nano Filter and Its Application for Inhalable PM Removal. *Polymers* **2023**, *15*, 1459. [CrossRef] [PubMed]
51. Wu, J.; Akampumuza, O.; Liu, P.; Quan, Z.; Zhang, H.; Qin, X.; Wang, R.; Yu, J. 3D structure design and simulation for efficient particles capture: The influence of nanofiber diameter and distribution. *Mater. Today Commun.* **2020**, *23*, 100897. [CrossRef]
52. Buivydiene, D.; Todea, A.M.; Asbach, C.; Krugly, E.; Martuzevicius, D.; Kliucininkas, L. Composite micro/nano fibrous air filter by simultaneous melt and solution electrospinning. *J. Aerosol Sci.* **2021**, *154*, 105754. [CrossRef]
53. Wang, H.; Bao, Y.; Yang, X.; Lan, X.; Guo, J.; Pan, Y.; Huang, W.; Tang, L.; Luo, Z.; Zhou, B.; et al. Study on Filtration Performance of PVDF/PUL Composite Air Filtration Membrane Based on Far-Field Electrospinning. *Polymers* **2022**, *14*, 3294. [CrossRef] [PubMed]
54. Kakoria, A.; Chandel, S.S.; Sinha-Ray, S. Novel supersonically solution blown nanofibers from waste PET bottle for PM0.1-2 filtration: From waste to pollution mitigation. *Polymer* **2021**, *234*, 124260. [CrossRef]
55. Fan, X.; Wang, Y.; Kong, L.; Fu, X.; Zheng, M.; Liu, T.; Zhong, W.-H.; Pan, S. A Nanoprotein-Functionalized Hierarchical Composite Air Filter. *ACS Sustain. Chem. Eng.* **2018**, *6*, 11606–11613. [CrossRef]
56. Chen, R.; Zhang, H.; Wang, M.; Zhang, X.; Gan, Z. Thermoplastic Polyurethane Nanofiber Membrane Based Air Filters for Efficient Removal of Ultrafine Particulate Matter PM0.1. *ACS Appl. Nano Mater.* **2021**, *4*, 182–189. [CrossRef]
57. Kadam, V.; Kyratzis, I.L.; Truong, Y.B.; Schutz, J.; Wang, L.; Padhye, R. Electrospun bilayer nanomembrane with hierarchical placement of bead-on-string and fibers for low resistance respiratory air filtration. *Sep. Purif. Technol.* **2019**, *224*, 247–254. [CrossRef]
58. Segovia, J.M.; Huang, C.-H.; Mamishev, M.; Yuan, N.; He, J.; Novosselov, I. Performance of Textile Mask Materials in Varied Humidity: Filtration Efficiency, Breathability, and Quality Factor. *Appl. Sci.* **2022**, *12*, 9360. [CrossRef]
59. Gao, Y.; Tian, E.; Zhang, Y.; Mo, J. Utilizing electrostatic effect in fibrous filters for efficient airborne particles removal: Principles, fabrication, and material properties. *Appl. Mater. Today* **2022**, *26*, 101369. [CrossRef]
60. Gao, H.; He, W.; Zhao, Y.B.; Opris, D.M.; Xu, G.; Wang, J. Electret mechanisms and kinetics of electrospun nanofiber membranes and lifetime in filtration applications in comparison with corona-charged membranes. *J. Membr. Sci.* **2020**, *600*, 117879. [CrossRef]
61. Liu, C.; Dai, Z.; He, B.; Ke, Q.F. The Effect of Temperature and Humidity on the Filtration Performance of Electret Melt-Blown Nonwovens. *Materials* **2020**, *13*, 4774. [CrossRef] [PubMed]
62. Lu, Y.; Li, S.; Chen, F.; Ma, H.; Gao, C.; Xue, L. Development of coin-shaped ZIF-7 functionalized superhydrophobic polysulfone composite foams for continuous removal of oily contaminants from water. *J. Hazard. Mater.* **2022**, *421*, 126788. [CrossRef] [PubMed]
63. Gu, J.; Zheng, M.; Zhu, T.; Wang, N.; Wang, L.; Yu, J.; Qin, X. Electrostatic-modulated interfacial crosslinking and waterborne emulsion coating toward waterproof, breathable, and antifouling fibrous membranes. *Chem. Eng. J.* **2023**, *454*, 140439. [CrossRef]
64. Huiming, X.; Gangjin, C.; Xumin, C.; Zhi, C. A Flexible Electret Membrane with Persistent Electrostatic Effect and Resistance to Harsh Environment for Energy Harvesting. *Sci. Rep.* **2017**, *7*, 8443. [CrossRef]
65. *GB 2626-2019*; Respiratory Protection—Non-Powered Air-Purifying Particle Respirator. China Standard Press: Beijing, China, 2019.
66. Fu, Q.; Liu, Y.; Liu, T.; Mo, J.; Zhang, W.; Zhang, S.; Luo, B.; Wang, J.; Qin, Y.; Wang, S.; et al. Air-permeable cellulosic triboelectric materials for self-powered healthcare products. *Nano Energy* **2022**, *102*, 107739. [CrossRef]
67. Zhang, X.; Wang, Y.; Liu, W.; Jin, X. Needle-punched electret air filters (NEAFs) with high filtration efficiency, low filtration resistance, and superior dust holding capacity. *Sep. Purif. Technol.* **2022**, *282*, 120146. [CrossRef]

polymers

MDPI

Article

Biodegradable Nanofiber/Metal–Organic Framework/Cotton Air Filtration Membranes Enabling Simultaneous Removal of Toxic Gases and Particulate Matter

Sujin Ryu [1,†], Doyeon Kim [1,2,†], Hyewon Lee [1], Yoonjin Kim [1], Youngbok Lee [2,3,4], Myungwoong Kim [5,*], Heedong Lee [1,*] and Hoik Lee [1,2,*]

1 Advanced Textile R&D Department, Research Institute of Convergence Technology, Korea Institute of Industrial Technology (KITECH), 143 Hanggaulro, Sangnok-gu, Ansan 15588, Republic of Korea; sjryu@kitech.re.kr (S.R.); dobbie@kitech.re.kr (D.K.)
2 HYU-KITECH Joint Department, Hanyang University, Ansan 15588, Republic of Korea; yblee@hanyang.ac.kr
3 Department of Applied Chemistry, Hanyang University, Ansan 15588, Republic of Korea
4 Department of Chemical and Molecular Engineering, Center for Bionano Intelligence Education and Research, Hanyang University, Ansan 15588, Republic of Korea
5 Department of Chemistry and Chemical Engineering, Inha University, Incheon 22212, Republic of Korea
* Correspondence: mkim233@inha.ac.kr (M.K.); lhd0121@kitech.re.kr (H.L.); hoik@kitech.re.kr (H.L.)
† These authors contributed equally to this work.

Abstract: The typical filters that protect us from harmful components, such as toxic gases and particulate matter (PM), are made from petroleum-based materials, which need to be replaced with other environmentally friendly materials. Herein, we demonstrate a route to fabricate biodegradable and dual-functional filtration membranes that effectively remove PM and toxic gases. The membrane was integrated using two layers: (i) cellulose-based nanofibers for PM filtration and (ii) metal–organic framework (MOF)-coated cotton fabric for removal of toxic gases. Zeolitic imidazolate framework (ZIF-8) was grown from the surface of the cotton fabric by the treatment of cotton fabric with an organic precursor solution and subsequent immersion in an inorganic precursor solution. Cellulose acetate nanofibers (NFs) were deposited on the MOF-coated cotton fabric via electrospinning. At the optimal thickness of the NF layer, the quality factor of 18.8×10^{-2} Pa^{-1} was achieved with a filtration efficiency of 93.1%, air permeability of 19.0 $cm^3/cm^2/s$, and pressure drop of 14.2 Pa. The membrane exhibits outstanding gas adsorption efficiencies (>99%) for H_2S, formaldehyde, and NH_3. The resulting membrane was highly biodegradable, with a weight loss of 62.5% after 45 days under standard test conditions. The proposed strategy should provide highly sustainable material platforms for practical multifunctional membranes in personal protective equipment.

Keywords: nanofiber; cellulose; biodegradable; metal organic frameworks; filtration

1. Introduction

Numerous air pollutants, such as hazardous chemicals and particles, have been released from industrial facilities and other human activities and may adversely affect the environment and human health [1]. The World Health Organization (WHO) has reported that approximately seven million people have died due to exposure to air pollution, suggesting that air pollution is a critical health risk [2]. In particular, particulate matter (PM) is currently considered a significant air pollutant that can potentially lead to respiratory disease [3,4]. Other hazardous contaminants are toxic gases that can eventually displace oxygen from the lungs or can damage the respiratory system, e.g., volatile organic compounds (VOCs), aldehydes, H_2S, NH_3, and so on [2,5,6]. In addition, exposure to air, including sources for infectious and major communicable diseases, such as COVID-19 is highly problematic. Therefore, protection from the above-mentioned pollutants requires the use of personal protective equipment (PPE), such as face masks that have become indispensable.

Membrane filters to block air pollutants are necessary for filter-type PPE, which is generally made from synthetic polymers, such as polypropylene (PP) and polyethylene terephthalate (PET). Although these synthetic polymers are highly processable and economical, their high stability has resulted in other environmental problems. For example, the mandatory use of face masks during the COVID-19 crisis has posed a strong challenge to plastic waste management and subsequent adverse environmental effects [7]. According to the WHO, the weight of PPE purchased worldwide due to COVID-19 in 1.5 years was approximately 87,000 tons, eventually ending in landfill [8]. Along with the dramatic increase in the demand for PPE, eco-friendly membrane filters with materials that can replace synthetic polymers are needed.

Cellulose is an abundant and biodegradable natural polymer material exhibiting outstanding mechanical, chemical, and biological properties and is advantageous in various applications, such as composite science [9], paper engineering [10,11], and medical engineering [12]. Cotton is a well-known cellulose-based fabric material utilized in clothing. Despite the advantages of cotton, it has not been used as a component for filtration devices because of its large pore size and incompatibility with the corona charging process, which is a typical method applied widely to PP non-woven filters for removing PM. To apply a conventional cotton fabric as a filtration media, nanofabrication can be an effective approach as it has been a workhorse in the field of air filtration [13,14]. In particular, electrospinning is a representative technique to produce ultra-fine polymeric nanofiber (NF) mat in a scale of from several tens to hundreds of nanometers [15], resulting in high surface area and a sub-micron sized pore structure, which is highly advantageous in various applications including filtration and biomedical engineering [16,17]. The effective combination, i.e., stacking NF on the cotton fabric, is feasible for improving the filtration performance against PM.

In addition, the chemical structure of the cellulose is beneficial to tailor its properties to achieve the target application because it has abundant multiple hydroxyl groups providing useful reactivity to incorporate various chemical functions for desirable properties [18,19]. The metal–organic framework (MOF) is a widely used material as an adsorbent for different gas molecules due to its high specific surface area and porosity [20]. Among the various MOF materials, the zeolitic imidazolate framework (ZIF) has a high potential for the removal or storage of specific gas molecules [21,22]. The incorporation of ZIF onto a cotton fabric possibly provides an effective route for MOF-incorporated cotton fabric that is applicable for removing toxic gases. Hydroxy groups in the cotton fabric can effectively interact with precursor molecules to synthesize ZIFs, allowing direct growth on the surface of the cotton fabric, eventually leading to a filtration membrane capable of adsorbing toxic gases.

The multifunctional fibrous membrane for removing particulate matter and toxic gases was demonstrated by employing simultaneous electrospinning/electrospray of poly(vinyl alcohol) and MOF onto PET nonwoven fabric [4]. However, it does not only require a complex co-electrospinning system, but also in the process, the adsorbed MOF particles can be desorbed in a variety of conditions. More importantly, the system is yet biodegradable, which is not preferred as a next-generation material platform. Herein, we report the fabrication route of a biodegradable cellulose-based filtration membrane via the direct growth of ZIF-8 from the surface of cotton fabric and subsequent electrospinning of cellulose acetate (CA) solution to form a CA NF/MOF-coated cotton fabric membrane that can filter PM and toxic gases simultaneously (Figure 1). The major components of the membrane are made from cellulose-based materials to ensure biodegradability. ZIF-8 was selected as a coating material on the cotton fabric to remove toxic gases, e.g., H_2S, formaldehyde, and NH_3 [22]. The filtration performance of the resulting membrane was assessed by controlling the amount of NF deposited on the MOF-coated cotton fabric, maximizing the quality factor towards the optimal filtration performance parameters. The multi-layered membrane effectively removed the three model toxic gas molecules. In addition, the membrane showed adequate biodegradability, highlighting its potential to replace the current petroleum-based polymer membrane filters.

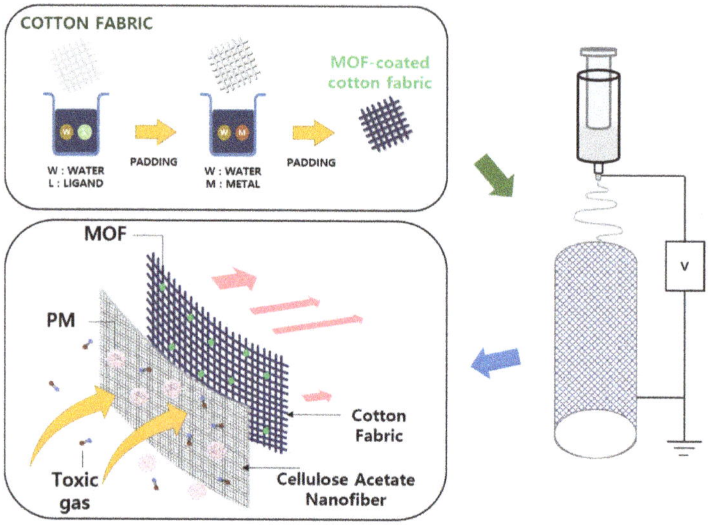

Figure 1. Schematic diagram displaying the fabrication of the multifunctional hybrid CA NF/MOF-coated cotton membrane for the simultaneous filtration of toxic gas molecules and PM.

2. Materials and Methods

2.1. Materials

Cellulose acetate (CA; $M_n \approx 50{,}000$ g/mol), cellulose (particle size ≈ 20 μm, as a standard material of biodegradability), *N,N*-dimethylformamide (DMF; 99.8%), acetone ($\geq$99.5%), zinc nitrate hexahydrate (98%), and 2-methylimidazole (99%) were purchased from Sigma-Aldrich Co., Ltd. (St. Louis, MO, USA). The cotton fabric (30 g/m^2) was purchased from Kolon Industries Inc. and was cut to a dimension of 20 cm $\times$ 28.5 cm for use in the fabrication of a filtration membrane. All chemical reagents were used as received.

2.2. Microscopic Characterization

The morphologies of the NFs on cotton fabrics were examined using field-emission electron microscopy (FE-SEM; SU8010, HITACHI, Tokyo, Japan).

2.3. Spectroscopic Characterization

X-ray diffraction (XRD) studies were carried out using a RINT2000 X-ray diffractometer (Rigaku, Tokyo, Japan) using a Cu Kα source. Fourier-transform infrared (FT-IR) spectra were recorded on Perkin Elmer Frontier in the wavenumber range of 400–4000cm^{-1} (Waltham, MA, USA).

2.4. Assessment of Air Permeability

Air permeability was estimated using FX 3300 Air Permeability Tester III (TEXTEST, Schwerzenbach, Switzerland) by measuring an air flow rate perpendicularly passing through the sample (effective area = 38 cm^2) under an air pressure difference of 125 Pa [23]. The particulate respiration filter test was performed using a TSI Automated Filter Tester 8130A (TSI, Shoreview, MN, USA) to examine the filtration efficiency and pressure drop using a sodium chloride challenge aerosol at a flow rate, i.e., a face velocity of 32 L/min. When the aerosol passed through the filter, the aerosol was measured simultaneously using a light scattering photometer. Five measurements were performed, and the average and the standard deviation values were reported. A filtration efficiency (η) that represents the filtration performance of the membrane was estimated using Equation (1), and a quality factor (Q_f) was obtained using Equation (2):

$$\eta = (C_{up} - C_{down})/C_{up} \quad (1)$$

where concentrations are upstream (C_{up}) and downstream (C_{down}).

$$Q_f = -\ln(1 - \eta)/\Delta P \quad (2)$$

where ΔP is the pressure drop across the filtration medium [24].

2.5. Evaluation of Gas Adsorption Capability

The adsorption capability for toxic gases was measured using a gas detector tube method. The membrane sample (10 cm × 10 cm) was placed in a chamber (2 L) filled with a gas of interest at 100 ppm. The gas concentration was measured using a valve for up to an hour. The rate of the gas concentration decrease (CDR) was determined using Equation (3):

$$\text{CDR } (\%) = [(C_i - C_f)/C_i] \times 100 \quad (3)$$

where C_i and C_f are concentrations at the initial and final states, respectively.

2.6. Assessment of Biodegradability

The biodegradability was assessed using the procedure and conditions following KS M ISO 14855-1. Standard compost (ABNEXO, Gyeonggi, Republic of Korea; dry solid content = 52.5%) was used as an inoculum. Cellulose was used as a standard material, and NF/MOF-coated cotton samples were used as test materials after cryo-milling to less than 250 μm. The mixture of standard compost with a moisture content of 50% (300 g), standard material (25 g), or test material (25 g) was placed in a composting container, followed by the measurement of CO_2 generation every 24 h for 45 days. Biodegradability (D_t) was calculated using Equation (4):

$$D_t = [((CO_2)_T - (CO_2)_B)/(CO_2)_{Th}] \times 100 \quad (4)$$

where the $(CO_2)_T$ and $(CO_2)_B$ are the cumulative amount of CO_2 generated from the composting container containing the test material and standard material, respectively, and $(CO_2)_{Th}$ is the theoretical amount of CO_2 generated from the test material.

2.7. Preparation of MOF-Coated Cotton Fabric

An aqueous solution of 2-methylimidazole (0.24 M) was prepared for use as a ligand solution. Cotton fabric was immersed in a ligand solution for 3 h at room temperature. The cotton fabric was applied to the padder (DL-2005, Daelim starlet, Siheung, Republic of Korea) at 0.5 bar and 1 m/min. The resulting cotton fabric was immersed in an aqueous solution of zinc nitrate hexahydrate (0.48 M) for 5 h, and dried under ambient conditions at room temperature for 24 h.

2.8. Electrospinning of CA NFs on Cotton Fabric

CA NFs were fabricated on the prepared MOF/cotton fabric using an electrospinning process. The electrospinning apparatus was equipped with a high-voltage power supply (NC-ESR100S, NanoNC, Seoul, Republic of Korea) as a source of the electric field, a plastic syringe with a metallic needle (25 gauge), and a cylinder-type metallic drum as a collector. A 19 wt% CA solution in a mixed solvent of DMF and acetone (4/6, *v*/*v*) was prepared via vigorous stirring at room temperature for 24 h. The solution was injected into the syringe and subjected to electrospinning at a voltage of 15 kV and a flow rate of 10 μL/min with a tip-to-collector distance of 15 cm. The cotton fabric was mounted on the metallic drum collector using scotch tape. The funnel collector rotated at the speed of 100 rpm. All electrospinning experiments were conducted at room temperature with a relative humidity of 35–40%.

3. Results and Discussion

3.1. Fabrication of MOF-Coated Cotton Fabric

A multifunctional biodegradable membrane filter was fabricated by stacking CA NFs onto the MOF-coated cotton fabric. The cotton fabric was used as a supporting material due to its outstanding mechanical properties, biodegradability, and inexpensiveness. More importantly, the abundant hydroxyl groups in the cellulose enable the formation of MOF structure on the surface of the cotton. The adsorption of the ligand readily occurs when the cotton fabric is immersed in the ligand solution because the hydroxyl group can effectively interact with 2-methylimidazole [25]. After immersion, the padding process was carried out to remove the excess ligands not adsorbed on the fabric surface. The resulting cotton fabric was finally immersed in a metal ion precursor solution to form ZIF-8 MOF particles. First, the effect of the variation in the amounts of reactants, i.e., Zn^{2+} and ligand, on the morphology of resulting MOF particles was assessed without cotton (Figure S1). Upon varying the [ligand] to [Zn^{2+}] ratio from 2.0 to 8.0, the morphology of the MOFs fabricated with a ratio of 2.0 led to 1–2 μm spherical particles. The morphology was changed to a disc-like or elliptical structure as the ratio increased [26]. In addition, at a ratio higher than 4, the size of disc-like particles gradually decreased, which is in good agreement with the literature [27]. For the applicability to remove gas species, [ligand]/[Zn^{2+}] was set to 2.0 to attain a small spherical MOF structure and maximize the adsorption capability and surface area. ZIF-8/cotton was fabricated at a [ligand]/[Zn^{2+}] ratio of 2.0, and the morphologies were examined (Figure 2). The surface of the pristine cotton fabric was very smooth (Figure 2a). In contrast, the MOF/cotton fabric exhibited some grains on its surface, indicating the formation of the MOF particles on the surface (Figure 2b). In addition, the resulting MOF/cotton fabric was washed sufficiently with ethanol, acetone, and water to remove weakly attached particles, and strongly attached MOF particles were observed on the surface (Figure 2c), which is of importance to ensure the stability of the ZIF-8 coating on the fabric.

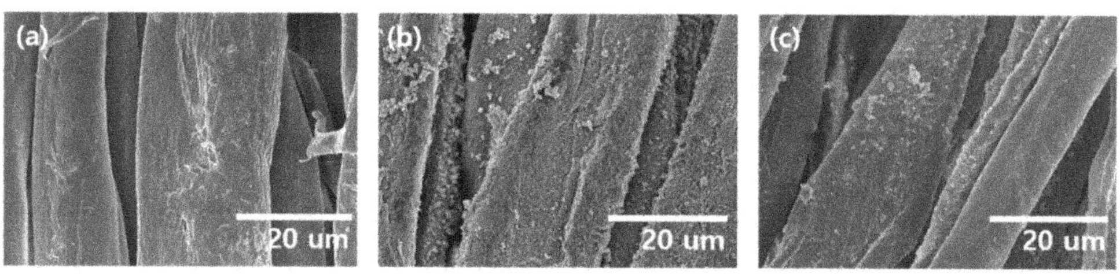

Figure 2. SEM images showing the morphology of (**a**) padded cotton and (**b**) ZIF-8-coated cotton before washing, and (**c**) ZIF-8-coated cotton after washing.

3.2. Electrospinning of CA Nanofiber on MOF-Coated Cotton Fabric

On the MOF-coated cotton membrane, a CA solution was electrospun at fixed concentration with different amounts of solution from 0.6 mL to 7.2 mL. Although the concentration affects the fiber diameter in the electrospinning process [28,29], in terms of filtration efficiency, the thickness of the nanofiber mat is more crucial [4,30]. Thus, we focused on the effect of the thickness of the nanofiber mat fabricated at the fixed concentration of the polymer solution. Figure 3 presents the morphologies of CA NFs on MOF/cotton fabrics, namely $NF_{0.6}$, $NF_{1.8}$, $NF_{3.6}$, $NF_{6.0}$, and $NF_{7.2}$, where the number indicates the volume (mL) of electrospun solution. All membranes exhibited a straight nanofibrous structure with high uniformity and smooth surface on cotton fabrics, suggesting that the amount of electrospun solution did not affect the morphology, as expected. The thicknesses of the resulting NF membranes were 0.222, 0.232, 0.253, 0.279, and 0.293 mm for $NF_{0.6}$, $NF_{1.8}$, $NF_{3.6}$, $NF_{6.0}$, and

$NF_{7.2}$, respectively. In $NF_{0.6}$, the underlying cotton fabric was observed readily even after NF was formed on its top (Figure 3a), suggesting that the NFs did not sufficiently cover the cotton fabric. In contrast, the NF samples fabricated with the solution with a volume higher than 1.8 mL fully covered the fabric, which was vital for the PM removal filter. The cross-sectional SEM image of $NF_{7.2}$ (Figure 3f) revealed a bilayer structure, confirming that NFs were deposited well on the cotton fabric without a gap between the two layers.

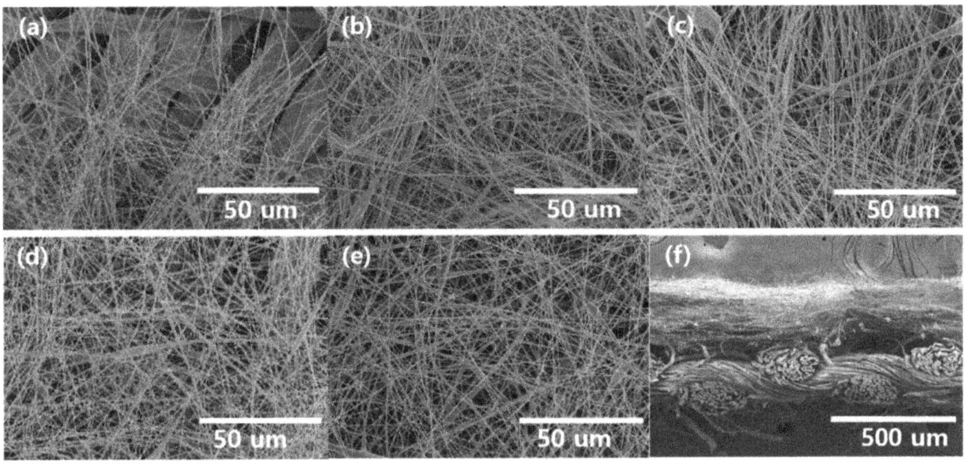

Figure 3. SEM images showing the morphology of (**a**) $NF_{0.6}$, (**b**) $NF_{1.8}$, (**c**) $NF_{3.6}$, (**d**) $NF_{6.0}$, (**e**) $NF_{7.2}$, and (**f**) cross-section SEM image of $NF_{7.2}$.

Commercial filters made from synthetic polymers, such as polyethylene (PE) and PP, can be used for filtration applications after treatment with corona charging, one of the most exploited techniques because of its convenience. Under humid or high-temperature conditions, however, the charges injected into the dielectric materials can be dissipated easily, resulting in a degradation of the durability of the resulting membrane [31]. Since cotton can effectively interact with water molecules, despite exhibiting excellent biodegradability, it cannot be used widely as a filter medium because of its high water content. In this perspective, electrospun CA NFs are beneficial as they can impart filtration capability and circumvent this issue. Nano-sized fibers should compensate for the disadvantages of cotton fabric by improving the filtration efficiency much higher than cotton fabrics because of its extremely small pores in the NF membrane. It should be noted that high filtration efficiency can also result in low air permeability and high-pressure drop, which are essential factors in filtration performance. Therefore, optimizing the filtration efficiency and air permeability is essential, which is feasible with the control of the thickness of the nanofiber mat, which is one of the most significant parameters to affect the performance of air filtration membranes.

3.3. Characterization of the Fabricated Filtration Membrane

The fabricated composite fibrous membranes were characterized by FT-IR spectroscopy and XRD (Figure 4). In the FT-IR spectrum of MOF (Figure 4a), the peak at $\approx$760 cm^{-1} was assigned to the stretching vibration mode of Zn-O in ZIF-8 [21]. The peaks at $\approx$1150 and $\approx$1310 cm^{-1} were assigned to the C–N bending mode of the imidazole group [21,32,33]. These characteristic peaks were barely observed in the spectra of MOF/cotton and NF/MOF/cotton membranes, which was expected, as the relative amount of MOF in the composite system is very small compared to cotton and NF; compared to cotton fabric, approximately 2.8 wt% of MOF particles were formed, resulting in a very low intensity of MOF in the FT-IR spectra. The peaks due to CA were readily observed in the composite fabric sample. The peaks at $\approx$1750 cm^{-1} and $\approx$1220 cm^{-1} were assigned to C=O and C–O bonds of a carbonyl group, which were not found in cotton and MOFs/cotton and only appeared in the CA NF/MOFs/cotton sample [18,19]. Given the configuration of the composite sample, the appearance of the characteristic peaks due to CA NF is acceptable. These results suggest that the CA NF had been deposited effectively on MOF-coated cotton fabric. The resulting sample was characterized by XRD (Figure 4b). In the pattern of only MOF (green trace), the XRD peaks of MOF were observed at 8.0°, 12.0°, 15.0°, and 17.6° 2θ [34,35]. In particular, the peak at 8.0° 2θ (011) was a sharp characteristic crystalline peak typically representing the sodalite structure of ZIF-8 with a body-centered cubic packing. The cotton sample did not exhibit an XRD peak at around 8.0°; some broad peaks at 14.75°, 16.48°, and 22.71° 2θ were observed, indicating the semicrystallinity of cotton fabric [36]. On the other hand, MOF/cotton and CA NF/MOF/cotton membranes exhibited an XRD peak at 8.0° 2θ, suggesting that ZIF-8 had grown on the cotton fabric, and the structure was fully preserved during the electrospinning process.

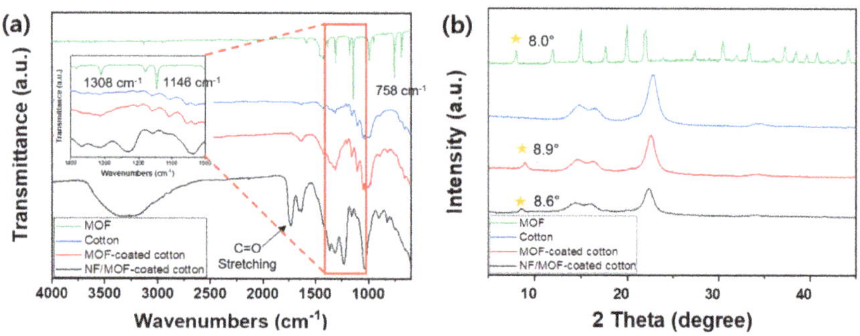

Figure 4. (**a**) FT-IR spectra and (**b**) XRD profiles of the ZIF-8 (green), cotton (blue), MOF-coated cotton (red), and NF/MOF-coated cotton (black) samples.

3.4. Air Filtration Performance of the Filtration Membrane

To optimize the filtration performance, we assessed the air permeability, filtration efficiency, pressure drop, and Q_f for cotton, MOF-coated cotton, and CA NF/MOF-coated cotton (Figure 5). In addition, the amount of the CA solution that was electrospun onto MOF-coated cotton fabric was varied because it determines the density of CA NF, which is directly related to the porosity of the NF membrane and, hence, the final filtration performance. In the case of MOF-coated cotton and $NF_{0.6}$/MOF-coated cotton membranes, air permeability and pressure drop did not change significantly compared to the pristine cotton fabric. The filtration efficiencies were 19.6, 21.3, and 54.6% for the cotton, MOF-coated cotton, and $NF_{0.6}$/MOF-coated cotton membrane, respectively (Figure 5a). These values are insufficient to filter PM effectively. These results suggest that the growth of ZIF-8 on the surface of cotton and the deposition of small amounts of NF did not result in high enough porosity for practical filtration performance. However, the $NF_{7.2}$/MOF-coated fabric showed a largely increased filtration efficiency of ≈99%, whereas the air permeability was reduced significantly from ≈23 to ≈10 $cm^3/cm^2/s$. In addition to these changes, a pressure drop increased gradually from ≈14 (NF1.8/MOF-coated cotton fabric) to ≈67 Pa ($NF_{7.2}$/MOF-coated fabric). Finding the optimal parameter, i.e., the amount of NF, is critical to attaining the membrane utilizable for PPE because these filtration performance parameters are in the trade-off relation. A quantitative evaluation of the filtration performance can be carried out by estimating Q_f, which is affected significantly by the thickness of the NF layer that can be tuned readily by controlling the amount of electrospun CA solution. The $NF_{1.8}$/MOF-coated cotton fabric membrane showed the highest Q_f value (18.8×10^{-2} Pa^{-1}) with an air permeability of ≈19 $cm^3/cm^2/s$ and pressure drop of ≈14 Pa, which are acceptable for the filtration medium in facemasks [37]. Recently, a dual-functional hybrid filtration system consisting of MOF and NF was reported for the removal of particulate matter and toxic gases in the air [4]. The study highlighted the importance of finding the most desirable NF, MOF, and non-woven fabric configurations for optimal filtration performance for the practical PPE application. The membrane in the current study also shows comparable air filtration performance, ensuring the effectiveness of the proposed fabrication approach. Therefore, the $NF_{1.8}$/MOF-coated cotton membrane exhibiting the highest Q_f was further assessed for its ability to remove toxic gases.

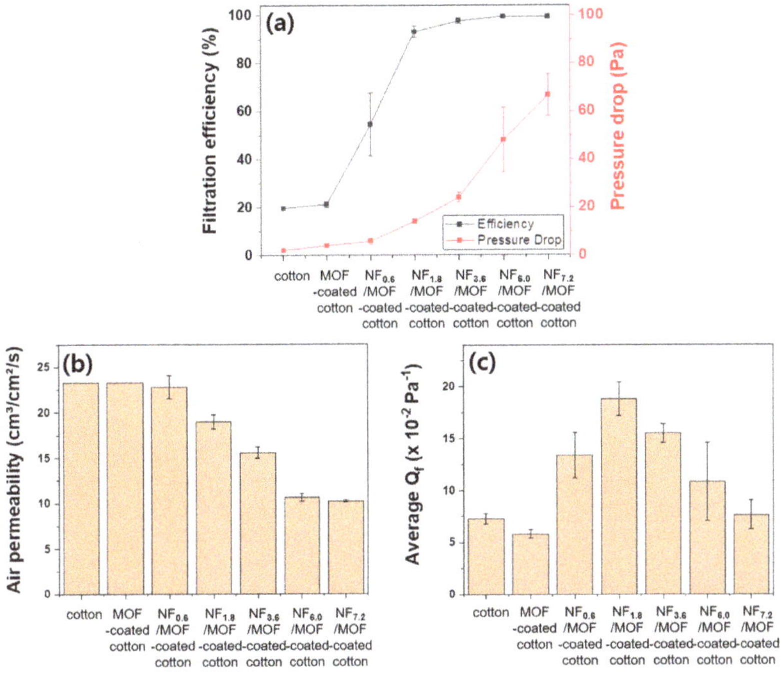

Figure 5. (**a**) Filtration efficiency (left y-axis, black) and pressure drop (right y-axis, red), (**b**) air permeability, and (**c**) average Q_f of cotton, MOF-coated cotton, and CA NF/MOF-coated cotton fabricated with different amounts of electrospun CA solutions.

3.5. Gas Adsorption Behavior of Membrane Filter

Three model gases, i.e., H_2S, formaldehyde, and NH_3, known as harmful gases representing acidic, neutral, and basic gases, were tested to assess the capability to remove toxic gases. Figure 6 shows the gas-capturing behaviors of the pristine cotton, MOF-coated cotton, and $NF_{1.8}$/MOF-coated cotton membranes. The CDR values after 60 min for pure cotton fabric showed ≈10%, ≈15%, and ≈30% for H_2S, formaldehyde, and NH_3, respectively. On the other hand, MOF-coated cotton fabric showed dramatically improved adsorption capability compared to pristine cotton fabric. MOF is highly effective in capturing these gases because of its active sites and large surface area [38]. In particular, ZIF-8, which was grown as an MOF on the cotton fabric, is used widely for adsorbing harmful gas species [39]. As expected, MOF-coated cotton and $NF_{1.8}$/MOF-coated cotton membranes exhibited a CDR of approximately 99% for all three gases after 60 min. These results strongly suggest that the direct growth of MOF-coated cotton fabric is highly effective for removing toxic gases, even with the NF layer imparting PM filtration capability.

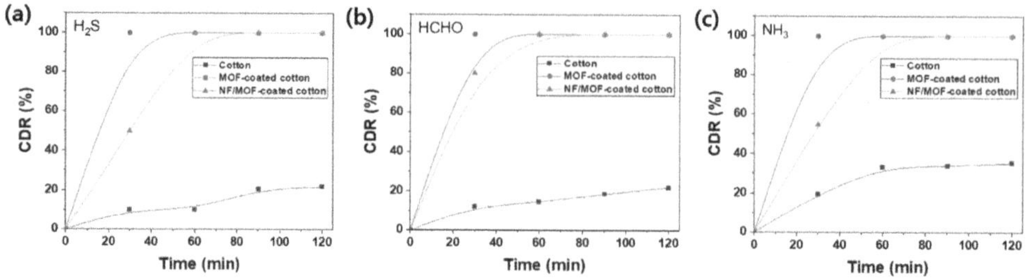

Figure 6. CDR as a function of exposure time for (**a**) H_2S, (**b**) formaldehyde, and (**c**) NH_3 measured with pristine cotton fabric (black), MOF-coated cotton fabric (red), and CA $NF_{1.8}$/MOF-coated cotton fabric membranes (blue).

3.6. Biodegradability of the Filtration Membrane

The majority of the $NF_{1.8}$/MOF-coated cotton membrane is a cellulose-based material because the calculated weight ratio of the components is 16.2:2.8:81.0 (NF, MOF, and cotton), suggesting that the hybrid membrane exhibits adequate biodegradability. The $NF_{1.8}$/MOF-coated cotton fabric showed comparable biodegradation behavior (Figure 7). The biodegradability of the membrane reached 56.9% after 30 days, converging to 62.5% after 45 days. According to the ASTM standard, a material that can be degraded by more than 60% (weight loss) within 180 months is biodegradable [40]. The cotton-based membrane filter is expected to exhibit biodegradability because cotton is readily degraded by microorganisms in the soil when landfilled. Venditti et al. reported that approximately 40% of cotton fabric was degraded after 40 days [41]. Li et al. reported that the degradation efficiency of the cotton fabric varied in the range of 50–77% after 90 days of treatment [42]. Our results agree well with the literature; therefore, the proposed route effectively fabricates a cellulose-based biodegradable dual-functional membrane to remove PM and toxic gases simultaneously.

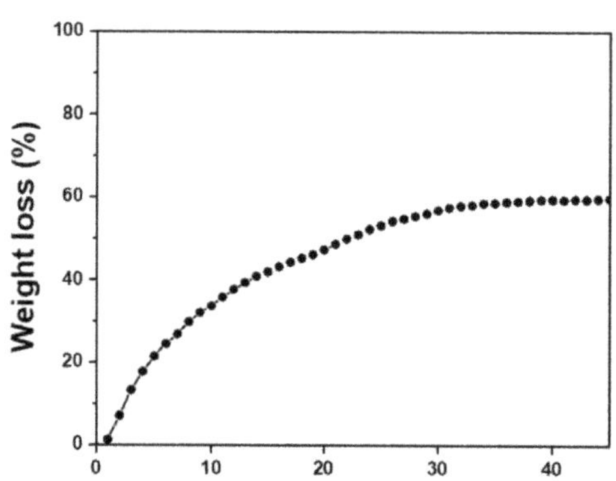

Figure 7. Biodegradability of the $NF_{1.8}$/MOF-coated cotton fabric membrane as a function of time.

4. Conclusions

A biodegradable dual-functional hybrid membrane was demonstrated via the route by combining the direct growth of MOF on cotton fabric and the subsequent electrospinning of a CA solution. The ZIF-8 was grown from the surface of a cotton fabric by treatment with an organic precursor solution and subsequent immersion in a metal ion solution. The CA NF was electrospun onto the MOF-coated cotton fabric with different amounts of CA solution to control the porosity of the membrane. The thickness of the NF layer largely affected the filtration performance of particulate matter, as indicated by the assessment of the filtration efficiency, pressure drop, and air permeability. In general, with a thicker NF layer, the filtration efficiency and the pressure drop tended to increase, even though the air permeability decreased. As all of these parameters are in the trade-off relationship: Q_f was assessed to find the optimal thickness of the NF layer; $NF_{1.8}$/MOF-coated cotton fabric exhibited the highest Q_f value (18.8×10^{-2} Pa^{-1}) with an air permeability of 19.0 $cm^3/cm^2/s$, a pressure drop of 14.2 Pa, and filtration efficiency of 93.1%. The MOF-coated cotton and $NF_{1.8}$/MOF-coated cotton membrane showed CDR values higher than 99% for H_2S, formaldehyde, and NH_3, confirming the achievement of the dual functions of toxic gas removal and the filtration of PM in the membrane system. Finally, the $NF_{1.8}$/MOF-coated cotton fabric membrane showed reasonable biodegradability with a weight loss of 62.5% after 45 days, which is acceptable to consider the system as a biodegradable material. The presented strategy combining the growth of MOF and electrospinning cellulose-based NF should offer an eco-friendly approach to achieve lightweight PPE filters that remove both PM and toxic gases simultaneously in a sustainable manner.

Supplementary Materials: The following supporting information can be downloaded at: https://www.mdpi.com/article/10.3390/polym15193965/s1, Figure S1: SEM images showing the morphology of the prepared ZIF-8 particles with different [ligand]/[Zn^{2+}] of (a) 2.0, (b) 4.0, (c) 6.0, and (d) 8.0.

Author Contributions: S.R. and D.K. contributed equally to this work. S.R. and D.K. performed fabrication processes and characterizations. H.L. (Hyewon Lee), Y.L. and Y.K. analyzed the experimental datasets. M.K., H.L. (Heedong Lee) and H.L. (Hoik Lee) conceived the concept, oversaw the study, and wrote the manuscript. All authors have read and agreed to the published version of the manuscript.

Funding: This study has been conducted with the support of the Korea Institute of Industrial Technology as "Development of fiber based technology for reduction of hazardous substances in the air (kitech EO-23-0005)", and the support of the National Research Foundation of Korea (NRF) grants (2022R1F1A1074129).

Institutional Review Board Statement: Not applicable.

Data Availability Statement: The datasets generated during and/or analyzed during the current study are available from the corresponding author on reasonable request.

Conflicts of Interest: The authors declare no conflict of interest.

References

1. Gu, H.; Yan, W.; Elahi, E.; Cao, Y. Air pollution risks human mental health: An implication of two-stages least squares estimation of interaction effects. *Environ. Sci. Pollut. Res.* **2020**, *27*, 2036–2043. [CrossRef]
2. World Health Organization. *Air Pollution: The Invisible Health Threat*; World Health Organization: Geneva, Switzerland, 2023.
3. Wu, S.; Deng, F.; Wei, H.; Huang, J.; Wang, X.; Hao, Y.; Zheng, C.; Qin, Y.; Lv, H.; Shima, M. Association of cardiopulmonary health effects with source-appointed ambient fine particulate in Beijing, China: A combined analysis from the Healthy Volunteer Natural Relocation (HVNR) study. *Environ. Sci. Technol.* **2014**, *48*, 3438–3448. [CrossRef]
4. Kim, D.; Kim, Y.; Kim, D.; Son, D.; Doh, S.J.; Kim, M.; Lee, H.; Yoon, K.R. Rational process design for facile fabrication of dual functional hybrid membrane of MOF and electrospun nanofiber towards high removal efficiency of $PM_{2.5}$ and toxic gases. *Macromol. Rapid Commun.* **2022**, *43*, 2100648. [CrossRef]
5. Kamal, M.S.; Razzak, S.A.; Hossain, M.M. Catalytic oxidation of volatile organic compounds (VOCs)—A review. *Atmos. Environ.* **2016**, *140*, 117–134. [CrossRef]
6. World Health Organization. What are the WHO air quality guidelines? In *Improving Health by Reducing Air Pollution*; World Health Organization: Geneva, Switzerland, 2021.
7. Benson, N.U.; Fred-Ahmadu, O.H.; Bassey, D.E.; Atayero, A.A. COVID-19 pandemic and emerging plastic-based personal protective equipment waste pollution and management in Africa. *J. Environ. Chem. Eng.* **2021**, *9*, 105222. [CrossRef]
8. World Health Organization. *Tonnes of COVID-19 Health Care Waste Expose Urgent Need to Improve Waste Management Systems*; World Health Organization: Geneva, Switzerland, 2022.
9. Venkatarajan, S.; Athijayamani, A. An overview on natural cellulose fiber reinforced polymer composites. *Mater. Today Proc.* **2021**, *37*, 3620–3624. [CrossRef]
10. Luo, H.; Xie, J.; Xiong, L.; Zhu, Y.; Yang, Z.; Wan, Y. Fabrication of flexible, ultra-strong, and highly conductive bacterial cellulose-based paper by engineering dispersion of graphene nanosheets. *Compos. Part B Eng.* **2019**, *162*, 484–490. [CrossRef]

11. Yao, Y.; Zeng, X.; Pan, G.; Sun, J.; Hu, J.; Huang, Y.; Sun, R.; Xu, J.-B.; Wong, C.-P. Interfacial engineering of silicon carbide nanowire/cellulose microcrystal paper toward high thermal conductivity. *ACS Appl. Mater. Interfaces* **2016**, *8*, 31248–31255. [CrossRef]
12. Islam, S.U.; Ul-Islam, M.; Ahsan, H.; Ahmed, M.B.; Shehzad, A.; Fatima, A.; Sonn, J.K.; Lee, Y.S. Potential applications of bacterial cellulose and its composites for cancer treatment. *Int. J. Biol. Macromol.* **2021**, *168*, 301–309. [CrossRef]
13. Woo, H.C.; Yoo, D.K.; Jhung, S.H. Highly improved performance of cotton air filters in particulate matter removal by the incorporation of metal–organic frameworks with functional groups capable of large charge separation. *ACS Appl. Mater. Interfaces* **2020**, *12*, 28885–28893. [CrossRef]
14. He, R.; Li, J.; Chen, M.; Zhang, S.; Cheng, Y.; Ning, X.; Wang, N. Tailoring moisture electroactive Ag/Zn@cotton coupled with electrospun PVDF/PS nanofibers for antimicrobial face masks. *J. Hazard. Mater.* **2022**, *428*, 128239. [CrossRef]
15. Xue, J.; Wu, T.; Dai, Y.; Xia, Y. Electrospinning and electrospun nanofibers: Methods, materials, and applications. *Chem. Rev.* **2019**, *119*, 5298–5415. [CrossRef]
16. Lu, T.; Cui, J.; Qu, Q.; Wang, Y.; Zhang, J.; Xiong, R.; Ma, W.; Huang, C. Multistructured electrospun nanofibers for air filtration: A review. *ACS Appl. Mater. Interfaces* **2021**, *13*, 23293–23313. [CrossRef]
17. Wu, S.; Dong, T.; Li, Y.; Sun, M.; Qi, Y.; Liu, J.; Kuss, M.A.; Chen, S.; Duan, B. State-of-the-art review of advanced electrospun nanofiber yarn-based textiles for biomedical applications. *Appl. Mater. Today* **2022**, *27*, 101473. [CrossRef]
18. An, S.; Jeon, B.; Bae, J.H.; Kim, I.S.; Paeng, K.; Kim, M.; Lee, H. Thiol-based chemistry as versatile routes for the effective functionalization of cellulose nanofibers. *Carbohydr. Polym.* **2019**, *226*, 115259. [CrossRef]
19. Choi, H.Y.; Bae, J.H.; Hasegawa, Y.; An, S.; Kim, I.S.; Lee, H.; Kim, M. Thiol-functionalized cellulose nanofiber membranes for the effective adsorption of heavy metal ions in water. *Carbohydr. Polym.* **2020**, *234*, 115881. [CrossRef]
20. Thorarinsdottir, A.E.; Harris, T.D. Metal–organic framework magnets. *Chem. Rev.* **2020**, *120*, 8716–8789. [CrossRef]
21. Huang, D.; Xin, Q.; Ni, Y.; Shuai, Y.; Wang, S.; Li, Y.; Ye, H.; Lin, L.; Ding, X.; Zhang, Y. Synergistic effects of zeolite imidazole framework@ graphene oxide composites in humidified mixed matrix membranes on CO_2 separation. *RSC Adv.* **2018**, *8*, 6099–6109. [CrossRef]
22. Jameh, A.A.; Mohammadi, T.; Bakhtiari, O.; Mahdyarfar, M. Synthesis and modification of Zeolitic Imidazolate Framework (ZIF-8) nanoparticles as highly efficient adsorbent for H_2S and CO_2 removal from natural gas. *J. Environ. Chem. Eng.* **2019**, *7*, 103058. [CrossRef]
23. *ASTM D737-18*; Standard Test Method for Air Permeability of Textile Fabrics. ASTM: West Conshohocken, PA, USA, 2023.
24. Huang, S.-H.; Chen, C.-W.; Kuo, Y.-M.; Lai, C.-Y.; McKay, R.; Chen, C.-C. Factors affecting filter penetration and quality factor of particulate respirators. *Aerosol Air Qual. Res.* **2013**, *13*, 162–171. [CrossRef]
25. Pradhan, S.; Behera, S.K.; Samal, S.K.; Panda, I.; Sahu, P.K.; Priyadarshini, S. Interaction between 2-methylimidazole and 1-butanol/1-octanol: Thermophysical and computational studies. *ChemistrySelect* **2023**, *8*, e202204931. [CrossRef]
26. Ramos, V.C.; Reyes, C.B.G.; García, G.M.; Quesada, M.I.S.; Barrero, F.J.M.-C.; Rábago, J.J.S.; Polo, M.S. ZIF-8 and its magnetic functionalization as vehicle for the transport and release of ciprofloxacin. *Pharmaceutics* **2022**, *14*, 2546. [CrossRef]
27. Watanabe, S.; Ohsaki, S.; Hanafusa, T.; Takada, K.; Tanaka, H.; Mae, K.; Miyahara, M.T. Synthesis of zeolitic imidazolate framework-8 particles of controlled sizes, shapes, and gate adsorption characteristics using a central collision-type microreactor. *Chem. Eng. J.* **2017**, *313*, 724–733. [CrossRef]
28. Lee, H.; An, S.; Kim, S.; Jeon, B.; Kim, M.; Kim, I.S. Readily functionalizable and stabilizable polymeric particles with controlled size and morphology by electrospray. *Sci. Rep.* **2018**, *8*, 15725. [CrossRef]
29. Lee, H.; Woo, J.; Son, D.; Kim, M.; Choi, W.I.; Sung, D. Electrospinning/electrospray of ferrocene containing copolymers to fabricate ROS-responsive particles and fibers. *Polymers* **2020**, *12*, 2520. [CrossRef]
30. Bian, Y.; Wang, S.; Zhang, L.; Chen, C. Influence of fiber diameter, filter thickness, and packing density on $PM_{2.5}$ removal efficiency of electrospun nanofiber air filters for indoor applications. *Build. Environ.* **2020**, *170*, 106628. [CrossRef]
31. Liu, T.; Cai, C.; Ma, R.; Deng, Y.; Tu, L.; Fan, Y.; Lu, D. Super-hydrophobic cellulose nanofiber air filter with highly efficient filtration and humidity resistance. *ACS Appl. Mater. Interfaces* **2021**, *13*, 24032–24041. [CrossRef]
32. Modi, A.; Jiang, Z.; Kasher, R. Hydrostable ZIF-8 layer on polyacrylonitrile membrane for efficient treatment of oilfield produced water. *Chem. Eng. J.* **2022**, *434*, 133513. [CrossRef]
33. Yang, D.; Li, Y.; Zhao, L.; Cheng, F.; Chang, L.; Wu, D. Constructing ZIF-8-decorated montmorillonite composite with charge neutralization effect and pore structure optimization for enhanced Pb^{2+} capture from water. *Chem. Eng. J.* **2023**, *466*, 143014. [CrossRef]
34. Schejn, A.; Balan, L.; Falk, V.; Aranda, L.; Medjahdi, G.; Schneider, R. Controlling ZIF-8 nano-and microcrystal formation and reactivity through zinc salt variations. *CrystEngComm* **2014**, *16*, 4493–4500. [CrossRef]
35. Park, K.S.; Ni, Z.; Côté, A.P.; Choi, J.Y.; Huang, R.; Uribe-Romo, F.J.; Chae, H.K.; O'Keeffe, M.; Yaghi, O.M. Exceptional chemical and thermal stability of zeolitic imidazolate frameworks. *Proc. Natl. Acad. Sci. USA* **2006**, *103*, 10186–10191. [CrossRef]
36. Cabrales, L.; Abidi, N.; Manciu, F. Characterization of developing cotton fibers by confocal Raman microscopy. *Fibers* **2014**, *2*, 285–294. [CrossRef]
37. Sousa-Pinto, B.; Fonte, A.P.; Lopes, A.A.; Oliveira, B.; Fonseca, J.A.; Costa-Pereira, A.; Correia, O. Face masks for community use: An awareness call to the differences in materials. *Respirology* **2020**, *25*, 894. [CrossRef]

38. Binaeian, E.; El-Sayed, E.-S.M.; Matikolaei, M.K.; Yuan, D. Experimental strategies on enhancing toxic gases uptake of metal–organic frameworks. *Coord. Chem. Rev.* **2021**, *430*, 213738. [CrossRef]
39. Rana, A.; Biswas, S. Electrophilicity modulated targeted luminescence of MOF-coated cotton composite for dual analyte detection in aqueous medium. *Inorg. Chem. Front.* **2023**, *10*, 2742–2753. [CrossRef]
40. *ASTM D6400-04*; Standard Specification for Compostable Plastics. ASTM: West Conshohocken, PA, USA, 1999.
41. Zambrano, M.C.; Pawlak, J.J.; Venditti, R.A. Effects of chemical and morphological structure on biodegradability of fibers, fabrics, and other polymeric materials. *BioResources* **2020**, *15*, 9786. [CrossRef]
42. Li, L.; Frey, M.; Browning, K.J. Biodegradability study on cotton and polyester fabrics. *J. Eng. Fibers Fabr.* **2010**, *5*, 155892501000500406. [CrossRef]

polymers

MDPI

Article

One-Dimensional Nickel Molybdate Nanostructures with Enhanced Supercapacitor Performance

Baodong Sun [1], Shaomin Wang [2] and Mingyi Zhang [2,*]

1 College of Teacher Education, Harbin Normal University, Harbin 150025, China
2 Key Laboratory for Photonic and Electronic Bandgap Materials, Ministry of Education, School of Physics and Electronic Engineering, Harbin Normal University, Harbin 150025, China
* Correspondence: zhangmingyi@hrbnu.edu.cn

Abstract: One-dimensional $NiMoO_4$ nanofibers were successfully prepared by electrospinning and high-temperature calcination. The supercapacitor performance tests were conducted on the prepared materials in a three-electrode system, and it was found that the calcination temperature during the preparation of the fibers seriously affects the final morphology and electrochemical performance of the obtained samples. The sample with a calcination temperature of 500 °C has better performance, its specific capacitance can reach 1947 F g^{-1}, and the retention rate is 82.35% after 3000 cycles of constant current charging–discharging. The improvement of electrochemical performance is primarily on account of the unique one-dimensional nanostructure of the material, which can both enhance the charge transfer efficiency and effectively increase the speed of electrolyte ion diffusion.

Keywords: $NiMoO_4$; supercapacitor; nanofiber; electrospinning

1. Introduction

In recent years, due to the consumption of fossil fuels and environmental damage, people have now feared that using non-renewable fossil energy will bring about air pollution and a rise in average global temperatures. Meanwhile, new energy has become the most urgent need for sustainable development of a global nature in the 21st century [1,2]. Over the past decade, the drive to replace fossil fuels with biofuels, wind power, tidal power, and energy storage has become more urgent [3,4]. At present, supercapacitors are one of the more important energy storage devices [5–7]. Because of its advantages of fast charging–discharging capability, high flexibility [8,9], and long cycle life, it can provide storage technology to help solve the energy crisis.

Transition metal oxide is a kind of supercapacitor capacitive material that has been widely studied [10–12]. This kind of material can have different oxidation states and high theoretical capacitance when the Faraday redox reaction occurs. At present, bimetallic transition metal oxides, for instance, Nickel–cobaltate ($NiCo_2O_4$), Nickel–manganese (Ni-MnBMO), Calcium–molybdate ($CaMoO_4$), Nickel–Tungstate($NiWO_4$) and Nickel–molybdate ($NiMoO_4$) have become a research hotspot in the field of new energy storage materials because of their outstanding electrochemical properties and superior electrical conductivity [13–18].

Among various available transition metal oxides, $NiMoO_4$ has attracted extensive attention due to its advantages such as good electrochemical conductivity, high theoretical capacitance value (2500 Fg^{-1}), high cost effectiveness, rapid redox activity, low cost, and non-toxicity [19–21]. At present, spinel structure oxide nickel molybdate has been widely used in lithium batteries and supercapacitors. For $NiMoO_4$ powder materials, the repeated ion embedding and extraction process will lead to volume expansion, and the loose connection between particles will also increase its own resistance, thus affecting its theoretical capacitance value cannot be maximized. How to further improve the capacitive properties of such materials is a meaningful research topic [22]. The above problems can be solved by constructing one-dimensional nanostructures to preserve the chemical properties of materials while ensuring the nanocrystallization of basic particles [23,24]. This kind of $NiWO_4$ nanofiber is expected to present several advantages: First of all, the ultra-long ordinate axis with 1D nanostructure can provide more efficient transport pathways for both electrons as well as ions with greater electrochemical reversibility and cycle stability. Secondly, the crosslinked network nanofibers composed of individual $NiWO_4$ nanoparticles can effectively decrease their surface energy. More importantly, the $NiWO_4$ nanofibers would possess a larger interspace and a doubled active surface area, resulting in enhanced charge transport and electrochemical performance.

In this work, one-dimensional $NiMoO_4$ nanofibers were prepared by simple electrospinning and calcination using nickel acetate and ammonium molybdate as reactants. The morphology and structure of nanofibers were controlled by a simple method of adjusting the calcination temperature. X-ray diffractometer (XRD), scanning electron microscope (SEM), transmission electron microscope (TEM), and X-ray photoelectron spectrometer (XPS) were used to characterize the morphology, structure, and chemical compositions of the prepared materials. The electrochemical properties of the materials were investigated through a three-electrode test system, and the corresponding structure–activity relationship was explained rationally.

2. Experiment

2.1. Chemicals

Polyacrylonitrile (PAN, Mw ~150,000) was purchased from Sigma-Aldrich Corporation (Shanghai, China). $Ni(CH_3COO)_2$, $(NH_4)_6Mo_7O_{24}\cdot 4H_2O$, and N, N-dimethylformamide (DMF) were purchased from Zhiyuan Reagent (Tianjin, China). All chemicals were used directly after purchase without any further purification.

2.2. $NiMoO_4$ Nanofibers' Synthesis Method

In a typical synthesis process, dissolve 0.5 g polyacrylonitrile (PAN) powder in 5 mL N, n-dimethylformamide (DMF) and stir the above mixture for 1 h. After that, 0.174 g of $Ni(CH_3COO)_2$ and 0.124 g of $(NH_4)_6Mo_7O_{24}\cdot 4H_2O$ were added to the above solution and stirred at room temperature for 12 h. Next, a well-mixed precursor solution was injected into the syringe with a plastic needle, and the distance between the syringe needle and the aluminum foil collector was controlled at 12 cm. The operating voltage was set to 6 kV and electrospinning under dry conditions for several hours to obtain light green nanofibers. The samples of $NiMoO_4$ nanofibers were synthesized by calcining the composite nanofibers at the temperatures of 450, 500, 600, and 700 °C and lasted for 2 h in air with a heating rate of 2 °C min^{-1}. As-obtained $NiMoO_4$ nanofibers were named NMO-450, NMO-500, NMO-600, and NMO-700 NFs, respectively.

2.3. Characterization

The structures of the prepared samples were collected by XRD (Rigaku D/max2600, Tokyo, Japan) which analyses with a Cu Kα radiation source (λ = 0.154178 nm), and the morphologies were characterized by SEM (SU70, Hitachi, Tokyo, Japan) and TEM (FEI, Tecnai TF20, Tokyo, Japan). The Brunauer Emmett Teller (BET) test was measured for the specific surface area of the samples. In addition, the chemical composition and element state of the samples were obtained from XPS.

2.4. Electrochemical Measurement Technology

The nature of the electrochemical property for as-prepared was investigated by a standard three-electrode system which was tested in 1 M KOH solution. Saturated calomel electrode (SCE) and Pt foil electrode with 1 cm^2 were regarded as the reference electrode and counter electrode, respectively. The electrode slurry was prepared by evenly mixing the active substance, acetylene black and polytetrafluoroethylene (PTFE) binder in a mortar at a weight ratio of 8:1:1. Then, the mixed electrode slurry was equably coated on 1 cm^2

nickel foam and dried under 60 °C vacuum for 12 h to obtain the working electrode. In this work, all the electrochemical data were tested by the electrochemical workstation (VMP3, France). The obtained data from the cyclic voltammetry (CV) test were collected by an optimal voltage range from 0 to 0.5 V at the scan rates for 5, 10, 20, 30, 40, and 50 mV s^{-1}. Galvanostatic charging–discharging (GCD) processes were attained at the current densities of 1, 2, 4, 6, 8, and 10 A g^{-1} with the voltage range 0~0.44 V, respectively, according to the following equation:

$$C_m = I \times \Delta t / m \times \Delta V \quad (1)$$

where C_m represents the specific capacitance (F g^{-1}), I is the discharge current (A), m is the mass of electrode material (mg), Δt refers to the discharge time (s), and ΔV is the voltage drop upon discharging (V).

3. Results and Discussion

The microstructure and morphological information were shown through SEM. Compared with the sample of NMO-450 (Figure 1a,b), NMO-600 (Figure 1e,f), and NMO-700 NFs (Figure 1g,h), the sample of NMO-500 NFs possess a well-organized 1D nanostructure as shown in Figure 1c,d. At 450 °C, the polymer template (PAN) served as a mighty carrier for the $NiMoO_4$ structure, while at 500 °C, PAN powder was completely removed with air. Subsequently, at 600 °C, the pores between the nanoparticles grew larger. More interestingly, at 700 °C, nanofibers still preserved a 1D structure, although the partial samples appeared an issue of local fracture. These results show that it is feasible to obtain $NiMoO_4$ with a special 1D structure through annealing the nanofibers at different high temperatures.

The crystal structures of the NMO-450, NMO-500, NMO-600, and NMO-700 NFs were represented by XRD patterns (Figure 2). The characteristic peaks which located at 23.36, 26.58, 27.31, 32.3, 34.18, 36.68, 43.8, and 47.4° belong to (0 2 −1), (2 2 0), (1 1 −2), (4 0 0), (2 2 2), (4 2 −2), (2 4 −1), and (2 0 −4) lattice planes of $NiMoO_4$ (JCPDS No. 45-0142), respectively. In the sample of NMO-450 NFs, the observed maxima are associated with the turbostratic structure typical of the char. Moreover, for the sample of NMO-600 and NMO-700 NFs, a few impurity peaks located at 37.03, 43.16, and 63.71° are assigned to (1 0 1), (0 1 2) and (1 1 0) lattice planes of NiO (JCPDS No. 44-1159). We believe that a pure phase NMO-500 NF with a one-dimensional nanostructure can be prepared under the appropriate temperature (500 °C) and the cooperative contribution of flexible electrospinning technology. These Mmonometallic oxides (NiO impurity) might be formed in a thermal oxidizing atmosphere at high temperatures. It can be seen from the XRD pattern that with the increase in calcination temperature, NiO becomes the main component of the sample. The electrochemical properties of the monometallic oxides are commonly lower than the bimetallic oxides, which hinder and impact the capacitive performances of pure $NiMoO_4$.

NMO-500 NFs' morphologies and structures could be further characterized by TEM and HRTEM analysis. The NMO-500 NFs exhibit obvious one-dimensional nanostructures, which are coated with a layer of interconnected nanoparticles on the nanofibers (Figure 3a). As a result, the NMO-500 NFs have a relatively rough surface. The lattice spacing of the NMO-500 NFs was observed to be 0.335 nm, which corresponded to the (2 2 0) lattice plane of the spinel $NiMoO_4$ (Figure 3b). Furthermore, the element surface distribution of Mo, Ni, and O elements for the as-fabricated samples significantly exhibit a high degree of overlap in the mapping (Figure 3c–e). This result strongly confirmed that the $NiMoO_4$ nanofibers were successfully prepared. And on top of all the above, the element composition was also outlined through the energy element analysis data using an energy dispersive spectrometer (EDS) (Figure 4).

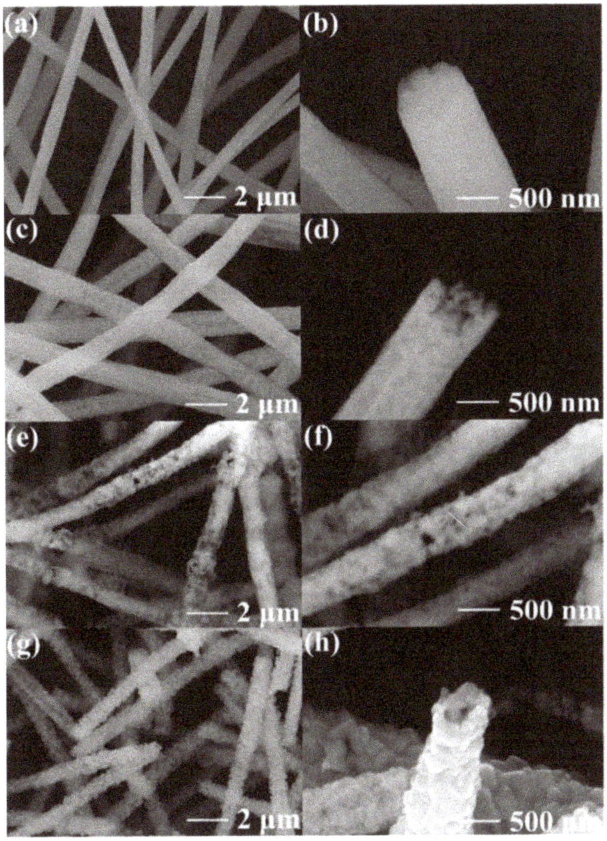

Figure 1. SEM images of (**a**) and (**b**) NMO-450, (**c**) and (**d**) NMO-500, (**e**) and (**f**) NMO-600, and (**g**) and (**h**) NMO-700 NFs under different magnifications.

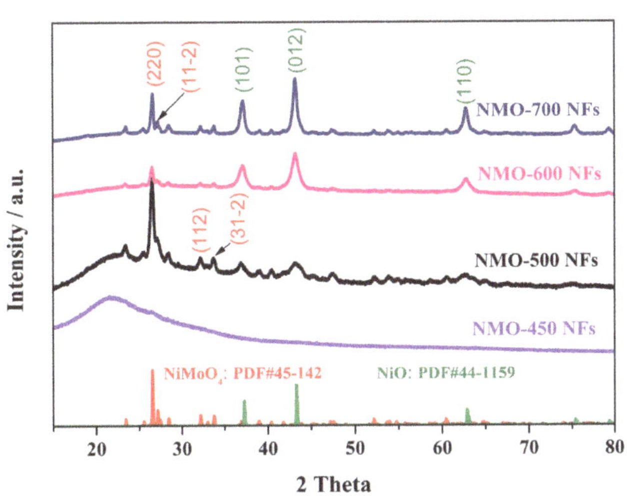

Figure 2. XRD patterns of NMO-450, NMO-500, NMO-600, and NMO-700 NFs.

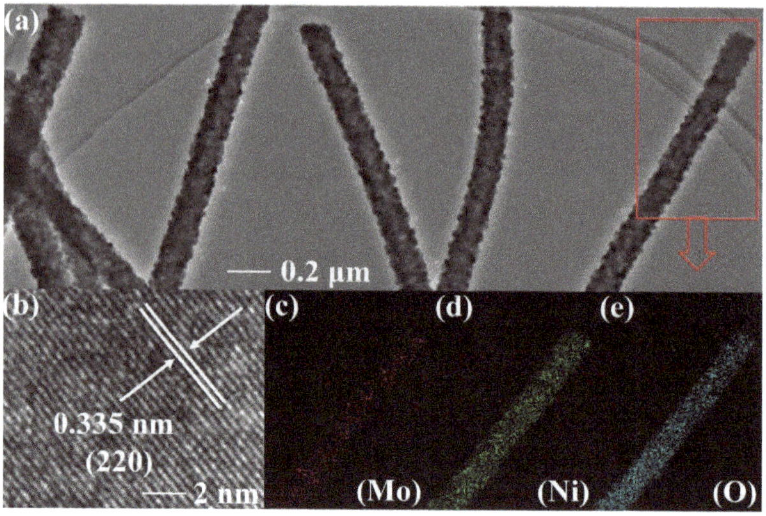

Figure 3. (**a**,**b**) TEM and HRTEM images of NMO-500 NFs, and (**c**–**e**) their EDX elemental mapping images.

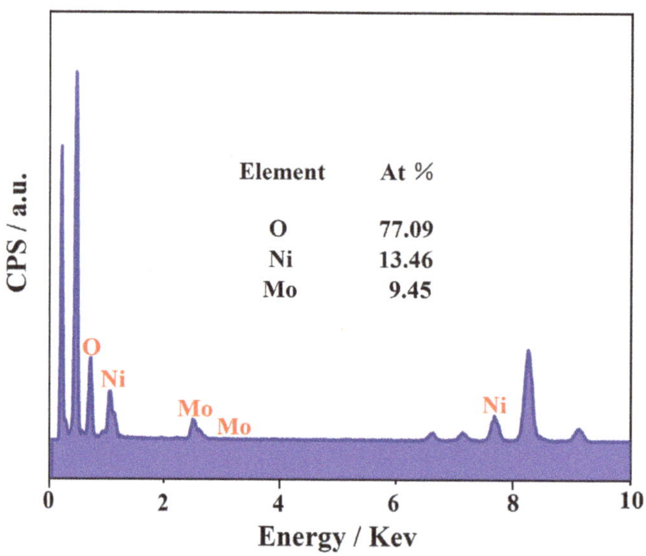

Figure 4. EDS of the NMO-500 NFs.

The BET-specific surface area and the pore size distribution of NMO-500 NFs were measured and analyzed via the N_2 adsorption and desorption test. The NMO-500 NFs are type-IV isotherm curves (Figure 5a). NMO-500 NFs' BET-specific surface area is apparently high (40.6 m^2 g^{-1}), which could perhaps be related to its one-dimensional structural characteristics and advantages. In addition, the pore size distribution of NMO-500 NFs has a pore distribution peak between 5 and 10 nm, indicating that the prepared sample is a mesoporous structure. (Figure 5b). This feasibly affords an advantageous path for the rapid diffusion of ions. Therefore, the NMO-500 NF, with its unique structural advantages, will provide pseudocapacitance to achieve the desired electrochemical characteristics of energy storage.

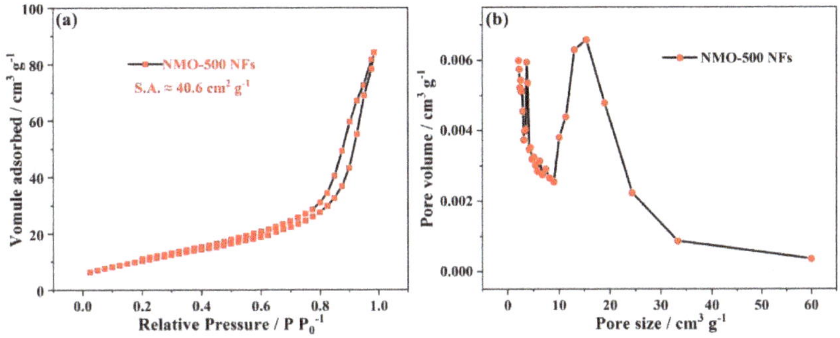

Figure 5. N_2 adsorption–desorption isotherm loop for (**a**) NMO-500 NFs and (**b**) the curves of pore size distribution.

In order to determine the surface electronic states and chemical compositions of the $NiMoO_4$ samples, XPS measurements were performed. XPS pattern spectra show the full spectrum of the NMO-500 NFs; there are four peaks located at 231.70, 284.00, 530.80, and 854.60 eV, matching with Mo 3d, C 1s, O 1s, and Ni 2p energy levels, respectively (Figure 6a). Among them, it could be deconvoluted as four peaks for Ni 2p core level spectra which peaks located at 855.90 and 861.50 eV (satellite peak) correspond to Ni $2p_{3/2}$ energy level, whereas the peaks located at 873.7 and 880 eV (satellite peak) ascribe to the Ni $2p_{1/2}$ energy level. As shown in Figure 6b, there is a binding energy gap of 17.80 eV between the main peaks of Ni $2p_{3/2}$ and Ni $2p_{1/2}$, which indicates that Ni^{2+} is an oxidation state. The Mo 3d energy level spectra show two peaks located at 232.00 and 235.10 eV which correspond to Mo $3d_{5/2}$ and Mo $3d_{3/2}$ (Figure 6c). And there is a binding energy gap of 3.10 eV for separated Mo 3d, which represents the oxidation state of Mo^{6+}. The peak of binding energy which is located at 530.79 eV belongs to the O 1s energy level of $NiMoO_4$ and the peak at 531.40 eV implies the generation of low coordination oxygen ions on the surface of the sample (Figure 6d). Thus, the XPS analysis results revealed the generation of the nanostructure $NiMoO_4$, and confirmed the formation of $NiMoO_4$ nanoparticles embedded in nanofibers, as already analyzed by the XRD pattern.

The CV curves of NMO-450 NFs, NMO-500 NFs, NMO-600 NFs, and NMO-700 NFs at the scan rates from 5 to 50 mV s^{-1} were shown in Figure 7a,c,e,g. The CV curves of NMO-500 NFs possessed symmetrical and stable redox current peaks with a small potential drift and good repeatability when the scan rate increased, and their integrated area without apparent deformation, suggesting that the NMO-500 NFs have a good electrochemical reversibility derived from the improved nanostructures for adapting the desired fast charge and discharge reactions.

Figure 7b,d,f,h show the GCD curves of NMO-450 NFs, NMO-500 NFs, NMO-600 NFs, and NMO-700 NFs at different current densities of 1, 2, 4, 6, 8, and 10 A g^{-1} with the voltage range from 0 to 0.44 V. Distinctly, with the gradual increase in the current density, the charging–discharging time was significantly reduced, which is because the ions could more fully diffuse deep into the electrode material at the lower current density and promote the specific capacitance. In addition, the slope of all GCD curves changes nonlinearly with the increase in charging–discharging time, which further elucidates the representative pseudo-capacitance performance generated by the electrochemical redox reaction of NMO-500 NFs at the electrode–electrolyte interface.

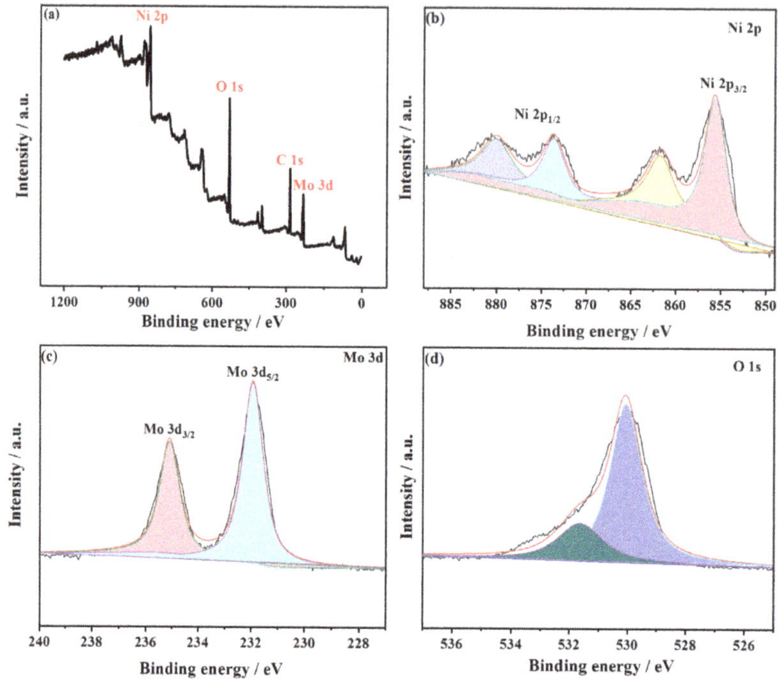

Figure 6. XPS characterization of NMO-500 NFs: (**a**) survey spectra, (**b**) nickel, (**c**) molybdenum, and (**d**) oxygen.

As demonstrated in Figure 8a,b, the prepared samples' electrochemical performances with well-designed nanostructures were estimated through combining the measured results of CV and GCD curves. The representative CV curves of NMO-450 NFs, NMO-500 NFs, NMO-600 NFs, and NMO-700 NFs which were supported at the scan rate of 5 mV s^{-1} were presented in Figure 8a. A pair of obviously and approximately symmetric sharp redox peaks in every CV curve were present, which indicated the role of the Faraday reaction (Ni^{2+}/Ni^{3+}) in charge storage. In addition, we found that the Mo element did not participate in the entire reaction and was still present in the sample for the form of polycation $MoO_4{}^{2-}$; meanwhile, there was no redox peak of Mo in the CV curve, which also confirmed the same results. It could only increase the conductivity of Nickel molybdate, thus improving the supernal specificity. The CV curve showed that the redox peaks remained symmetrical and also retained their shape, even at higher scanning rates, which indicated the reversibility of the reaction and the glorious rate capability of the electrode material. It was significantly more outstanding than those of NMO-450 NFs, NMO-600 NFs, and NMO-700 NFs for the electrochemical response current density and CV curve of NMO-500 NFs, and combined with XRD pattern analysis, which may be due to the presence of impurities (NiO) in the sample. In addition, the charging–discharging curves of a series of samples when the current density is 1 A g^{-1} were shown in Figure 8b. The curves of NMO-450, NMO-500, NMO-600, and NMO-700 had representative asymmetry and good reversibility, showing Faraday pseudocapacitance behavior. The discrepant specific capacitance for different electrode materials can be calculated by Equation (1). Notably, at a current density of 1 A g^{-1}, the NMO-500 HNFs were more efficient such as showing a longer discharge time and a higher 1947 F g^{-1} specific capacitance than the NMO-450 NFs (328 F g^{-1}), NMO-600 NFs (290 F g^{-1}), and NMO-700 NFs (63 F g^{-1}). These results were corresponded with the analysis of CV curves. It is well known that in reversible or quasi-reversible reactions, the peak current is proportional to the square root of the scanning rate ($\nu^{1/2}$). And, the first-order linear constant K of the ionic diffusion rate can effectively reflect the diffusion and transport rate of ionic electrolyte during the electrochemical reaction. NMO-500 NFs (K_b~450.5) presented a higher value of K than that of NMO-450 NFs (K_a~404.7), NMO-600 NFs (K_c~371.6), and NMO-700 NFs (K_d~253.7), which suggested that the one-dimensional nanostructure has more advantages than bulk nanostructure, particularly during the ion transport process (Figure 8c). More significantly, NMO-500 NFs manifested an apparent pseudocapacitive property, which may include the diffusion-controlled insertion behavior and surface capacitive effects. Furthermore, according to the GCD curves in Figure 8d, the variation law of specific capacitance can be calculated accurately. When the current density increases from 1 to 10 A g^{-1}, the specific capacitance of the NMO-450 NFs, NMO-500 NFs, NMO-600 NFs, and NMO-700 NFs decreases significantly. This phenomenon is possible given the limitation of electrolyte diffusion in conjunction and contact resistance with the electrochemical reaction process, resulting in a decrease in capacitance value as the current density increases. Meanwhile, this demonstrates that NMO-500 NFs have a higher specific capacitance and significant rate capability than most pure $NiMoO_4$ or $NiMoO_4$ matrix composites through different experimental parameters, as shown in Table 1.

Table 1. Specific capacitance and cycle life of the NMO-500 NFs electrode material for comparison with as-reported related electrodes.

Electrode Material	Electrolyte	Specific Capacitance	Cycle Life	Ref.
$NiMoO_4$ nanospheres	3 M KOH	974.4 F g^{-1} (at 1 A g^{-1})	75% (2000th at 5 A g^{-1})	[14]
$NiMoO_4$	2 M NaOH	392.53 F g^{-1}	87.14% (1000th at 5 A g^{-1})	[17]
$NiMoO_4$ nanorods	3 M KOH	672 F g^{-1} (at 4 A g^{-1})	72% (1000th at 1 A g^{-1})	[25]
$NiMoO_4$	6 M KOH	594 F g^{-1} (at 1 A g^{-1})	56% (1000th at 1 A g^{-1})	[26]
$NiMoO_4$/MWCNTs	3 M KOH	805 F g^{-1} (at 1 A g^{-1})	66.7% (1000th at 1 A g^{-1})	[27]
$NiMoO_4$/$CoMoO_4$ nanorods	1 M KOH	1445 F g^{-1} (at 1 A g^{-1})	78.8% (3000th at 10 A g^{-1})	[28]
$NiMoO_4$/3D-rGO (II)	3 M KOH	932 F g^{-1} (at 1 A g^{-1})	76% (500th at 1 A g^{-1})	[29]
MnO_2/$NiMoO_4$ nanostructure	5 M KOH	918 F g^{-1} (at 1 A g^{-1})	80% (10000th at 5 A g^{-1})	[30]
$NiCo_2O_4$@$NiWO_4$ core–shell nanowire	6 M KOH	1384 F g^{-1} (at 1 A g^{-1})	87.6% (6000th at 5 A g^{-1})	[31]
Amorphous $NiWO_4$	2 M KOH	586.2 F g^{-1} (at 0.5 A g^{-1})	91.4% (5000th at 2A g^{-1})	[32]
$NiMoO_4$ NFs	1 M KOH	1947 F g^{-1} (at 1 A g^{-1})	82.35% (3000th at 6 A g^{-1})	this work

Furthermore, the Faraday reaction on the electrode surface was evaluated by EIS. The Nyquist plots of NMO-450 NFs, NMO-500 NFs, NMO-600 NFs, and NMO-700 show the standard shape in the high-frequency zone (Figure 9). In addition, the equivalent series resistance (Rs) can be calculated by interception of the Z′-axis, and the diameter distance in the impedance spectrum matches the transfer charge resistance (Rct). It is clearly shown that the electrochemical impedance spectroscopy of the NMO-500 NFs showed the lowest values of Rs and Rct among these catalysts.

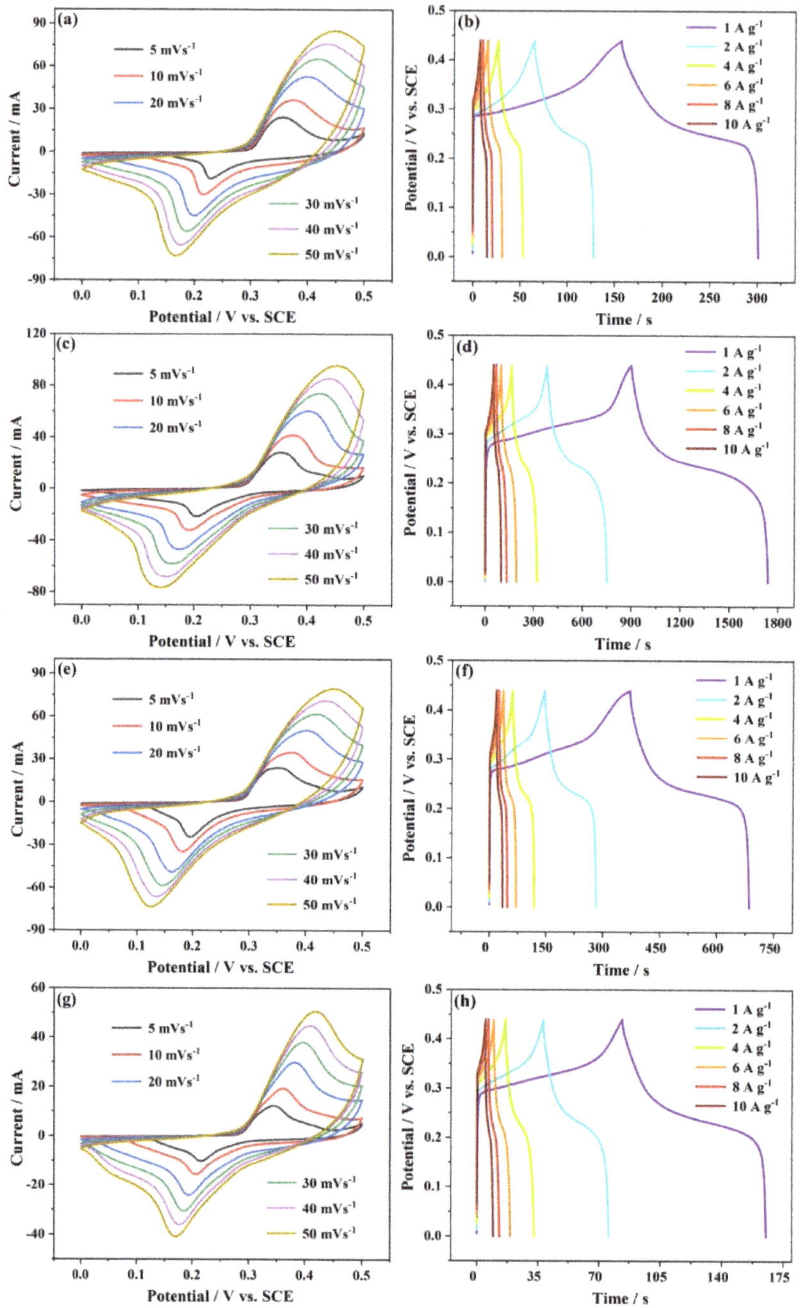

Figure 7. CV curves of (**a**) NMO-450 NFs, (**c**) NMO500 NFs (**e**) NMO-600 NFs, and (**g**) NMO-700 NFs at different scan rates in 1 M KOH electrolyte; GCD curves of (**b**) NMO-450 NFs, (**d**) NMO-500 NFs, (**f**) NMO-600 NFs, and (**h**) NMO-700 NFs at different current densities in 1 M KOH electrolyte.

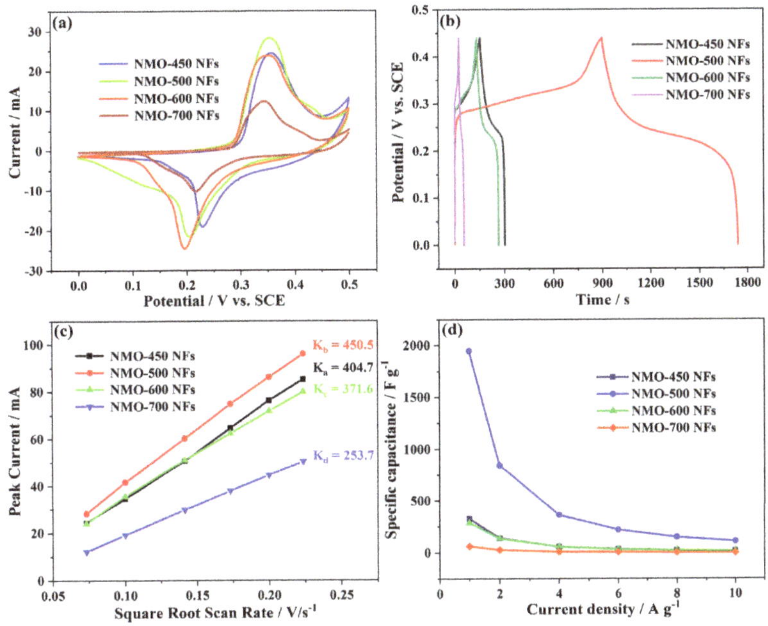

Figure 8. (**a**) CV curves of NMO-450 NFs, NMO-500 NFs, NMO-600 NFs, and NMO-700 NFs at 5 mV s^{-1} in 1 M KOH electrolyte. (**b**) GCD curves of NMO-450 NFs, NMO-500 NFs, NMO-600 NFs, and NMO-700 NFs at 1 A g^{-1} in 1 M KOH electrolyte. (**c**) Linear relationship of peak current vs. square root of scan rates of the NMO-450 NFs, NMO-500 NFs, NMO-600 NFs, and NMO-700 NFs electrode materials. (**d**) Mass-specific capacitances of pristine NMO-450 NFs, NMO-500 NFs, NMO-600 NFs, and NMO-700 NFs at different current densities.

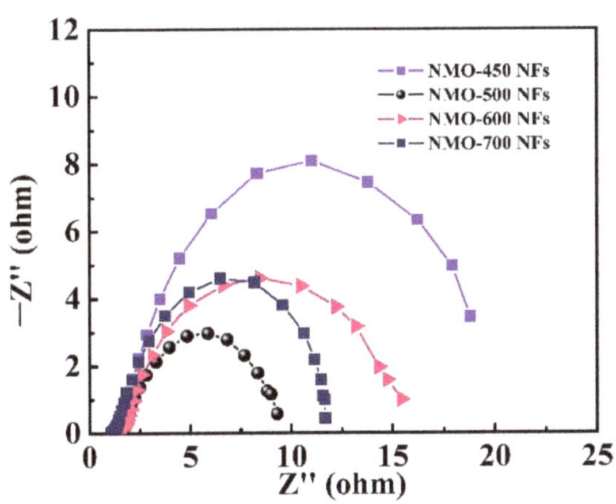

Figure 9. Nyquist plots of the NMO-450 NFs, NMO-500 NFs, NMO-600 NFs, and NMO-700 NFs.

For evaluating the cycling properties of the electrodes, the values of capacitance retention were calculated by charging–discharging tests at the current density of 6 A g^{-1} (Figure 10). The NMO-500 NFs still maintained an excellent capacitance retention rate (~82.35%) over 3000 cycles at high current density (6 A g^{-1}). In addition, as shown in the inset of Figure 9, the typical GCD curves of NMO-500 NFs still revealed a linear correlation of potential–time after the 1st, 800th, 1600th, 2400th, and 3000th circles, respectively. These results indicated the NMO-500 NFs with unique 1D nanostructures show excellent long-term electrochemical stability and characteristics of ideal capacitors. Therefore, it may be due to its distinct nanostructure that leads to improved cyclic stability, which not only increases the contact area of electrons and ions but also shortens the path between the electrolyte and active materials.

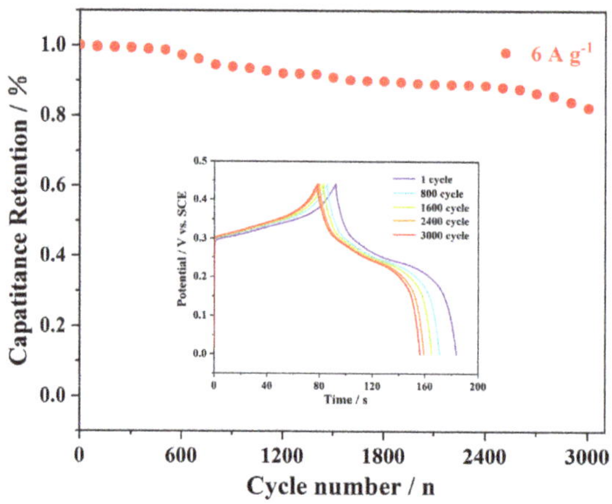

Figure 10. Cycling stability of NMO-500 NFs at a current density of 6 A g^{-1} for 3000 cycles, and (inset) the galvanostatic charge/discharge curves of the NMO-500 NFs over the 1st, 800th, 1600th, 2400th and 3000th cycles at a current density of 6 A g^{-1}.

4. Conclusions

In this work, one-dimensional nanostructured molybdenum nickel oxide ($NiMoO_4$) electrode with excellent electrochemical properties was constructed at an optimal temperature (500 °C) using an innovative electrospinning technique. $NiMoO_4$ nanofibers exhibit excellent specific capacitance (1947 F g^{-1}) and outstanding capacitance retention (82.35%) undergoing the 3000 cycles under 6 A g^{-1} conditions. This result can be attributed to their unique 1D nanostructure, which eliminated the negative effect of the powder agglomerates and exhibited good crystallinity at an appropriate temperature. This shows that it can be widely used in various portable electronic devices and provides a reference for the preparation of $NiMoO_4$-based electrode materials with excellent electrochemical properties and good cycle stability.

Author Contributions: B.S.: Conceptualization, methodology, validation, investigation, resources; S.W.: Writing—original draft, visualization; M.Z.: conceptualization, validation, writing—review and editing, supervision, funding acquisition. All authors have read and agreed to the published version of the manuscript.

Funding: The authors are grateful to be supported in part by the National Natural Science Foundation of China (No. 51872068).

Institutional Review Board Statement: Not applicable.

Data Availability Statement: All the relevant data are included in this published article.

Conflicts of Interest: The authors declare no conflict of interest.

References

1. Zardkhoshoui, A.M.; Davarani, S.S.H. Formation of Graphene-wrapped Multi-shelled $NiGa_2O_4$ Hollow Spheres and Graphene-wrapped Yolk-shell $NiFe_2O_4$ Hollow Spheres Derived from Metal-organic Frameworks for High-performance Hybrid Supercapacitors. *Nanoscale* **2020**, *12*, 1643–1656. [CrossRef] [PubMed]
2. Mohammadi Zardkhoshoui, A.; Hosseiny Davarani, S.S.; Ashtiani, M.M.; Sarparast, M. High-Performance Energy Storage Device Based on Triple-Shelled Cobalt Gallium Oxide Hollow Spheres and Graphene Wrapped Copper Iron Disulfide Porous Spheres. *ACS Sustain. Chem. Eng.* **2019**, *7*, 7908. [CrossRef]
3. Dubal, D.P.; Chodankar, N.R.; Kim, D.H.; Gomez-Romero, P. Towards Flexible Solid-state Supercapacitors for Smart and Wearable Electronics. *Chem. Soc. Rev.* **2018**, *47*, 2065–2129. [CrossRef] [PubMed]
4. Chen, D.; Jiang, K.; Huang, T.; Shen, G. Recent Advances in Fiber Supercapacitors, Materials, Device Configurations, and Applications. *Adv. Mater.* **2020**, *32*, 1901806. [CrossRef] [PubMed]
5. Simon, P.; Gogotsi, Y. Perspectives for Electrochemical Capacitors and Related Devices. *Nat. Mater.* **2020**, *19*, 1151–1163. [CrossRef] [PubMed]
6. Bi, Z.; Kong, Q.; Cao, Y.; Sun, G.; Su, F.; Wei, X.; Li, X.; Ahmad, A.; Xie, L.; Chen, C.M. Biomass-derived Porous Carbon Materials with Different Dimensions for Supercapacitor Electrodes: A Review. *J. Mater. Chem. A* **2019**, *7*, 16028–16045. [CrossRef]
7. Liu, P.; Liu, W.; Huang, Y.; Li, P.; Yan, J.; Liu, K. Mesoporous Hollow Carbon Spheres Boosted, Integrated High Performance Aqueous Zn-Ion Energy Storage. *Energy Storage Mater.* **2020**, *25*, 858–865. [CrossRef]
8. Abdollahifar, M.; Huang, S.S.; Lin, Y.H.; Lin, Y.C.; Shih, B.Y.; Sheu, H.S.; Liao, Y.-F.; Wu, N.L. High-performance Carbon-coated $ZnMn_2O_4$ Nanocrystallite Supercapacitors with Tailored Microstructures Enabled by a Novel Solution Combustion Method. *J. Power Sources* **2018**, *378*, 90–97. [CrossRef]
9. Zhang, X.; Sui, Y.; Wei, F.; Qi, J.; Meng, Q.; Ren, Y.; He, Y. Self-supported 3D Layered Zinc/nickel Metal-organic-framework with Enhanced Performance for Supercapacitors. *J. Mater. Sci. Mater. Electron.* **2019**, *30*, 18101–18110. [CrossRef]
10. Lu, X.; Jia, W.; Chai, H.; Hu, J.; Wang, S.; Cao, Y. Solid-State Chemical Fabrication of One-Dimensional Mesoporous β-Nickel Molybdate Nanorods as Remarkable Electrode Material for Supercapacitors. *J. Colloid Interface Sci.* **2018**, *5*, 322–331. [CrossRef]
11. Zhong, Y.; Shi, T.; Huang, Y.; Cheng, S.; Liao, G.; Tang, Z. One-step Synthesis of Porous Carbon Derived from Starch For all-carbon Binder-free High-rate Supercapacitor. *Electrochim. Acta* **2018**, *269*, 676–685. [CrossRef]
12. Minakshi, M.; Mitchell, D.R.G.; Jones, R.T.; Pramanik, N.C.; Jean-Fulcrand, A.; Garnweitner, G. A Hybrid Electrochemical Energy Storage Device Using Sustainable Electrode Materials. *Chemistry Select* **2020**, *5*, 1597–1606. [CrossRef]
13. Zhang, G.; Lou, X.W. General solution growth of mesoporous $NiCo_2O_4$ nanosheets on various conductive substrates as high-performance electrodes for supercapacitors. *Adv. Mater.* **2013**, *25*, 976–979. [CrossRef] [PubMed]
14. Cai, D.; Wang, D.; Liu, B.; Wang, Y.; Liu, Y.; Wang, L.; Li, H.; Huang, H.; Li, Q.; Wang, T. Comparison of the electrochemicalperformance of $NiMoO_4$ nanorods and hierarchical nanospheres for supercapacitor applications. *ACS Appl. Mater. Interfaces* **2013**, *5*, 12905–12910. [CrossRef]
15. Tang, X.; Zhang, B.; Lui, Y.H.; Hu, S. Ni-Mn bimetallic oxide nanosheets as high-performance electrode materials for asymmetric supercapacitors. *J. Energy Storage* **2019**, *25*, 100897. [CrossRef]
16. Minakshi, M.; Mitchell, D.R.G.; Baur, C.; Chable, J.; Barlow, A.J.; Fichtner, M.; Banerjee, A.; Chakraborty, S.; Ahuja, R. Phase evolution in calcium molybdate nanoparticles as a function of synthesis temperature and its electrochemical effect on energy storage. *Nanoscale Adv.* **2019**, *1*, 565–580. [CrossRef]
17. Sharma, P.; Minakshi, M.; Whale, J.; Jean-Fulcrand, A.; Garnweitner, G. Effect of the Anionic Counterpart: Molybdate vs. Tungstate in Energy Storage for Pseudo-Capacitor Applications. *Nanomaterials* **2021**, *11*, 580. [CrossRef]
18. Sharma, P.; Minakshi Sundaram, M.; Watcharatharapong, T.; Laird, D.; Euchner, H.; Ahuja, R. Zn Metal Atom Doping on the Surface Plane of One-Dimesional $NiMoO_4$ Nanorods with Improved Redox Chemistry. *ACS Appl. Mater. Interfaces* **2020**, *12*, 44815–44829. [CrossRef]
19. Chen, Z.Y.; He, B.; Yan, D.; Yu, X.F.; Li, W.C. Peapod-like MnO@Hollow Carbon Nanofibers Film as Self-standing Electrode for Li-Ion Capacitors with Enhanced Rate Capacity. *J. Power Sources* **2020**, *472*, 228501. [CrossRef]
20. Gao, S.; Sui, Y.; Wei, F.; Qi, J.; Meng, Q.; He, Y. Facile Synthesis of Cuboid Ni-MoF for High-performance Supercapacitors. *J. Mater. Sci.* **2018**, *53*, 6807–6818. [CrossRef]
21. Liu, Y.; Liu, Q.; Wang, L.; Yang, X.; Yang, W.; Zheng, J.; Hou, H. Advanced Supercapacitors Based on Porous Hollow Carbon Nanofiber Electrodes with High Specific Capacitance and Large Energy Density. *ACS Appl. Mater. Interfaces* **2020**, *12*, 4777–4786. [CrossRef] [PubMed]
22. Tao, X.; Wang, J.; Liu, C.; Wang, H.; Yao, H.; Zheng, G.; Seh, Z.W.; Cai, Q.; Li, W.; Zhou, G.; et al. Balancing Surface Adsorption and Diffusion of Lithium-polysulfides on Nonconductive Oxides for Lithium-sulfur Battery Design. *Nat. Chem.* **2016**, *7*, 11203. [CrossRef] [PubMed]
23. Çıplak, Z.; Yıldız, A.; Yıldız, N. Green Preparation of Ternary Reduced Graphene Oxide-au@polyaniline Nanocomposite for Supercapacitor Application. *J. Energy Storage* **2020**, *32*, 101846. [CrossRef]
24. Wang, Y.; Xie, W.; Li, D.; Han, P.; Shi, L.; Luo, Y.; Cong, G.; Li, C.; Yu, J.; Zhu, C.; et al. One-pot Synthesis of Hierarchical Co1-x $S/NC@MoS_2/C$ Hollow Nanofibers Based on One-dimensional Metal Coordination Polymers for Enhanced Lithium and Sodium-ion Storage. *Sci. Bull.* **2020**, *65*, 1460–1469. [CrossRef] [PubMed]

25. Zhang, Y.; Chang, C.R.; Gao, H.L.; Wang, S.W.; Yan, J.; Gao, K.Z.; Jia, X.D.; Luo, H.W.; Fang, H.; Zhang, A.Q.; et al. High-performance supercapacitor electrodes based on $NiMoO_4$ nanorods. *J. Mater. Res.* **2019**, *34*, 2435–2444. [CrossRef]
26. Neeraj, N.S.; Mordina, B.; Srivastava, A.K.; Mukhopadhyay, K.; Prasad, N.E. Impact of Process Conditions on the Electrochemical Performances of $NiMoO_4$ Nanorods and Activated Carbon-Based Asymmetric Supercapacitor. *Appl. Surf. Sci.* **2019**, *473*, 807–819. [CrossRef]
27. Zhang, Y.; Chang, C.R.; Jia, X.D.; Huo, Q.Y.; Gao, H.L.; Yan, J.; Zhang, A.Q.; Ru, Y.; Mei, H.X.; Gao, K.Z.; et al. Morphology-Dependent $NiMoO_4$/Carbon Composites for High Performance Supercapacitors. *Inorg. Chem. Commun.* **2020**, *111*, 107631. [CrossRef]
28. Nti, F.; Anang, D.A.; Han, J.I. Facilely Synthesized $NiMoO_4/CoMoO_4$ Nanorods as Electrode Material for High Performance Supercapacitor. *J. Alloys Compd.* **2018**, *742*, 342–350. [CrossRef]
29. Arshadi Rastabi, S.; Sarraf Mamoory, R.; Dabir, F.; Blomquist, N.; Phadatare, M.; Olin, H. Synthesis of $NiMoO_4$/3D-rGO Nanocomposite in Alkaline Environments for Supercapacitor Electrodes. *Crystals* **2019**, *9*, 31. [CrossRef]
30. Chu, Y.; Xiong, S.; Li, B.; Qian, Y.; Xi, B. Designed Formation of MnO_2@NiO/$NiMoO_4$ Nanowires@Nanosheets Hierarchical Structures with Enhanced Pseudocapacitive Properties. *ChemElectroChem* **2016**, *3*, 1347–1353. [CrossRef]
31. Chen, S.; Yang, G.; Jia, Y.; Zheng, H. Three-dimensional $NiCo_2O_4$@$NiWO_4$ core–shell nanowire arrays for high performance supercapacitors. *J. Mater. Chem. A* **2017**, *5*, 1028–1034. [CrossRef]
32. Niu, L.; Li, Z.; Xu, Y.; Sun, J.; Hong, W.; Liu, X.; Wang, J.; Yang, S. Simple synthesis of amorphous $NiWO_4$ nanostructure and its application as a novel cathode material for asymmetric supercapacitors. *ACS Appl. Mater. Interfaces* **2013**, *5*, 8044–8052. [CrossRef] [PubMed]

polymers

MDPI

Review

Potential of Chitosan/Gelatin-Based Nanofibers in Delivering Drugs for the Management of Varied Complications: A Review

Popat Mohite [1], Abhijeet Puri [1], Shubham Munde [1], Roshan Dave [1], Showkhiya Khan [1], Riteshkumar Patil [1], Anil Kumar Singh [2], Pratchaya Tipduangta [3], Sudarshan Singh [3,4,*] and Chuda Chittasupho [3,*]

1 AETs St. John Institute of Pharmacy and Research, Palghar 401404, Maharashtra, India; mohitepb@gmail.com (P.M.); abhijeetp@sjipr.edu.in (A.P.); shubhamvmunde@gmail.com (S.M.); 122roshan4001@sjipr.edu.in (R.D.); showkhiya@outlook.com (S.K.); riteshkumar280201@gmail.com (R.P.)
2 United Institute of Pharmacy, Prayagraj 211010, Uttar Pradesh, India; singhanil2682@gmail.com
3 Department of Pharmaceutical Sciences, Faculty of Pharmacy, Chiang Mai University, Chiang Mai 50200, Thailand; pratchaya.t@cmu.ac.th
4 Office of Research Administration, Chiang Mai University, Chiang Mai 50200, Thailand
* Correspondence: sudarshan.s@cmu.ac.th (S.S.); chuda.c@cmu.ac.th (C.C.)

Abstract: Drug delivery systems have revolutionized traditional drug administration methods by addressing various challenges, such as enhancing drug solubility, prolonging effectiveness, minimizing adverse effects, and preserving potency. Nanotechnology-based drug delivery systems, particularly nanoparticles (NPs) and nanofibers (NFs), have emerged as promising solutions for biomedicine delivery. NFs, with their ability to mimic the porous and fibrous structures of biological tissues, have garnered significant interest in drug-delivering applications. Biopolymers such as gelatin (Ge) and chitosan (CH) have gained much more attention due to their biocompatibility, biodegradability, and versatility in biomedical applications. CH exhibits exceptional biocompatibility, anti-bacterial activity, and wound healing capabilities, whereas Ge provides good biocompatibility and cell adhesion properties. Ge/CH-based NFs stimulate cellular connections and facilitate tissue regeneration owing to their structural resemblance to the extracellular matrix. This review explores the additive methods of preparation, including electrospinning, force pinning, and template synthesis, focusing on electrospinning and the factors influencing the fiber structure. The properties of Ge and CH, their role in drug release, formulation strategies, and characterization techniques for electrospun fibers are discussed. Furthermore, this review addresses applications in delivering active moieties in the management of orthopedics and wound healing with regulatory considerations, along with challenges related to them. Thus, the review aims to provide a comprehensive overview of the potential of Ge/CH-based NFs for drug delivery and biomedical applications.

Keywords: biopolymers; gelatin; chitosan; electrospinning; nanofibers; nanotechnology

1. Introduction

A drug delivery system (DDS) is a formulation or device that facilitates the administration or release of a drug substance into the body, enhancing its effectiveness and safety by regulating the timing, location, and rate of drug release. This involves administering the therapeutic product, releasing its active ingredients, and transporting them across biological membranes to the target site [1]. Novel DDSs have transformed traditional drug administration methods by addressing a range of challenges. These systems enhance the solubility of drugs, prolong their effectiveness, minimize their adverse effects, and preserve their potency. Currently, DDS significantly improves the availability of drugs in the body, enhances their absorption, regulates the release of drugs to maintain consistent levels, and targets specific cells to reduce side effects [2]. Nanotechnology-based DDS is a growing field in medicine delivery that addresses various challenges. These include enhancing bioavailability in the face of low solubility, addressing shortcomings in intestinal absorption mechanisms through degradation, improving delivery precision to target sites, enhancing therapeutic efficacy, minimizing side effects, and mitigating fluctuations in drug levels within the bloodstream [3]. NPs are a class of materials in nanotechnology, characterized by particle sizes ranging from 10 to 100 nm. These materials exhibit diverse structural dimensions, including zero- and one-dimensional configurations. NPs are composed of a core, shell, and surface layer that can be modified with various functional groups and possess unique properties. Their applications span numerous fields, with particular promise in the biomedical domain, including drug delivery, chemical and biological sensing, gas sensing, and carbon dioxide capture [4]. Various classes of nanomaterials have been used in the medical field. These include dendrimers, metallic NPs, carbon nanomaterials (such as nanorods, nanowires, tubes, fibers, and platelets) [5,6], magnetic nanomaterials, semiconductor nanomaterials [7], hydrogel [8], nanocomposites [9], biodegradable polymers, liposomes [10], and polymer nanocomposites [11]. Each class offers unique properties that can be harnessed for medical applications such as drug delivery, imaging, and tissue engineering (TE) [12,13]. NFs stand out among the different nanomaterials because they mimic the porous and fibrous structures in biological tissues, making them valuable for multifaceted applications [14]. NFs have garnered significant interest in delivering medicine due to their unique properties like mechanical characteristics of polymeric NFs, such as the modulus, shear modulus, and tensile strength, which improve as the fiber diameter decreases and are used to govern the behavior of cells and provide vigor to endure the forces of the cell cytoskeleton [15,16].

Biopolymers are emerging as promising candidates for the fabrication of NFs due to their biocompatibility, biodegradability, and versatility in various biomedical applications [17]. Ge and CH have gained significant attention among various biopolymers due to their unique and functional properties, which are essential for NF development. CH is an excellent option for biomedical applications such as TE scaffolds, wound dressings, and DDS due to its exceptional biocompatibility, anti-bacterial activity, and wound healing capabilities [18]. CH is produced by the deacetylation of chitin [19], while Ge, a denatured form of collagen derived from animal connective tissues, provides good biocompatibility and cell adhesion capabilities [20]. These properties are essential for TE and regenerative medicine. Owing to their structural resemblance to the extracellular matrix (ECM) [21], Ge-based NFs stimulate cellular connections and facilitate tissue regeneration. The tuneable mechanical characteristics of Ge-based NF can be controlled by modifying the electrospinning process settings. This allows the fabrication of scaffolds with customized stiffness and elasticity that resemble natural tissues. Furthermore, the versatility and application of Ge in the fabrication of controlled-release DDS and wound healing therapies are expanded by its susceptibility to chemical alterations and mixing with other polymers [22]. GE, a protein-rich in glycine, proline, and hydroxyproline, has various applications, such as in plasma extrusion, wound dressing, and adhesives, and as an absorbent pad during surgery [23]. According to research, it has been claimed that GE can activate macrophages, can exhibit a strong hemostatic effect, and does not induce antigenicity [24]. The present review explores additive preparation methods, including techniques such as electrospinning, force pinning, and template synthesis. Moreover, the review also focuses on electrospinning and discusses the process, recent advancements, and factors influencing the fiber structure. Additionally, the review covers polymer properties, the role of GE/CH in drug release, formulation strategies, and characterization techniques, focusing on fabrication using electrospun technology. In addition, applications in DDSs, orthopedics, wound healing, and regulatory considerations are addressed, along with challenges related to polymers.

2. Nanofibers

Nanofibers are ultrathin fibers, ranging from 1 to 1000 nm in diameter, produced using polymers. Although according to ASTM (2012) and ISO/ASTM standards, a nanomaterial is defined as having at least one dimension in the 1–100 nm range, many studies classify fibers with diameters exceeding 100 nm as nanofibers due to their nanoscale properties and functional advantages [25]. Using polymeric fibers and controlled-release administration methods, medications can be administered once or twice daily, enhancing patient compliance and preventing toxic plasma peaks resulting from frequent dosing of immediate-release formulations. These small NFs offer several benefits, such as exceptional stability, targeted delivery, high drug-loading capacity, increased surface area, reduced toxicity, enhanced mechanical properties, and suitability for delivering heat-sensitive drugs [26] (Figure 1). Electrospinning began over four centuries ago with William Gilbert's work around 1600, pioneering this technique. Key developments included Louis Schwabe's silk spinning method in 1845, Hughes and Chambers' first carbon NF patent in 1889, and John Francis Cooley's 1902 patent for the first electrospinning machine. In 1938, "Petryanov filters" were developed using electrospun fibers, and hollow graphitic carbon fibers were invented in 1952. Doshi and Reneker popularized "electrospinning" in 1995. This process is preferred for producing NFs with high surface-area-to-volume ratios and numerous pores [27].

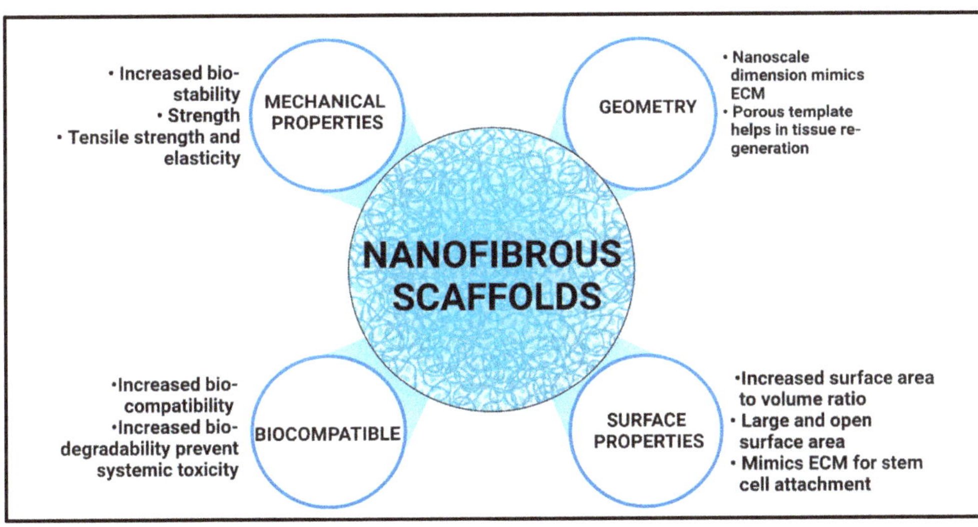

Figure 1. Illustration for ideal properties of NFs.

Nanofiber Fabrication Techniques

NFs are fabricated using various methods (Figure 2), each offering distinct advantages and applications. Electrospinning is the most widely used method whereas other notable techniques include phase separation, where a polymer solution undergoes gelation and phase extraction to form NFs, and template synthesis, which utilizes porous membranes to guide the formation of fibers. Also, self-assembly involves molecules' spontaneous organization into fiber structures, although it typically requires complex procedures and has low productivity. Techniques for fabricating NFs are divided into several categories, including physical, chemical, and biological procedures, as well as top-down and bottom-up approaches. The top-down method reduces complicated materials to micro- or nanoscale particles, whereas the bottom-up method assembles small particles, usually between 1 and 10 nm, to form new bulk materials [28]. Physical techniques use mechanical energy or high-energy radiation to develop NFs. In contrast, chemical techniques, part of the bottom-up approach, depend on specific conditions to help atoms or ions combine to form clusters of single NPs. Environmentally friendly biological methods develop NFs by treating source materials with enzymes or microorganisms [29].

Electrospinning: It is a prominent technique that relies on parameters such as voltage, viscosity, solvent choice, and tip-to-collector distance. It is widely employed in the development of NFs from polymers and the incorporation of NPs. For example, polyurethane, polyacrylonitrile, and polyvinylidene fluoride yield NFs with diameters of approximately 150 nm, requiring a voltage of approximately 11 kV. Similarly, using acetone as a solvent and combining cellulose acetate with essential oils result in NFs that can be produced ranging between 700 and 1500 nm in size, necessitating a higher voltage of 120 kV [30]. Biologically generated NFs, such as eugenol and GE, are fabricated using acetic acid, ethanol, and water solvents. Their diameters range from 157 nm to 293 nm at a voltage of 13 kV. Moreover, NFs fabricated using GE produce a diameter of approximately 422 nm when mixed with polyethylene oxide, silica, and ciprofloxacin [31].

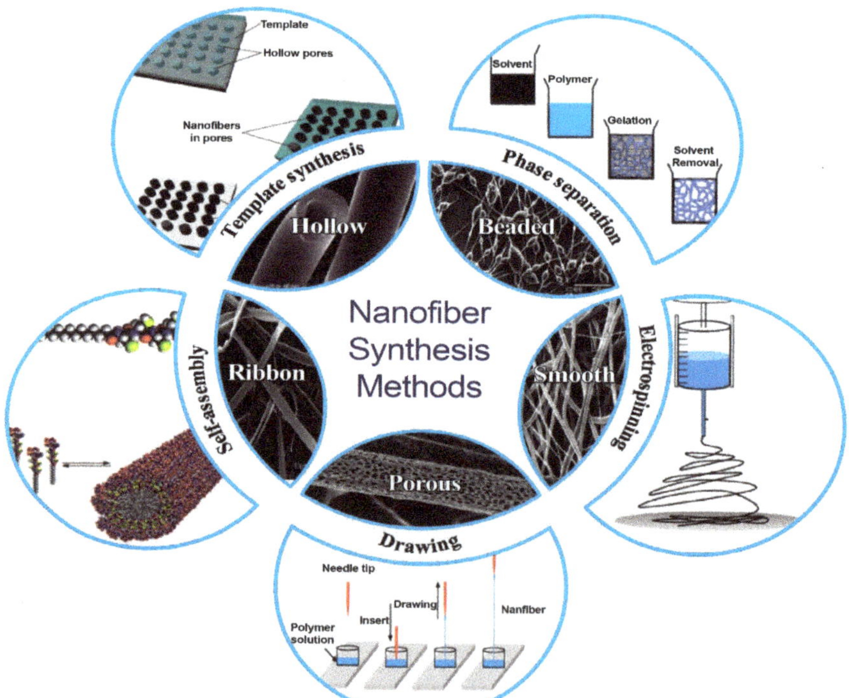

Figure 2. Illustration demonstrating comparative analysis of different NF fabrication techniques. Reproduced with permission from [32]. Copyright Clearance Center's RightsLink service.

The drawing process can result in the development of fibers that are long and exhibit a diameter of several nanometers. Each NF is drawn using a micropipette during solvent evaporation and at the moment of solidification. Cooling and solvent evaporation are the two processes that cause freezing [32].

The template synthesis technique uses different solvents and templates, such as cellulose or a phenylalanine–phenylalanine–aspartic acid tripeptide, to produce NFs with regulated pore sizes and shapes.

Self-assembly: Small molecules are used as the fundamental building blocks in the self-assembly method [32].

Phase-separation-mediated NF growth demonstrates a separation of phase due to physical incompatibility. In this technique, a gel is prepared by lowering the temperature of the polymer solution developed by mixing a polymer material with an appropriate solvent. After producing the gel, it is immersed in another solvent that causes the gel to separate from the initial solvent, and in this way, the separation occurs between the two phases. The resulting polymer structures typically exhibit pores with a sponge-like morphology and spherical pores [32].

Force spinning or centrifugal spinning, also known as rotary or rotational jet spinning, has been extensively developed and utilized in fabricating NFs. The primary distinction between force spinning and regular electrospinning is that the first uses a centrifugal force instead of an electric field, which is used in electrospinning. In this method, a liquid jet is ejected from the nozzle tip when the centrifugal force exceeds the surface tension and the spinning head containing the polymer solution reaches a crucial rotational speed. Subsequently, the jets are stretched, and the solvent evaporates, causing the deposition of NFs over the collector. The temperature, collecting system (e.g., collection distance), nozzle configuration, rotating speed of the heated structure, and other parameters determine the geometry and morphology of the centrifugally spun NFs [33], as shown in Figure 3. Many conductive and nonconductive materials can be spun using centrifugal spinning techniques. However, certain solid materials do not require chemical pretreatment before spinning after melting [34]. Centrifugal spinning eliminates the safety problems associated with electrospinning because it does not require high voltage. Moreover, force spinning helps to reduce typical electrospinning constraints, such as high electric field requirements, limited productivity, conductivity difficulties, and expensive production costs. Compared with electrospinning, centrifugal spinning yields NFs using polymers at higher concentrations, low solvent usage, and reduced production costs. Nevertheless, a notable drawback of centrifugal spinning is that fiber quality and productivity are profoundly influenced by the material properties and spinneret design [35]. Direct current and alternating current are two possible power source types. The electrospinning process can be broadly described as follows [36]: (i) charging the liquid droplet and developing a cone-shaped or Taylor cone-shaped jet, and the charged jet extends in a straight line; (ii) the application of a high-voltage electric field, ejection, and the formation of a polymer jet; (iii) thinning the jet in the presence of an electric field and producing an electrical bending instability, also called whipping instability; (iv) solidify and collect the jet as a solid fiber on a grounded collector [36].

CH and Ge NFs are versatile materials that can be used in various applications, including wound healing, drug delivery, tissue engineering, etc. Concise electrospinning is a process that involves applying a voltage to stretch polymer solutions into a fine fiber form. The flow rate and distance between the needle and collector are crucial parameters while fabricating fiber with uniformity. Additionally, environmental conditions such as temperature, humidity, and airflow also impact the process. Therefore, the proper control of these factors is essential for producing smooth, defect-free nanofibers, whereas proper temperature and humidity control are vital for consistency and reproducibility. Airflow around the setup affects fiber deposition patterns. Post-treatment and crosslinking are essential steps for stabilizing nanofibers, enhancing mechanical strength, water resistance, and durability [37].

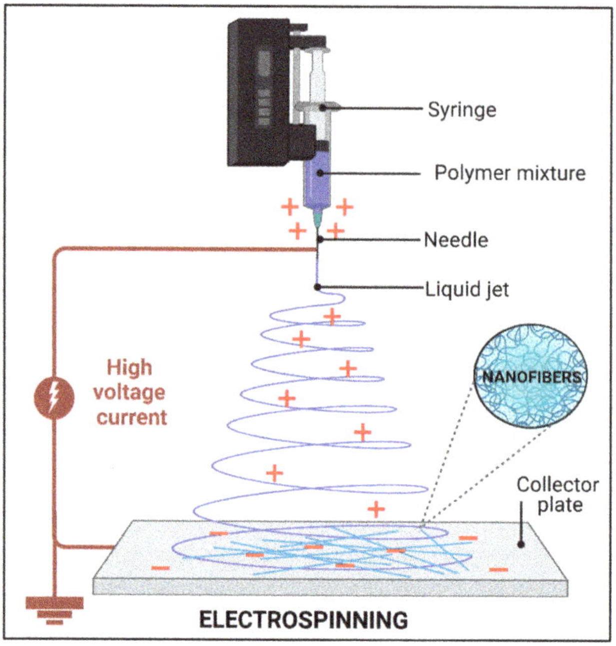

Figure 3. Illustration for fabrication of NFs using electrospinning.

CH is a rigid polysaccharide [38,39] that requires blending with flexible polymers like polyethylene oxide or polyvinyl alcohol to improve its spinnability and mechanical strength [11,40]. The critical electrospinning process involves parameters like polymer concentration, solution viscosity, conductivity, and environmental conditions. Moreover, the process involves optimizing the voltage applied and flow rate of the polymer solution to avoid bead formation and ensure smooth, continuous fibers [41]. CH, a polymeric biomaterial aligned with rigid D-glucosamine repeat units, is difficult to convert into a submicron-sized fibrous form due to its low solubility in water and organic solvents. However, modified chitosan demonstrates improvement in solubility when hydrogen bonds form water molecules, which are prevented by electrostatic repulsive interactions between positive ammonium groups. Acetic acid is a common pH adjustment reagent, and pure chitosan can be electrospun using a solvent with a high concentration. However, reducing pH using acids can make chitosan solutions more viscous, affecting spinnability. The electrospinning technique for chitosan NFs further faces challenges like limited solvent availability, ultimately affecting the quality and yield [42]. Aniseiei et al. fabricated CH-based NFs using HFIP as a solvent due to its low boiling point and ability to interrupt hydrogen bonding networks. Chitosan solutions were successfully electrospun at 0.4% concentration, producing amorphous nanofibers with a mean diameter of 42 ± 15 nm. The experimental results of the study revealed that fibers stick together under slightly acidic conditions, highlighting the challenge of managing solvent quantities during the electrospinning process [43].

Ge is another versatile polymeric material suitable for various applications [44]. However, their electrospinning process presents challenges due to their hydrophilic nature and tendency to dissolve or degrade in aqueous environments. Solution parameters, such as polymer concentration, molecular weight, solvent choice, and blend ratio, play a crucial role in the fabrication of nanofiber using Ge [45]. Ge is often combined with synthetic polymers or additives to improve fiber stability and mechanical performance. Blend ratios affect fiber properties, with higher proportions increasing bioactivity but reducing stability. Process parameters, such as applied voltage, flow rate, and needle-to-collector distance, also play a pivotal role in controlling fiber morphology and uniformity, while environmental factors, such as temperature, humidity, and airflow, influence the electrospinning process [45]. Moreover, the temperature affects the viscosity of the spinning solution and the evaporation rate of the solvent. In contrast, humidity plays a critical role in determining fiber morphology, with high humidity promoting the condensation of solvent vapors on the fiber surface, while controlled humidity ensures uniform and smooth fibers. Airflow around the electrospinning setup affects fiber deposition patterns and alignment. Post-spinning treatments, such as chemical crosslinking, are vital for enhancing the mechanical stability and water resistance of gelatin nanofibers. However, chemical crosslinkers may introduce cytotoxicity, requiring thorough removal of residual agents to maintain biocompatibility. Physical crosslinking methods, such as heat treatment or UV irradiation, offer residue-free alternatives but may compromise fiber morphology or mechanical properties [46].

3. Advancements in Electrospinning

Advancements in the fabrication of NFs through electrospinning technology have significantly enhanced the capabilities and applications of these materials, particularly in three-dimensional (3D) and four-dimensional (4D) cases using techniques such as chaotic electrospinning, which allows the development of multilayered structures with improved surface area and functionality, making them suitable for applications like energy storage area and biomedical scaffolds. Additionally, four-dimensional electrospinning incorporates time as a variable, enabling the fabrication of response to external stimuli, thus broadening their potential uses in innovative materials and DDS. Additionally, techniques such as near-field electrospinning have also emerged, which provide precise control over fiber deposition, allowing for the development of aligned fibers with specific topologies that enhance mechanical properties and cellular interactions. These innovations address traditional challenges in NF production, such as chaotic fiber deposition and low yield, and pave the way for more sophisticated applications in fields ranging from medicine to energy storage.

The electrospinning technique is now capable of more than two-dimensional fiber deposition owing to recent advancements. The development of 3D and 4D electrospinning methods has brought about revolutionary changes in several sectors, such as drug delivery [47], filtration [48], TE [48], and energy storage [49]. The engineering and science of sophisticated materials have been completely transformed using 3D printing. Rapid prototyping, often known as additive manufacturing technology or 3D printing, is a simple method for fabricating 3D products/materials that have been computer-designed. The software opens the door to innovation and intricate product design with ground-breaking functionality that will completely change the manufacturing landscape. Nonetheless, two significant drawbacks of 3D printing are its limited resolution and sluggish pace of production [50]. The term "4D printing" refers to the process of 3D printing, which exhibits the ability to alter shape over time or in response to external stimuli. These shape-changing materials have applications in medication delivery, biomedical devices, soft robotic systems, and intelligent textiles. They can also adapt to changing environmental factors, including heat, light, moisture, and pH. This revolutionary new technology will make the goods of the future possible, although 4D printing at the nanoscale has not yet been investigated [51]. Electrospinning is a highly adaptable method that can quickly fabricate 3D/4D printing solutions at the nanoscale by processing melts, solutions, or suspensions into continuous NFs using a high voltage. A revolutionary nano-printing technique called 3D/4D electrospinning combines the benefits of electrospinning with 3D printing technology [52].

Additionally, 3D/4D structures at the nanoscale provide new prospects for material development. Fibers are deposited through three-dimensional structures using several nozzles or platforms in 3D electrospinning. This method makes it possible to precisely regulate the orientation and distribution of fibers while fabricating intricate scaffolds. However, 4D electrospinning adds a temporal dimension that permits dynamic modifications to the shape of the fiber or other characteristics after deposition. This potential invents new opportunities for developing responsive fabrics, programmed medication-release systems, and shape-memory materials. Enhancement of functionality and adaptation to different physiological or environmental conditions is possible with 4D electrospinning through the incorporation of stimuli-responsive materials or controlled degradation processes [53].

4. Biocompatible and Biodegradable Polymers for Fabrication of Nanofibers

Biocompatible and biodegradable polymers play a crucial role in the fabrication of NFS, which are increasingly utilized in various applications, such as TE, drug delivery, and environmental remediation, as they interact harmoniously with biological systems. Different polymers have been employed in the fabrication of NFs using electrospinning. Natural polymers are collagen, alginate, Ge, CH, etc., whereas synthetic polymers, including polycaprolactone (PCL), polylactic acid [54], poly(lactic-co-glycolic acid) (PLGA), etc., offer advantages such as tuneable mechanical properties and ease of processing, making them suitable for diverse applications (Figure 4). Their selection is based on the type of targeted action and the method of preparation. Herein, we only focus on using natural polymers such as GE and GE for the fabrication of NFs, as they are biocompatible and quickly degrade when they come in contact with biological tissue.

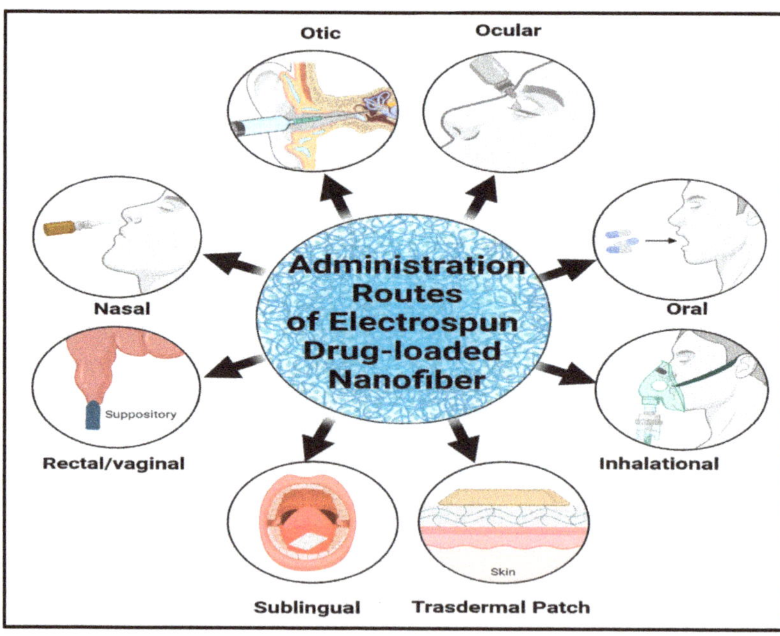

Figure 4. Common administration routes of drug-loaded GE/CH nanofibers.

4.1. Gelatin

GE (Figure 5) is a water-soluble biopolymer versatile protein produced through collagen hydrolysis. Its biodegradability, bio-affinity, formability, and cost effectiveness make it a valuable material in various fields, including the biomedical and environmental fields [55]. GE can be facilitated in various shapes, including NPs, microparticles, films, hydrogels, and NFs. Owing to their low electrospinnability, suitable solvents like 2,2,2-trifluoroethanol, acetic acid, and formic acid are used [56]. Mixtures of GE and synthetic and natural polymers offer desirable biocompatibility and bio-mechanical characteristics. Electrospun Ge-NFs are ideal for drug delivery, wound dressings, and cell and tissue culture. GE is preliminarily found in plant connective tissues and has a significant role in preserving the integrity of connective tissues, including bones, corneas, cartilage, and tendons. Approximately 80–90% of the collagen in the body is made up of the three most common forms: type I, type II, and type III [57]. Three helical polypeptide chains, each measuring around 1.5 nm in diameter and 300 nm in length, make up collagen type I [58]. Two α-chains and one β-chain form the typical triple-helix structure of type I collagen are the molecules in the middle regions, and the polypeptide architectures of the three spiral chains of collagen have limited immunogenicity and antigenicity [59]. The limited antigenicity of collagen has led to a significant decrease in its use in biological applications and its involvement in the manufacturing of NFs [60].

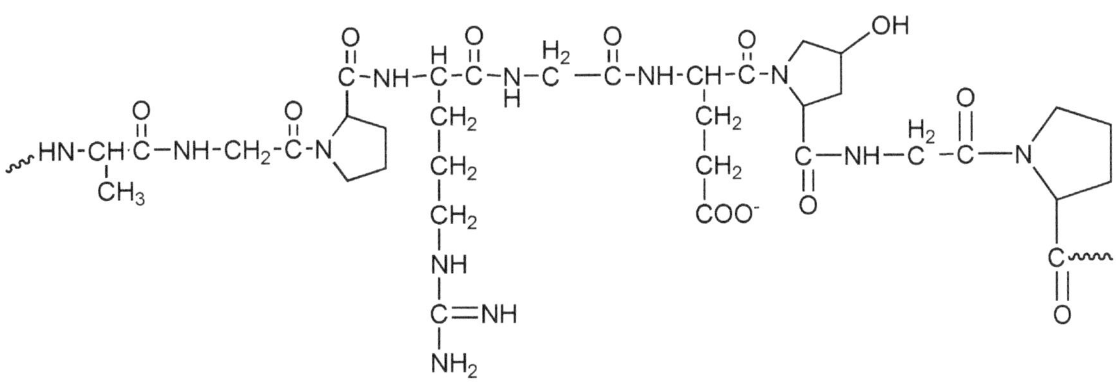

Figure 5. Structure of gelatin.

4.2. Chitosan

CH (Figure 6), which is thought to be the second most abundant biopolymer after cellulose, is obtained from natural sources that can be fungal, aquatic, or terrestrial. Chitin is the starting material obtained after chemical and enzymatic treatments, and GE is obtained by deacetylating chitin [61]. The semi-crystalline substance known as GE is composed of N-acetyl-d-glucosamine residues. GE molecules are composed of hydroxyl (–OH) and amino (–NH_2) groups [62]. This particular polymer is cationic, which aids in constructing ECM components by drawing negatively charged ions. It also exhibits biopotential activities, improves mechanical strength, and strengthens bonds by causing structural changes. GE, a cationic biopolymer, is problematic to electrospin because of its rigid D-glucosamine repeat unit, high crystallinity, and hydrogen bonding ability. For electrospun defect-free fibers, the polymer concentration must be at least 2 to 2.5 times the entanglement concentration [63]. However, even moderate concentrations can become too viscous to overcome the electric field, making electrospinning impossible. The rheology of the solutions also affects the electrospinning process. To improve electrospunability, researchers have blended GE with other polymers [64]. A comparative feature of GE and CH is presented in Table 1, collected from various previously published data [65–75].

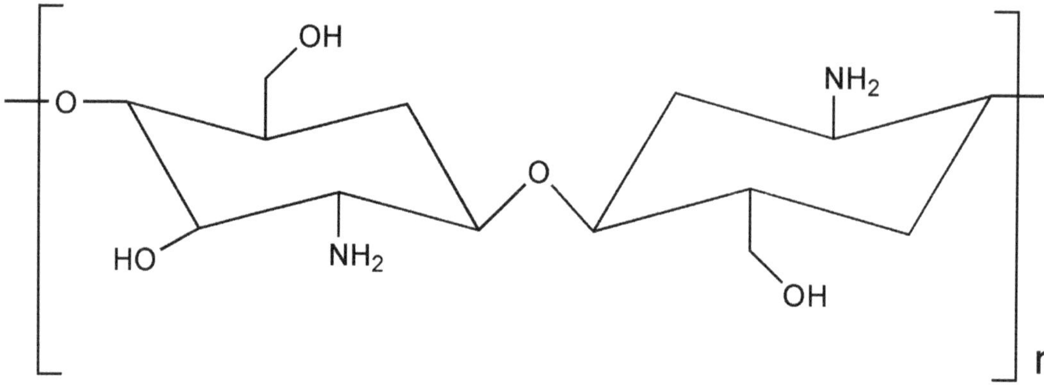

Figure 6. Structure of chitosan.

Table 1. Comparison of GE and chitosan as biopolymers for fabrication of nanofibers (NFs) for topical drug delivery.

Property	Gelatin	Chitosan
Source	Derived from collagen (animal-based)	Derived from chitin
Fabrication method	Electrospinning	Electrospinning
NF diameter (nm)	~100–360 nm	~50–250 nm
Surface morphology	Smooth, slightly porous	Smooth, uniform
Mechanical strength	Lower compared to CH	Moderate
Drug-loading efficiency	Moderate (better for hydrophobic drugs)	High (especially for hydrophilic drugs)
Biocompatibility	High, non-toxic for topical use	High, suitable for skin applications
Anti-bacterial properties	Limited anti-bacterial effect, but supports healing	Anti-bacterial activity (due to positive charge)

5. Formulation Strategies for Polymer-Based Electrospun Drug Delivery Systems

Polymer-based NFs are widely used owing to their unique properties, such as a high surface-area-to-volume ratio, tuneable porosity, and ability to encapsulate a wide range of therapeutic applications [76]. To ensure controlled release, improved stability, and therapeutic efficacy, formulating an effective NF-based DDS involves intricate formulation strategies such as polymer selection, a drug encapsulation method, an optimization process, and material parameters [77].

Synthetic polymers (PLGA and PCL) and natural polymers (GE and CH) are used in controlled-release applications owing to their tuneable degradation rate, safety profile, compatibility with surfactants or co-solvents, and TE applications [78]. Moreover, the polymer ratio is another crucial aspect to be considered, as changing the concentration of polymers may affect the NF structure and release pattern. Vu et al. successfully fabricated polyvinyl alcohol (PVA)–CH NFs to understand the effect of polymers using a solution containing 5% PVA, 5% CH, and 60% CH_3COOH, with diameters ranging from 77 to 292 nm and an average diameter of 153 nm, suitable for drug integration [79].

The second parameter is drug incorporation: drugs or bioactive compounds can be added to the polymer either in melted or in solution form using various techniques before electrospinning. Drugs are dispersed throughout the polymer solution by solvent mixing, sonication, or stirring for physical encapsulation. Both hydrophilic and hydrophobic medications can be used with this technique; yet, the NFs may distribute pharmaceuticals unevenly [79]. Chemical conjugation, on the other hand, entails covalently joining medications or functional groups to polymer chains to improve stability and allow regulated release. This technique guarantees precise control over drug-loading and release kinetics; compatibility between drug and polymer chemistry is necessary [80]. Combining polymers of different features using blend electrospinning can produce synergistic advantages, including improved drug release patterns, biodegradation, and mechanical strength. Additionally, the polymer ratio influences the physical and chemical interactions between the drug molecules and the polymer matrix, which substantially impacts the entrapment efficiency of the electrospun NFs [81]. The polymer's solubility, the spinning solution's viscosity, electrospinning parameters, and the compatibility of the drug with the polymers all affect the entrapment efficiency. To achieve maximum drug loading, entrapment efficiency can be optimized by combining polymers with diverse properties (e.g., hydrophilic and hydrophobic). Hydrophilic polymers have been observed to interact favorably with hydrophilic medicines through hydrogen bonding or electrostatic interactions, leading to improved drug encapsulation inside the NF matrix. On the other hand, owing to enhanced compatibility and less drug outflow during electrospinning, hydrophobic medicines may show better entrapment efficiencies when electrospun with hydrophobic polymers [81,82]. The third parameter is controlled release and the release behavior of drugs from electrospun NFs, which is closely intertwined with the entrapment efficiency and polymer ratio. A higher entrapment efficiency typically results in more significant drug loading within the NFs, directly influencing the initial burst release and sustained release kinetics over time [83]. The polymer ratio controls factors such as the drug encapsulation efficiency, release kinetics, and drug liberation mechanism, all of which are directly related to the release of pharmaceuticals from the electrospun NFs. Furthermore, the drug release profile is regulated by modifying the polymer composition, including the polymer degradation rate, swelling behavior, and diffusion paths within the NF matrix [84].

6. Characterization Techniques Involved with Polymeric Electrospun Nanofibers

The physical, chemical, and structural characteristics of polymeric electrospun NFs must be evaluated to determine their applicability in various applications, including filtration, drug delivery, and TE. As described below, the electrospun nanofibers are characterized through primary advanced instrumental techniques.

6.1. Microscopic Analysis for Fiber Morphology Assessment

Scanning electron microscopy [SEM] and transmission electron microscopy [TEM] are used to visualize the morphology and surface topology of NFs. SEM allows the examination of cross-sectional views and enables analyses of the fiber structure, layering, and defects, whereas TEM provides high-resolution images and detailed imaging of NF structures at the nanoscale to elucidate the interaction between polymers and drug molecules within the NF matrix [85]. A microscopic examination using SEM for NF illustration is presented in Figure 7.

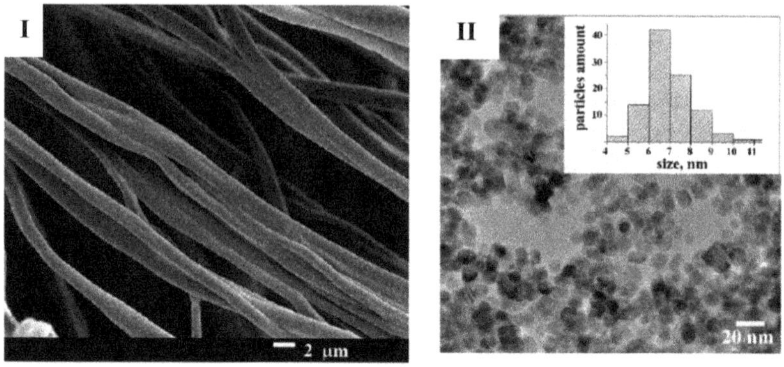

Figure 7. Morphological illustration of GE NFs (**I**) and TEM microscopy image of ZnO-NP-coated NFs with particle size distribution (**II**). Reproduced with permission from [86] under CCBY.

6.2. Evaluation of Mechanical Properties of Electrospun Nanofibers

The mechanical properties such as tensile strength, elasticity, and flexibility of the NFs are assessed using the tensile testing method following AST standards. This test measures the force required to stretch or deform the NFs, providing quantitative data on mechanical behavior under physiological or operational conditions, as these are necessary for TE scaffolds and wound dressings [87].

6.3. Material Interactions Through Spectroscopic Techniques

Various spectral/functional techniques have been employed to study the interactions between polymers and drug molecules between ECM and NFs, while FTIR has been used to analyze the chemical and molecular structures. FTIR spectra generally detect changes in the polymer conformation and crystallinity induced during electrospinning [88,89]. On the other hand, X-ray diffraction is used to investigate the crystalline structure and phase transition of the NFs. It provides information on the size, direction, and degree of crystallinity of crystallites by identifying the crystalline peaks corresponding to polymer chains or embedded compounds. XRD studies enabled understanding of the changes in NF structure under mechanical stress or environmental circumstances, managing drug release processes driven by crystalline domains, and optimizing polymer blends [90]. A thermogravimetric analysis and differential scanning calorimetry are two thermal characterization methods used to assess the stability, phase transitions, and thermal behavior of electrospun NFs. Differential scanning calorimetry sheds light on the mobility of polymer chains and thermal stability by measuring the variations in heat flow caused by melting, crystallization, or glass transition temperatures. By calculating the weight loss and breakdown temperatures of NFs, the thermogravimetric analysis evaluates their degradation kinetics and appropriateness for specific uses that require regulated or heat-resistant degradation [91].

7. Characterization Techniques Involved with Polymeric Electrospun Nanofibers

Biomimetic applications of polymer-based electrospun NF materials, utilizing nature-inspired design principles, offer advanced functional materials with unique properties, such as a high surface-area-to-volume ratio and mechanical flexibility.

7.1. Transdermal Drug Delivery Systems

Transdermal DDS (TDDS) offers a non-invasive, patient-friendly method of delivering therapeutic agents through the skin for localized or systemic effects. This research represents a significant gap in pharmaceutical technology [92]. Owing to their distinct structural and functional characteristics, NFs, especially electrospun NFs, have emerged as intriguing platforms for improving TDDS. The use of NFs in TDDS is examined in this article, which also discusses current developments, difficulties, benefits, and possible future paths. To produce systemic circulation or specific local effects, transdermal medication delivery entails the introduction of medicinal substances through the stratum corneum, the outermost layer of the skin. The inherent impermeability of the skin restricts the administration of hydrophilic medications and large molecules, which is why conventional transdermal patches and topical formulations rely on passive diffusion through the skin barrier. Overcoming these obstacles while preserving therapeutic efficacy and patient compliance is a significant problem in transdermal drug delivery. NFs are widely used to deliver drugs directly to the skin surface in dermatological conditions such as wound healing, tissue regeneration, psoriasis, and other fungal infections [92]. NF patches are commonly used to deliver hormones or vaccines to the skin. Du et al. fabricated propolis extract-enriched silk fibroin (SF)/Ge NFs as wound dressings. These findings revealed that both cytocompatibility and hemocompatibility were suitable for the SF/Ge-1%-propolis extract. Furthermore, it has the potential to enhance fibroblast cell motility significantly. The SF/Ge-1%-propolis extract was applied to a mouse model of full-thickness skin defects (Figure 8) [93].

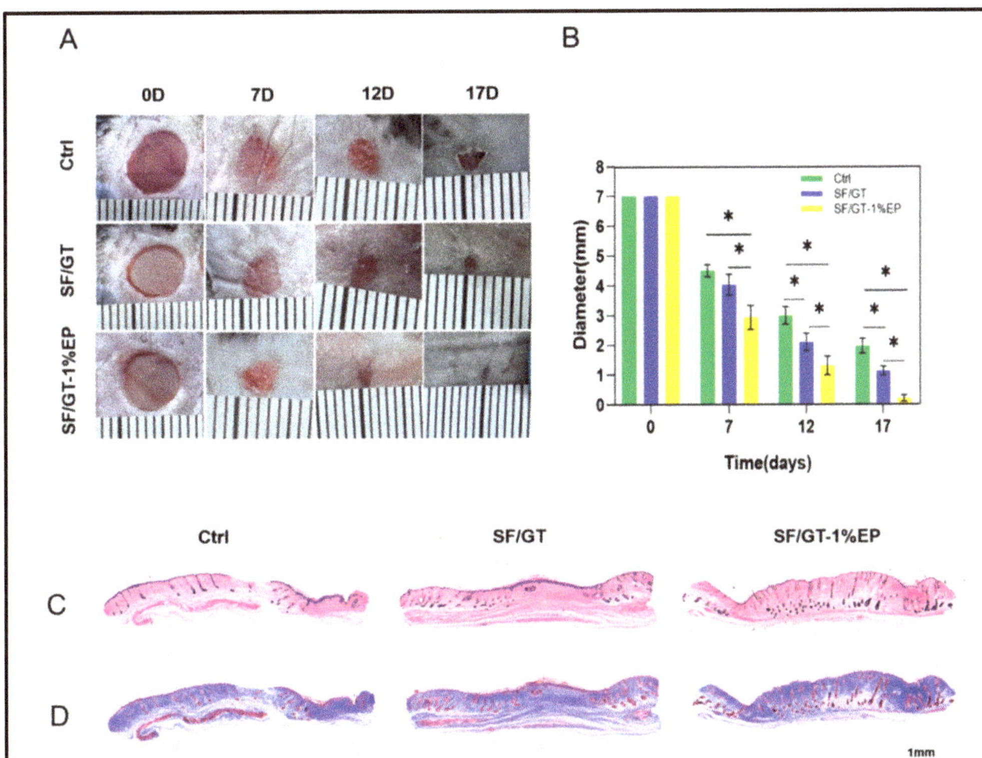

Figure 8. In vivo wound healing results. Macroscopic photographs of wounds at 0, 7, 12, and 17 days after a full-thickness skin wound was created (**A**). Wound diameters (0, 7, 12, and 17 days) of different groups (control, SF/GT, and SF/GT-1%-propolis extract) (**B**). Histology analyses of the wound area sections at different groups 17 days after the operation (control, SF/GT, and SF/GT-1%-propolis extract) (**C**,**D**). Reproduced with permission from [93] under CCBY.

Goudarzi et al. fabricated GE-PVA NFs for skin TE, and their findings revealed excellent cell viability and cell attachment of NFs for mouse fibroblast cells according to the MTT assay, which indicated that the use of both GE and GE complexes exhibit potent activity [94]. Doostmohammadi et al. used a histopathological analysis to examine the in vivo wound healing potential of tellurium NPs embedded in PCL and Ge-NFs (electrospun). This study was conducted on wounds in Wistar rats. Scaffolds incorporating NPs demonstrated the highest healing, scoring 15 out of 19. Furthermore, the scaffold reduced edema and inflammation at the injury site and had favorable effects on collagen horizontalization and synthesis in a dose-dependent manner. Their healing action was verified by the data, which can be helpful for wound dressing [95].

7.2. Orthopedic and Wound Healing Applications

Originally, wound dressings and fracture healing were fabricated using natural substances such as plant fibers and animal fats, which were used to cover the injured area. This has progressed to the point where healing applicants can now be made from artificial materials using various cutting-edge technologies. Enhanced antimicrobial properties, quick hemostasis, and tissue repair are essential for contemporary wound healing (Figure 9) and other orthopedic treatments. Wound healing has the advantage of developing nanofibrous membranes with fast hemostasis and anti-bacterial properties using electrospinning techniques. Anti-bacterial NFs have been prepared using conventional anti-bacterial ingredients, such as metal oxide NPs, triclosan, antibiotics, biguanides, quaternary ammonium compounds, and silver NPs, using the electrospinning method for fast wound healing with less toxicity and side effects [96]. In clinical studies, diabetic foot ulcers have been identified as a prevalent and severe form of chronic wounds. Diabetic feet can result from ischemia, neuropathy, or neurobiochemical illnesses. A diabetic foot causes ulcers and infections or necessitates foot amputation [97]. By 2030, diabetes is expected to affect 552 million people globally, and 25% of them are likely to develop foot ulcers [98]. Castings and orthopedics are more expensive than patient compliance. Fiber scaffolds fabricated by electrospinning devices can provide structural and morphological clues for the adhesion of various types of cells in wounds and serve as templates for wound tissue regeneration. Microstructures similar to ECM proteins and electrospun fiber scaffolds play an essential role in maintaining cell growth and infiltration and are, therefore, suitable for use in TE [99]. Polymeric electrospun NFs are among the most thoroughly studied nanostructures for orthopedic TE applications. The interconnected porous structure of nanofibrous scaffolds offers a large surface area for cell adhesion and room for the passage of nutrients [100]. In addition to providing mechanical support, applying NFs as delivery systems for bioactive compounds at the wound site offers two of three domains to initiate and accelerate wound healing, which may lead to effective tissue repair or regeneration [101]. Electrospun NFs can reduce tissue reactivity, reduce inflammatory responses, and provide a native habitat for cells at the wound site. They also resemble collagen fibers in the ECM [102]. Poly(ethylene oxide), used for a variety of biomedical applications such as cartilage TE and wound dressing, can facilitate the electrospinning of GE in aqueous solvents and eliminate the possible toxic effect of organic solvent residues in the final electrospun mat [103]. The effectiveness of polymeric NF scaffolds in promoting osteoblast adhesion and proliferation was shown by Li et al., who demonstrated their capacity to direct the orientation of cell cytoskeletal proteins and offer a large surface area for cell adhesion and proliferation [104]. Chu et al. reported a hepatocyte nanofibrous scaffold composed of PCL–ethyl ethylene phosphate containing a galactose ligand, which can aid cell attachment, spheroid formation, and functional maintenance of hepatocytes [105]. In another study, Linn and co-workers fabricated NFs using alone PVA, and GE, and a blend of PVA with Ge for the TE, considering processing parameters such as the concentration of polymers, electrical field, and tip-to-collector distance. The results in NF diameter via morphology for alone PVA, and GE, and a blend of PVA with Ge ranged between 50 and 150 nm. At the same time, an X-ray diffraction analysis demonstrated the existence of the crystalline nature of NFs. Meanwhile, thermal and tensile strength tests suggested improved physical properties without significant chemical interaction during processing, suggesting great promise for use in TE applications [106] (Figure 10). In a study by electrospun hydroxyapatite-containing CH-NFs crosslinked with genipin for bone, TE was tested for physical structure and mechanical properties with a consideration that NFs' composite may represent a similar microenvironment with the promotion of osteoblast differentiation and maturation. The resulting NFs showed a gain in diameter from 227 nm to 335 after genipin was added in CH-NFs with remarkable mechanical strength via Young's modulus of 142 Mpa, similar to natural periosteum. Moreover, the expression and enzymatic activity of alkaline phosphates were 2.4-fold higher for genipin-incorporated CH-NFs in cell culture, compared to CH NFs alone, suggesting that crosslinking CH with hydroxyapatite facilitates the proliferation, differentiation, and maturation of osteoblast-like cells [107] (Figure 11).

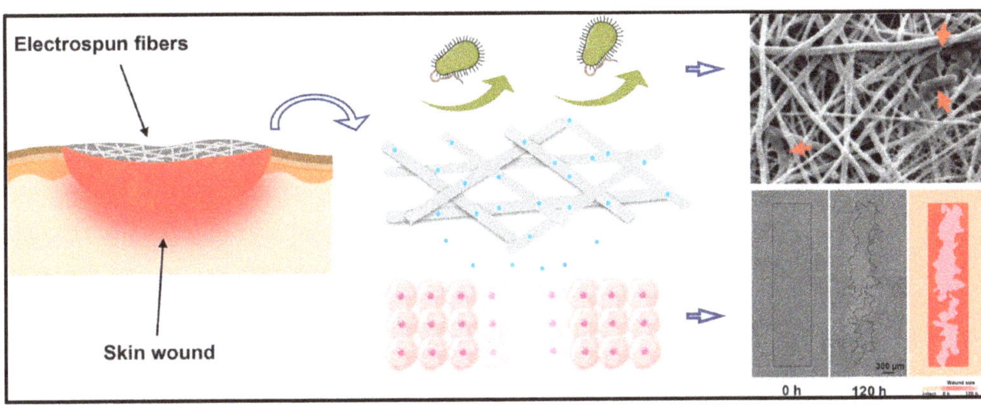

Figure 9. Illustration for NFs and their role in wound healing. Reproduced with permission from [108] under Rightlinks Copyright Clearance Center.

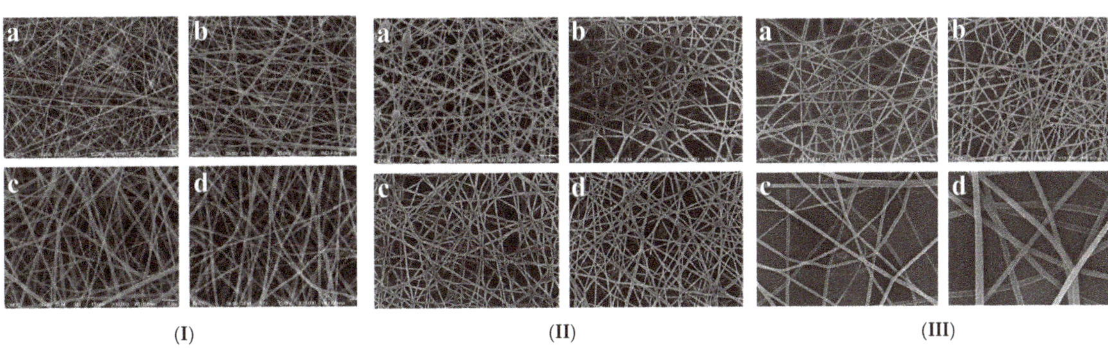

Figure 10. Scan electron microscopy images of PVA, GE, and their blend at ratio of **(I-a)** 0/10 [109], **(I-b)** 2/8, **(I-c)** 6/4, and **(I-d)** 8/2. Scan electron microscopy images of PVA-GE blend at ratio of 2/8 with applied voltages of **(II-a)** 18 kV, **(II-b)** 20 kV, **(II-c)** 22 kV, and **(II-d)** 24 kV at tip–target distance of 10 cm. Scan electron microscopy images of PVA-GE blend at ratio of 2/8 showing variation in diameter distribution with tip–target distance at **(III-a)** 7 cm, **(III-b)** 10 cm, **(III-c)** 15 cm, and **(III-d)** 20 cm at 22 kV. Reproduced with permission from [106] under Rightlinks Copyright Clearance Center.

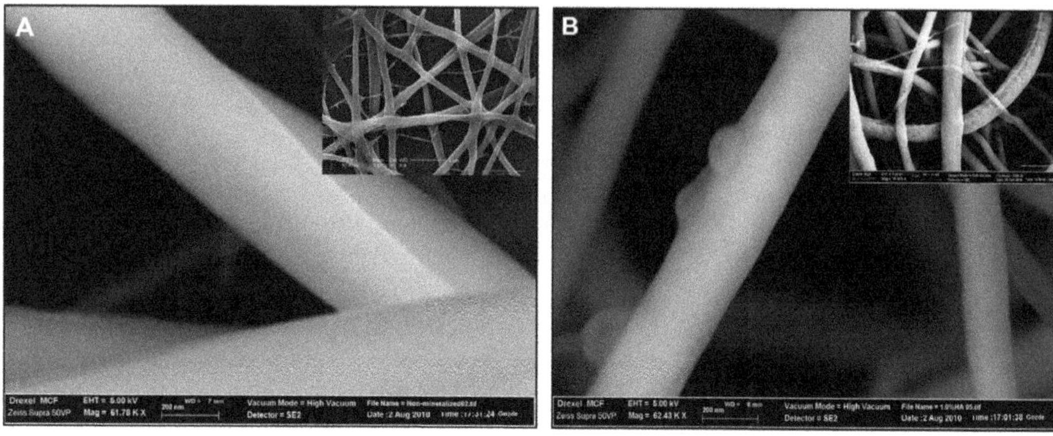

Figure 11. Scan electron image of 0.1% genipin crosslinked with chitosan electrospun scaffold mats: 0.1% genipin crosslinked with 1.0% hydroxyapatite (**A**) and 0.1% genipin crosslinked with 7.0% CH (**B**) at scale bar of 200 nm. Reproduced with permission from [107] under Rightlinks Copyright Clearance Center.

7.3. Central Nervous System-Targeted Drug Delivery

Traumatic and vascular injuries to the central nervous system are major concerns for international health. The delivery of medications into the brain for the treatment of central nervous system (CNS) disorders and their repercussions, including encephalitis, multiple sclerosis, and Alzheimer's disease, has become extremely difficult due to permeable barriers. Current pharmacological therapies are ineffective for treating CNS damage. Therefore, the use of nanocarrier-based drug delivery has become more popular as a novel and promising approach to neurological protection and treatment as a result of developments in nanotechnology and medicine [110]. Electrospinning offers the ability to adjust the micro/nano-architecture regarding fiber diameter, surface porosity, and orientation through process parameter control. Electrospun fiber scaffolds are a biomaterial technique that can direct axonal regeneration through spinal cord injury. When an electrospun fiber scaffold is implanted following spinal cord injury, it will change the phenotype of cells to best support functional recovery and have physical properties that allow for quick, targeted tissue regeneration. The behavior of CNS cells in vitro and in vivo is significantly influenced by fiber diameter, alignment, and density [111]. Dose, control, and site specificity are necessary to develop a revolutionary medication delivery system. These systems have been created using a variety of physical, chemical, mechanical, and biological signals, among which light has been used mainly in the last ten years. Peripheral nerve injury, spinal cord injury, and traumatic brain injury are examples of nervous system injuries that pose serious risks to patient health [112]. Polymeric electrospun materials are adaptable to various treatments, including proteins and small compounds. This makes it possible to implement customized therapies to reduce inflammation and hasten neuro-regeneration at the injury site [113].

Chemotherapeutics can be delivered to the tumor site at a high concentration and with prolonged release via electrospun NFs, reducing systemic exposure and toxicity. Aligned electrospun fibers can replicate the topographical structures of the tumor microenvironment encountered by invasive GBM cells in vitro, which can forecast cell activity in vivo and provide non-invasive treatment methods for malignant gliomas [114]. Bini et al. attempted to develop porous nanofibrous scaffolds using electrospinning. This method forms nanometer-diameter polymer fibers by applying a strong electric field to a stream of polymer fluid. Their study predicted the orientation and proliferation of nerve stem cells on PLGA, suggesting that it is a viable substrate for nerve tissue creation [115]. Yang developed brain and spinal cord neurites, especially those derived from in vitro-cultured neurons, guided in their interaction using two-dimensional nanofibrous scaffolds. They employed 3D nanofibrous scaffolds in brain and spinal cord injury models, both in vitro and in vivo, to create a favorable microenvironment for cell adhesion and contact guidance or structural bridges for axons to govern reconnection. They rebuild glial circuitry over the "lesion gaps" left by CNS injuries and direct neurite outgrowth, two potential uses for nanofibrous scaffolds. They are hypothesized to be composed of biodegradable and biocompatible materials [116]. Electrospun nanofibrous assemblies exhibit distinctive characteristics, including an elevated specific surface area, high porosity, and structural resemblance to the extracellular matrix. Owing to these qualities, they are beneficial in many domains, such as medication release and TE [117].

7.4. Vaginal-Targeted Drug Delivery

Drug-loaded electrospun NFs represent a novel approach to drug delivery via the vaginal mucosa, which may prove advantageous in situations where local drug delivery is necessary. The versatility of these NFs, high loading capacity, high mucoadhesive strength, typical softness, targeted drug delivery in the vagina, decreased toxicity to other organs, and lack of sharp corners that suit lesions with easy-to-use properties contribute to their benefits [118]. The ability of NFs to administer drugs directly to the vaginal mucosa is a noteworthy benefit. This feature is helpful in the treatment of disorders, such as infections, menstrual problems, and vaginal atrophy. By reducing systemic drug requirements, this localized therapy method lowers the risk of systemic adverse effects and improves the therapeutic outcomes. The capacity of NF-based systems to contour the vaginal membrane and provide effective action is another benefit [119]. Vidyavaridhi et al. fabricated luliconazole-loaded mucoadhesive PCL-GE NFs for anticandidal activity. The findings showed that the Fickian release mechanism allowed the release of luliconazole from LCZ-loaded NFs. When luliconazole was entrapped in NFs, as opposed to a drug suspension, penetration in the vaginal mucosa increased. Anticandidal action was demonstrated by in silico activity and microbiological assays [120]. Nematpur et al. fabricated clotrimazole-loaded NF films for vaginitis, and comparative studies were performed to assess their antifungal activity. This finding reveals that clotrimazole-loaded NFs exhibit higher antifungal activity than films [121]. Ball et al. developed a hollow tube-shaped NF suitable for a vaginal tampon applicator using a two-axis mandrel electrospinning rig for fiber collection [122]. Ge-PCL smooth NFs (Figure 12) fortified with *Myrtle berries*, which are a plethora of bioactive reservoirs and exhibit antioxidant, antiseptic, and anti-inflammatory activity, tested against a series of microbial flora showed significant capability in counteracting that microbial proliferation, suggesting the potential of *Myrtle berries* for future vaginal infection management [123].

The human immunodeficiency virus (HIV) still affects the health of millions of people worldwide despite numerous advancements in modern medicine, and significant efforts have been made to develop strategies to either prevent infection or remove the virus after it has already occurred. Antiviral medications are easily incorporated within electrospun cellulose acetate phthalate fibers during the electrospinning process and can be used as a preventative measure against HIV [124]. In a small-scale study, Agrawal et al. examined the therapeutic effectiveness of electrospun NFs loaded with cisplatin and a GE composite for the local treatment of cervical tumors in mice released for up to one month. When compared to ordinary medication, the prepared NFs demonstrated a lower percentage of cell survival, stronger mucoadhesive strength, and superior antitumor activity [125]. Herpes simplex virus infections caused by HSV-1 (oral herpes) and HSV-2 (genital herpes) were examined against electrospun NFs. This was achieved by encapsulating acyclovir in polymeric NFs for controlled-release delivery via intravaginal delivery [126]. The production of NFs with desired shapes provides an advantage for easy vaginal administration [127].

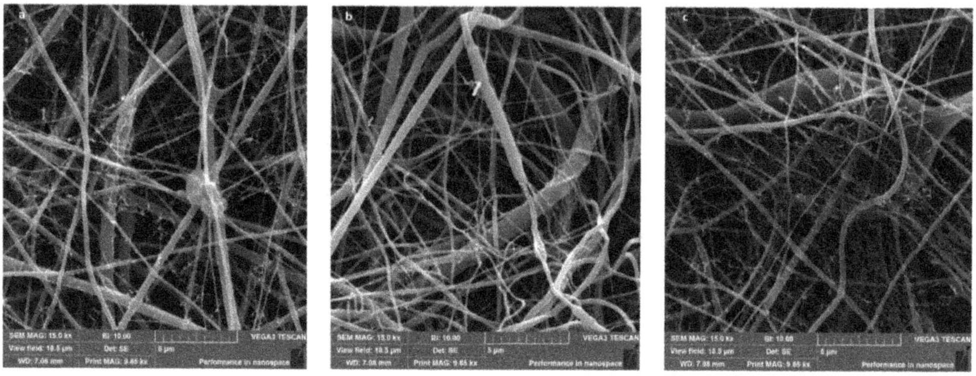

Figure 12. Scan electron microscopy image of Ge-PCL NFs fortified with *Myrtle berries* fruit (**a**), seeds (**b**), and leaves (**c**). Reproduced with permission from [123] under CCBY-4.0.

7.5. Ocular-Targeted Drug Delivery

Since their inception as DDS, NFs have been utilized for ocular drug targeting. The materials employed in their fabrication enhance drug penetration and prolong the contact time with ocular tissues, demonstrating excellent biocompatibility. Furthermore, their extended-release profile decreases the dosing frequency [127]. Researchers have demonstrated the effective utilization of NF-incorporated drugs for the treatment of various ocular diseases and disorders (Figure 13). Cejkova et al. developed and utilized cyclosporine A NFs for treating alkali-injured corneas and demonstrated superior efficacy in suppressing inflammation and neovascularization compared with eye drops. The prolonged contact time of drug-loaded NFs with the injured eye has been suggested to be a key factor for their enhanced effectiveness [128]. Similarly, Gagandeep et al. reported the treatment of glaucoma with NFs of timolol maleate and dorzolamide hydrochloride, which showed a significant reduction in intraocular pressure compared to commercial eye drops [129]. In another study, Hsu et al. developed polycaprolactone NFs and microbeads for the extended release of dexamethasone, aiming to reduce the frequency of administration and associated side effects [130]. These studies illustrate the potential of NF-based DDS to improve the treatment outcomes for various eye disorders. Mirzaeei and co-workers developed modified electrospun-based NFs for enhanced ocular residence of ofloxacin using a CH/PVA blend as a hydrophilic composition and Eudragit RL100/PVA crosslinked using glutaraldehyde for the management of infectious conjunctivitis. The average diameter obtained for both types of NFs was around 123 nm and 159 nm when NFs were fortified with ofloxacin. The release content of ofloxacin from NFs was tested for antimicrobial potency against *Staphylococcus aureus* and *Escherichia coli*, demonstrating good activity. Moreover, in vivo studies suggested that NF inserts released the drug for 96 h with $AUC_{0\text{-}96}$ of 9–20-fold, which was higher compared to that of the alone ofloxacin drug solution, suggesting the overall utility of NFs in ocular DDS (Figure 14) [131]. In another study, electrospun NFs of Ge crosslinked with hyaluronic acid through vapor of glutaraldehyde were fabricated for delivering human corneal mesenchymal stromal cells for the management of deep corneal injuries. The developed NFs exhibited excellent mechanics and transmittance with the improved proliferation of corneal mesenchymal stromal cells through a scratch migration assay after 48 h. Moreover, corneal wound healing and stromal α-smooth muscle actin lowered expression compared to those untreated keratectomy and corneal mesenchymal stromal cell groups, suggesting the therapeutic potential of NFs for the stromal regeneration and management of deep corneal defects [132].

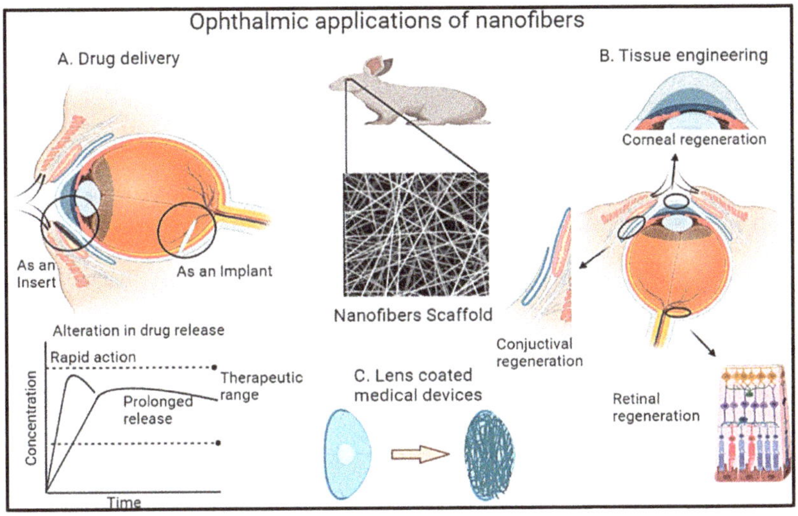

Figure 13. Illustration of ophthalmic applications of NFs.

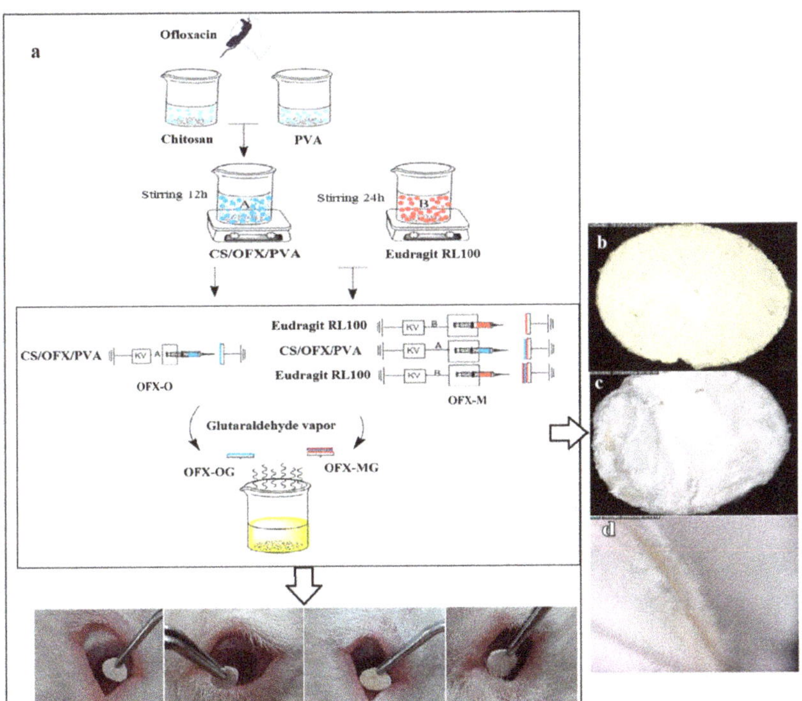

Figure 14. Illustration demonstrating fabrication of modified electrospun-based NFs (**a**) for enhanced ocular residence of ofloxacin using CH/PVA blends (**b**–**d**) as hydrophilic composition and Eudragit RL100/PVA crosslinked using glutaraldehyde for management of infectious conjunctivitis. Reproduced with permission from [131] under CCBY-4.0.

8. Challenges and Key Considerations in the Fabrication of Nanofibers

Below are important factors affecting the shape and structure of fibers in the context of biopolymer-based nanofibers, as well as information on how they are prepared and manufactured additively.

Properties of the polymer solution: The viscosity, concentration, solvent type, and molecular weight of the polymer solution have a significant impact on the morphology of the fiber. Thicker fibers are often produced at higher polymer concentrations and viscosities, whereas finer fibers are produced at lower concentrations. Fiber diameter and bead formation are influenced by solvent volatility; smoother fibers usually result from slower evaporation rates [133].

Electric field parameters: In electrospinning, variables such as the voltage, flow rate, and separation between the spinneret and collector are essential. Higher flow rates result in bead production, whereas higher voltages produce more stretching and thinner fibers. The fiber diameter and alignment are influenced by the distance between the spinneret and collector; closer distances result in better-aligned fibers [134].

Environmental conditions: Temperature and humidity affect fiber production. Low humidity can result in thicker fibers and quicker solvent evaporation, whereas high humidity can cause beads to develop and reduce the fiber diameter. Temperature affects the solvent evaporation rates and viscosity of the polymer solutions, which in turn affects the fiber shape [135].

Additives and crosslinking agents: The characteristics of the fiber can be altered by the addition of additives or crosslinking agents to the polymer solution. While crosslinking agents increase fiber stability and resistance to degradation, additives, such as surfactants or NPs, can improve mechanical qualities or introduce functions [136].

Biopolymer-based nanofibers: Biopolymers have adjustable characteristics, biocompatibility, and biodegradability, making them advantageous for therapeutic applications. Among their many applications are biosensors, scaffolds for TE, wound dressings, and DDS. Utilizing the distinct characteristics of biopolymers, including GE, collagen, alginate, and cellulose, scientists may create customized NF platforms that target certain therapeutic requirements [137].

9. Regulatory Consideration and Patents

The commercial and regulatory landscapes surrounding electrospinning have expanded in recent years. The relevance of electrospinning depends on the composition of the NFs. Many studies have been conducted on crosslinking treatments and the use of toxic solvents that are not FDA-approved. Operating conditions have encouraged its use, and it is a cutting-edge technology used in several industries, including the food and pharmaceutical industries. The primary issue is the scarcity of solvents that can dissolve both manufactured and natural polymers. The low volatility of certain solvents employed in the production of multiaxial fibers is a concern, and the slow production rate of 0.0010.1 g/h per spinneret in large-scale electrospinning is another issue [138]. International directives such as those established by the Food and Agriculture Organization (FAO) and the United Nations Environment Program (UNEP), which linked strategic principles with technological advancements in NF technology by incorporating the production of "green," environmentally friendly NFs using an electrospinning method, facilitated the formulation of national and regional policies [139]. Currently, there are few formal regulatory guidance documents and unclear regulations regarding nanomedicines. A few factors that could be significant from a regulatory standpoint should be considered in the early stages of research when creating NFs as a possible medical device. The materials utilized must first meet the safety regulations. Selecting a polymer that is deemed safe and has received official approval from authorities is advised. Validation can be used to show that the electrospinning procedure and apparatus are appropriate for accurately controlled and repeatable production. Nozzle-based electrospinning is superior to nozzle-free electrospinning in this respect because the latter produces a non-uniform fiber diameter owing to simultaneous jets [139]. Polymers that are both biocompatible and biodegradable are typically preferred. Using electrospinning to innovate an existing approved product could be a wise move. To close endonasal surgical defects, Campbell et al. synthesized NFs from an FDA-approved cyanoacrylate polymer and compared them with Adherus®, an FDA-approved common dural sealant [140]. In general, medical devices are governed by a clearly defined regulatory framework, but the regulation of those devices that contain nanomaterials is not. It is generally accepted that medical devices that employ nanomaterials should be categorized as having a higher risk. Therefore, caution should be exercised. While the EU's directives and applicable ISO standards are undoubtedly helpful starting points, interacting with the authorities on particular matters may also be beneficial. Following this classification, adherence to class norms is required [141]. Nonetheless, owing to its potential for use in biomedical and other nanotechnical applications, it has recently garnered considerable attention. The alteration of the electrospinning system configurations and the impact of the process parameters on the fibers, as well as how they are used for local chemotherapy, describe their use in drug administration, including carrier materials, loaded medicines, and their release kinetics. Currently, most research on the release of anti-bacterial agents and medications (such as psychotropic and antineoplastic agents) has been conducted in vitro. Comprehensive systemic investigations are required prior to considering any clinical commercialization in vivo, particularly those pertaining to the kinetics and dynamics of drug release, the impact of drug dosage and release kinetics on therapeutic efficacy, and the biodistribution of the released pharmaceuticals. Extensive research is necessary to determine the hazardous impact of polymeric carriers, as well as their distribution and removal mechanisms.

10. Conclusions and Future Perspective

The utilization of GE/CH-based NFs in drug delivery holds great significance for advancing biomedical applications. These biopolymers offer unique advantages including biocompatibility, biodegradability, and the ability to mimic the extracellular matrix, making them ideal candidates for tissue regeneration, wound healing, and drug delivery. Among other methods, electrospinning has proven to be effective in fabricating NFs with controlled morphology, enabling precise wound healing properties coupled with GE/CH cell adhesion and regenerative properties. However, challenges such as the optimization of fiber structures, drug-loading efficiency, and stability still remain. Despite these challenges, the potential of GE/CH-based NFs to improve drug delivery, minimize adverse effects, and enhance patient outcomes continues to drive research and innovation in this field. Furthermore, advancements in the widespread application of CH/Ge-based NFs in diverse medical fields may ultimately contribute to more effective biomedical applications. The CH/Ge-based NFs indicate the aforementioned widespread applicability due to their biocompatibility and biodegradability in nature, which ultimately makes them the most applied polymeric materials, compared to synthetic polymers.

The future of CH/Ge-based NFs in drug delivery holds great promise, particularly in enhancing targeted therapy for various complications. Advances in functionalization techniques, such as incorporating specific targeting ligands and stimuli-responsive elements, can improve the precision and control of drug release. The further optimization of drug-loading capacity and release profiles is crucial to enhance their therapeutic efficacy. Ensuring biocompatibility and safety through comprehensive studies will be vital for clinical application. Additionally, the potential for multi-drug delivery systems could offer innovative solutions for complex diseases. Scalability and cost-effective production methods will be key to making these systems commercially viable, while the combination with other nanomaterials may further enhance their therapeutic potential. Ultimately, successful clinical translation and regulatory approval will be pivotal in realizing the full potential of these nanofibers for managing a range of health complications.

Author Contributions: Conceptualization, P.M., S.S. and C.C.; methodology, A.P.; software, A.P.; validation, P.T., S.S., P.M. and C.C.; formal analysis, P.T., A.K.S. and C.C.; investigation, S.M., R.D., S.K. and R.P.; resources, P.M.; data curation, S.M., R.D., S.K. and R.P.; writing—original draft preparation, S.M., R.D., S.K. and R.P.; writing—review and editing, S.S., P.M. and C.C.; visualization, A.K.S.; supervision, C.C. and P.M.; project administration, C.C. All authors have read and agreed to the published version of the manuscript.

Funding: This research received no external funding.

Institutional Review Board Statement: Not applicable.

Data Availability Statement: Data can be made available by corresponding authors upon request.

Acknowledgments: This work was partially supported by the CMU Proactive Researcher Scheme (2023), Chiang Mai University, for Sudarshan Singh.

Conflicts of Interest: The authors declare no conflicts of interest.

References

1. Jain, K.K. Drug Delivery Systems—An Overview. In *Drug Delivery Systems*; Jain, K.K., Ed.; Humana Press: Totowa, NJ, USA, 2008; pp. 1–50.
2. Ezike, T.C.; Okpala, U.S.; Onoja, U.L.; Nwike, C.P.; Ezeako, E.C.; Okpara, O.J.; Okoroafor, C.C.; Eze, S.C.; Kalu, O.L.; Odoh, E.C.; et al. Advances in drug delivery systems, challenges and future directions. *Heliyon* **2023**, *9*, e17488. [CrossRef]
3. Jacob, J.; Haponiuk, J.T.; Thomas, S.; Gopi, S. Biopolymer based nanomaterials in drug delivery systems: A review. *Mater. Today Chem.* **2018**, *9*, 43–55. [CrossRef]
4. Sahu, T.; Ratre, Y.K.; Chauhan, S.; Bhaskar, L.V.K.S.; Nair, M.P.; Verma, H.K. Nanotechnology based drug delivery system: Current strategies and emerging therapeutic potential for medical science. *J. Drug Deliv. Sci. Technol.* **2021**, *63*, 102487. [CrossRef]
5. Guleria, M.; Malhan, A.; Singh, S.; Chidrawar, V.R.; Nagime, P.V. Navigating the Potential of Natural Products through Nano-enabled Drug Deliveries in Mitigation of Cancer: A Review. *Nano Life* **2025**, *15*, 2430006. [CrossRef]
6. Malhan, A.; Guleria, M.; Das, U.; Majee, S.B.; Singh, S.; Nagime, P.V.; Syukri, D.M.; Chidrawar, V.R.; Arania, R.; Shafi, S. Navigating the future of infectious diseases management through carbon nanotubes: A review. *Nano LIFE* **2025**, 2530001. [CrossRef]
7. Nagime, P.V.; Shaikh, N.M.; Shaikh, S.B.; Lokhande, C.D.; Patil, V.V.; Shafi, S.; Syukri, D.M.; Chidrawar, V.R.; Kumar, A.; Singh, S. Facile synthesis of silver nanoparticles using Calotropis procera leaves: Unraveling biological and electrochemical potentials. *Discov. Nano* **2024**, *19*, 139. [CrossRef]
8. Singh, S.; Syukri, D.M.; Ushir, Y.V.; Mishra, A.; Ontong, J.C.; Nwabor, O.F.; Darekar, S.M.; Samee, W.; Chidrawar, V.R.; Chittasupho, C. Post-operative Wound Healing Efficacy of Eucalyptus Camaldulensis Phenolic-rich Extracts Incorporated Hydrogel With Enhanced Antioxidant, Antibacterial, and Anti-inflammatory Activities. *J. Polym. Environ.* **2024**, *33*, 814–839. [CrossRef]
9. Eze, F.N.; Eze, R.C.; Singh, S.; Okpara, K.E. Fabrication of a versatile and efficient ultraviolet blocking biodegradable composite film consisting of Tara gum/PVA/Riceberry phenolics reinforced with biogenic riceberry phenolic-rich extract-nano-silver. *Int. J. Biol. Macromol.* **2024**, *278*, 134914. [CrossRef]
10. Ontong, J.C.; Singh, S.; Siriyong, T.; Voravuthikunchai, S.P. Transferosomes stabilized hydrogel incorporated rhodomyrtone-rich extract from Rhodomyrtus tomentosa leaf fortified with phosphatidylcholine for the management of skin and soft-tissue infections. *Biotechnol. Lett.* **2024**, *46*, 127–142. [CrossRef]
11. Singh, S.; Nwabor, O.F.; Sukri, D.M.; Wunnoo, S.; Dumjun, K.; Lethongkam, S.; Kusolphat, P.; Hemtanon, N.; Klinprathum, K.; Sunghan, J.; et al. Poly (vinyl alcohol) copolymerized with xanthan gum/hypromellose/sodium carboxymethyl cellulose dermal dressings functionalized with biogenic nanostructured materials for antibacterial and wound healing application. *Int. J. Biol. Macromol.* **2022**, *216*, 235–250. [CrossRef]
12. Datta, D.; Prajapati, B.; Jethva, H.; Agrawal, K.; Singh, S.; Prajapati, B.G. Value-Added Nanocellulose Valorized from Fruit Peel Waste for Potential Dermal Wound Healing and Tissue Regenerative Applications. *Regen. Eng. Transl. Med.* **2024**. [CrossRef]

13. Singh, S.A.; Vellapandian, C.; Shah, D.D.; Jayeoye, T.J.; Chorawala, M.R.; Singh, S.; Prajapati, B.G. Valorised Calcium-Rich Biomass from Fish Waste and Eggshells in the Fabrication of Antibacterial Scaffold for Wound Healing Applications: A Review. *Waste Biomass Valorization* **2024**, *15*, 1917–1941. [CrossRef]
14. Basu, B.; Dutta, A.; Ash, D.; Garala, K.; Singh, S.; Prajapati, B.G. Recent Advancements in the Diagnosis and Management of Cancer Using Biomaterials-Fabricated Nanofibers: A Review. *Curr. Med. Chem.* **2024**, *31*, 1–25. [CrossRef]
15. Leung, V.; Ko, F. Biomedical applications of nanofibers. *Polym. Adv. Technol.* **2011**, *22*, 350–365. [CrossRef]
16. Lasenko, I.; Sanchaniya, J.V.; Kanukuntla, S.P.; Ladani, Y.; Viluma-Gudmona, A.; Kononova, O.; Lusis, V.; Tipans, I.; Selga, T. The Mechanical Properties of Nanocomposites Reinforced with PA6 Electrospun Nanofibers. *Polymers* **2023**, *15*, 673. [CrossRef] [PubMed]
17. Zhou, J.; Li, X.; Zhang, Z.; Hou, T.; Xu, J.; Wang, Y.; Ye, H.; Yang, B. Bio-based and bio-degradable nanofiber materials: A sustainable platform for energy, environmental, and biomedical applications. *Chem. Eng. J.* **2024**, *491*, 152105. [CrossRef]
18. Rajinikanth, B.S.; Rajkumar, D.S.R.; K, K.; Vijayaragavan, V. Chitosan-Based Biomaterial in Wound Healing: A Review. *Cureus* **2024**, *16*, e55193. [CrossRef]
19. Mathaba, M.; Daramola, M.O. Effect of Chitosan's Degree of Deacetylation on the Performance of PES Membrane Infused with Chitosan during AMD Treatment. *Membranes* **2020**, *10*, 52. [CrossRef] [PubMed]
20. Naomi, R.; Bahari, H.; Ridzuan, P.M.; Othman, F. Natural-Based Biomaterial for Skin Wound Healing (Gelatin vs. Collagen): Expert Review. *Polymers* **2021**, *13*, 2319. [CrossRef] [PubMed]
21. Revete, A.; Aparicio, A.; Cisterna, B.A.; Revete, J.; Luis, L.; Ibarra, E.; Segura González, E.A.; Molino, J.; Reginensi, D. Advancements in the Use of Hydrogels for Regenerative Medicine: Properties and Biomedical Applications. *Int. J. Biomater.* **2022**, *2022*, 3606765. [CrossRef]
22. Vasita, R.; Katti, D.S. Nanofibers and their applications in tissue engineering. *Int. J. Nanomed.* **2006**, *1*, 15–30. [CrossRef]
23. Ndlovu, S.P.; Ngece, K.; Alven, S.; Aderibigbe, B.A. Gelatin-Based Hybrid Scaffolds: Promising Wound Dressings. *Polymers* **2021**, *13*, 2959. [CrossRef] [PubMed]
24. Dhandayuthapani, B.; Krishnan, U.M.; Sethuraman, S. Fabrication and characterization of chitosan-gelatin blend nanofibers for skin tissue engineering. *J. Biomed. Mater. Res. Part B Appl. Biomater.* **2010**, *94B*, 264–272. [CrossRef] [PubMed]
25. Barhoum, A.; Bechelany, M.; Hamdy Makhlouf, A.S. *Handbook of Nanofibers*; Springer: Berlin/Heidelberg, Germany, 2019.
26. Kalayil, N.; Budar, A.A.; Dave, R.K. Nanofibers for Drug Delivery: Design and Fabrication Strategies. *BIO Integr.* **2024**, *5*, 978. [CrossRef]
27. Hiwrale, A.; Bharati, S.; Pingale, P.; Rajput, A. Nanofibers: A current era in drug delivery system. *Heliyon* **2023**, *9*, e18917. [CrossRef]
28. Anusiya, G.; Jaiganesh, R. A review on fabrication methods of nanofibers and a special focus on application of cellulose nanofibers. *Carbohydr. Polym. Technol. Appl.* **2022**, *4*, 100262. [CrossRef]
29. Spizzirri, U.G.; Aiello, F.; Carullo, G.; Facente, A.; Restuccia, D. Nanotechnologies: An Innovative Tool to Release Natural Extracts with Antimicrobial Properties. *Pharmaceutics* **2021**, *13*, 230. [CrossRef]
30. Wen, P.; Zhu, D.-H.; Feng, K.; Liu, F.-J.; Lou, W.-Y.; Li, N.; Zong, M.-H.; Wu, H. Fabrication of electrospun polylactic acid nanofilm incorporating cinnamon essential oil/β-cyclodextrin inclusion complex for antimicrobial packaging. *Food Chem.* **2016**, *196*, 996–1004. [CrossRef] [PubMed]
31. Li, H.; Williams, G.R.; Wu, J.; Lv, Y.; Sun, X.; Wu, H.; Zhu, L.-M. Thermosensitive nanofibers loaded with ciprofloxacin as antibacterial wound dressing materials. *Int. J. Pharm.* **2017**, *517*, 135–147. [CrossRef] [PubMed]
32. Chamanehpour, E.; Thouti, S.; Rubahn, H.-G.; Dolatshahi-Pirouz, A.; Mishra, Y.K. Smart Nanofibers: Synthesis, Properties, and Scopes in Future Advanced Technologies. *Adv. Mater. Technol.* **2024**, *9*, 2301392. [CrossRef]
33. Zhang, X.; Lu, Y. Centrifugal Spinning: An Alternative Approach to Fabricate Nanofibers at High Speed and Low Cost. *Polym. Rev.* **2014**, *54*, 677–701. [CrossRef]
34. Zhao, Y.; Qiu, Y.; Wang, H.; Chen, Y.; Jin, S.; Chen, S. Preparation of Nanofibers with Renewable Polymers and Their Application in Wound Dressing. *Int. J. Polym. Sci.* **2016**, *2016*, 4672839. [CrossRef]
35. Luo, C.J.; Stoyanov, S.D.; Stride, E.; Pelan, E.; Edirisinghe, M. Electrospinning versus fibre production methods: From specifics to technological convergence. *Chem. Soc. Rev.* **2012**, *41*, 4708–4735. [CrossRef]
36. Xue, J.; Wu, T.; Dai, Y.; Xia, Y. Electrospinning and Electrospun Nanofibers: Methods, Materials, and Applications. *Chem. Rev.* **2019**, *119*, 5298–5415. [CrossRef]
37. Ibrahim, M.A.; Alhalafi, M.H.; Emam, E.M.; Ibrahim, H.; Mosaad, R.M. A Review of Chitosan and Chitosan Nanofiber: Preparation, Characterization, and Its Potential Applications. *Polymers* **2023**, *15*, 2820. [CrossRef]
38. Nwabor, O.F.; Singh, S.; Paosen, S.; Vongkamjan, K.; Voravuthikunchai, S.P. Enhancement of food shelf life with polyvinyl alcohol-chitosan nanocomposite films from bioactive Eucalyptus leaf extracts. *Food Biosci.* **2020**, *36*, 100609. [CrossRef]

39. Singh, S.; Nwabor, O.F.; Syukri, D.M.; Voravuthikunchai, S.P. Chitosan-poly(vinyl alcohol) intelligent films fortified with anthocyanins isolated from Clitoria ternatea and Carissa carandas for monitoring beverage freshness. *Int. J. Biol. Macromol.* **2021**, *182*, 1015–1025. [CrossRef] [PubMed]
40. Singh, S.; Supaweera, N.; Nwabor, O.F.; Chaichompoo, W.; Suksamrarn, A.; Chittasupho, C.; Chunglok, W. Poly (vinyl alcohol)-gelatin-sericin copolymerized film fortified with vesicle-entrapped demethoxycurcumin/bisdemethoxycurcumin for improved stability, antibacterial, anti-inflammatory, and skin tissue regeneration. *Int. J. Biol. Macromol.* **2024**, *258*, 129071. [CrossRef] [PubMed]
41. Ahmadi Bonakdar, M.; Rodrigue, D. Electrospinning: Processes, Structures, and Materials. *Macromol* **2024**, *4*, 58–103. [CrossRef]
42. Tamilarasi, G.P.; Sabarees, G.; Manikandan, K.; Gouthaman, S.; Alagarsamy, V.; Solomon, V.R. Advances in electrospun chitosan nanofiber biomaterials for biomedical applications. *Mater. Adv.* **2023**, *4*, 3114–3139. [CrossRef]
43. Denis, P.; Wrzecionek, M.; Gadomska-Gajadhur, A.; Sajkiewicz, P. Poly(Glycerol Sebacate)–Poly(l-Lactide) Nonwovens. Towards Attractive Electrospun Material for Tissue Engineering. *Polymers* **2019**, *11*, 2113. [CrossRef]
44. Singh, S.; Supaweera, N.; Nwabor, O.F.; Yusakul, G.; Chaichompoo, W.; Suksamrarn, A.; Panpipat, W.; Chunglok, W. Polymeric scaffold integrated with nanovesicle-entrapped curcuminoids for enhanced therapeutic efficacy. *Nanomedicine* **2024**, *19*, 1313–1329. [CrossRef]
45. El-Seedi, H.R.; Said, N.S.; Yosri, N.; Hawash, H.B.; El-Sherif, D.M.; Abouzid, M.; Abdel-Daim, M.M.; Yaseen, M.; Omar, H.; Shou, Q.; et al. Gelatin nanofibers: Recent insights in synthesis, bio-medical applications and limitations. *Heliyon* **2023**, *9*, e16228. [CrossRef]
46. Huang, Z.-M.; Zhang, Y.Z.; Ramakrishna, S.; Lim, C.T. Electrospinning and mechanical characterization of gelatin nanofibers. *Polymer* **2004**, *45*, 5361–5368. [CrossRef]
47. Baig, A.; Barse, R.; Paryekar, A.; Jagtap, V. Evolution of pharmaceuticals using 3D and 4D printing. *Intell. Pharm.* **2024**, *2*, 804–813. [CrossRef]
48. Mahmood, A.; Akram, T.; Shenggui, C.; Chen, H. Revolutionizing manufacturing: A review of 4D printing materials, stimuli, and cutting-edge applications. *Compos. Part B Eng.* **2023**, *266*, 110952. [CrossRef]
49. Hu, J.; Zhang, S.; Tang, B. Rational design of nanomaterials for high energy density dielectric capacitors via electrospinning. *Energy Storage Mater.* **2021**, *37*, 530–555. [CrossRef]
50. Truby, R.L.; Lewis, J.A. Printing soft matter in three dimensions. *Nature* **2016**, *540*, 371–378. [CrossRef]
51. Tibbits, S. 4D Printing: Multi-Material Shape Change. *Archit. Des.* **2014**, *84*, 116–121. [CrossRef]
52. Radacsi, N.; Nuansing, W. Chapter 7—Fabrication of 3D and 4D polymer micro- and nanostructures based on electrospinning. In *3D and 4D Printing of Polymer Nanocomposite Materials*; Sadasivuni, K.K., Deshmukh, K., Almaadeed, M.A., Eds.; Elsevier: Amsterdam, The Netherlands, 2020; pp. 191–229.
53. Abadi, B.; Goshtasbi, N.; Bolourian, S.; Tahsili, J.; Adeli-Sardou, M.; Forootanfar, H. Electrospun hybrid nanofibers: Fabrication, characterization, and biomedical applications. *Front. Bioeng. Biotechnol.* **2022**, *10*, 986975. [CrossRef]
54. Park, Y.R.; Sultan, M.T.; Park, H.J.; Lee, J.M.; Ju, H.W.; Lee, O.J.; Lee, D.J.; Kaplan, D.L.; Park, C.H. NF-κB signaling is key in the wound healing processes of silk fibroin. *Acta Biomater.* **2018**, *67*, 183–195. [CrossRef]
55. Salunke, M.; Viswalingam, V.; Shinde, V. Electrospun nanofibers of carbohydrate polymers for neoteric wound dressing: A review. *Int. J. Polym. Mater. Polym. Biomater.* **2024**, *74*, 103–127. [CrossRef]
56. Li, T.; Sun, M.; Wu, S. State-of-the-Art Review of Electrospun Gelatin-Based Nanofiber Dressings for Wound Healing Applications. *Nanomaterials* **2022**, *12*, 784. [CrossRef] [PubMed]
57. Amirrah, I.N.; Lokanathan, Y.; Zulkiflee, I.; Wee, M.; Motta, A.; Fauzi, M.B. A Comprehensive Review on Collagen Type I Development of Biomaterials for Tissue Engineering: From Biosynthesis to Bioscaffold. *Biomedicines* **2022**, *10*, 2307. [CrossRef] [PubMed]
58. Shoulders, M.D.; Raines, R.T. Collagen structure and stability. *Annu. Rev. Biochem.* **2009**, *78*, 929–958. [CrossRef] [PubMed]
59. Boyle, A.L. 3—Applications of de novo designed peptides. In *Peptide Applications in Biomedicine, Biotechnology and Bioengineering*; Koutsopoulos, S., Ed.; Woodhead Publishing: Sawston, UK, 2018; pp. 51–86.
60. Sabarees, G.; Vishvaja, S.; Raghuraman, S.; Velmurugan, V.; Alagarsamy, V.; Raja Solomon, V.; Padmini Tamilarasi, G. Collagen-based nanofibers: Revolutionizing therapeutics for impaired wound healing. *Int. J. Polym. Mater. Polym. Biomater.* **2024**, 1–29. [CrossRef]
61. Mohite, P.; Rahayu, P.; Munde, S.; Ade, N.; Chidrawar, V.R.; Singh, S.; Jayeoye, T.J.; Prajapati, B.G.; Bhattacharya, S.; Patel, R.J. Chitosan-Based Hydrogel in the Management of Dermal Infections: A Review. *Gels* **2023**, *9*, 594. [CrossRef] [PubMed]
62. Mohite, P.; Shah, S.R.; Singh, S.; Rajput, T.; Munde, S.; Ade, N.; Prajapati, B.G.; Paliwal, H.; Mori, D.D.; Dudhrejiya, A.V. Chitosan and chito-oligosaccharide: A versatile biopolymer with endless grafting possibilities for multifarious applications. *Front. Bioeng. Biotechnol.* **2023**, *11*, 1190879. [CrossRef]

63. Huang, B.; Xiao, D.; Tan, B.; Xiao, H.; Wang, J.; Yin, J.; Duan, J.; Huang, R.; Yang, C.; Yin, Y. Chitosan Oligosaccharide Reduces Intestinal Inflammation That Involves Calcium-Sensing Receptor (CaSR) Activation in Lipopolysaccharide (LPS)-Challenged Piglets. *J. Agric. Food Chem.* **2016**, *64*, 245–252. [CrossRef]
64. Senthil Muthu Kumar, T.; Senthil Kumar, K.; Rajini, N.; Siengchin, S.; Ayrilmis, N.; Varada Rajulu, A. A comprehensive review of electrospun nanofibers: Food and packaging perspective. *Compos. Part B Eng.* **2019**, *175*, 107074. [CrossRef]
65. Rather, J.A.; Akhter, N.; Ashraf, Q.S.; Mir, S.A.; Makroo, H.A.; Majid, D.; Barba, F.J.; Khaneghah, A.M.; Dar, B.N. A comprehensive review on gelatin: Understanding impact of the sources, extraction methods, and modifications on potential packaging applications. *Food Packag. Shelf Life* **2022**, *34*, 100945. [CrossRef]
66. Pellis, A.; Guebitz, G.M.; Nyanhongo, G.S. Chitosan: Sources, Processing and Modification Techniques. *Gels* **2022**, *8*, 393. [CrossRef]
67. Chi, H.Y.; Chang, N.Y.; Li, C.; Chan, V.; Hsieh, J.H.; Tsai, Y.-H.; Lin, T. Fabrication of Gelatin Nanofibers by Electrospinning—Mixture of Gelatin and Polyvinyl Alcohol. *Polymers* **2022**, *14*, 2610. [CrossRef] [PubMed]
68. Cui, C.; Sun, S.; Wu, S.; Chen, S.; Ma, J.; Zhou, F. Electrospun chitosan nanofibers for wound healing application. *Eng. Regen.* **2021**, *2*, 82–90. [CrossRef]
69. Cai, Z.X.; Mo, X.M.; Zhang, K.H.; Fan, L.P.; Yin, A.L.; He, C.L.; Wang, H.S. Fabrication of chitosan/silk fibroin composite nanofibers for wound-dressing applications. *Int. J. Mol. Sci.* **2010**, *11*, 3529–3539. [CrossRef] [PubMed]
70. Ki, C.S.; Baek, D.H.; Gang, K.D.; Lee, K.H.; Um, I.C.; Park, Y.H. Characterization of gelatin nanofiber prepared from gelatin–formic acid solution. *Polymer* **2005**, *46*, 5094–5102. [CrossRef]
71. Khanzada, H.; Kumpikaitė, E. Anti-bacterial nanofibers and their biomedical applications—A review. *J. Text. Inst.* **2024**, *116*, 309–327. [CrossRef]
72. Alotaibi, B.S.; Khan, A.K.; Kharaba, Z.; Yasin, H.; Yasmin, R.; Ijaz, M.; Khan, M.; Murtaza, G. Development of Poly(vinyl alcohol)–Chitosan Composite Nanofibers for Dual Drug Therapy of Wounds. *ACS Omega* **2024**, *9*, 12825–12834. [CrossRef] [PubMed]
73. Alyamani, A.A.; Al-Musawi, M.H.; Albukhaty, S.; Sulaiman, G.M.; Ibrahim, K.M.; Ahmed, E.M.; Jabir, M.S.; Al-Karagoly, H.; Aljahmany, A.A.; Mohammed, M.K.A. Electrospun Polycaprolactone/Chitosan Nanofibers Containing *Cordia myxa* Fruit Extract as Potential Biocompatible Antibacterial Wound Dressings. *Molecules* **2023**, *28*, 2501. [CrossRef] [PubMed]
74. Gulzar, S.; Tagrida, M.; Nilsuwan, K.; Prodpran, T.; Benjakul, S. Electrospinning of gelatin/chitosan nanofibers incorporated with tannic acid and chitooligosaccharides on polylactic acid film: Characteristics and bioactivities. *Food Hydrocoll.* **2022**, *133*, 107916. [CrossRef]
75. Arkoun, M.; Daigle, F.; Heuzey, M.C.; Ajji, A. Antibacterial electrospun chitosan-based nanofibers: A bacterial membrane perforator. *Food Sci. Nutr.* **2017**, *5*, 865–874. [CrossRef] [PubMed]
76. Zahmatkeshan, M.; Adel, M.; Bahrami, S.; Esmaeili, F.; Rezayat, S.M.; Saeedi, Y.; Mehravi, B.; Jameie, S.B.; Ashtari, K. Polymer Based Nanofibers: Preparation, Fabrication, and Applications. In *Handbook of Nanofibers*; Barhoum, A., Bechelany, M., Makhlouf, A., Eds.; Springer International Publishing: Cham, Switzerland, 2018; pp. 1–47.
77. Adepu, S.; Ramakrishna, S. Controlled Drug Delivery Systems: Current Status and Future Directions. *Molecules* **2021**, *26*, 5905. [CrossRef] [PubMed]
78. Rocha, C.V.; Gonçalves, V.; da Silva, M.C.; Bañobre-López, M.; Gallo, J. PLGA-Based Composites for Various Biomedical Applications. *Int. J. Mol. Sci.* **2022**, *23*, 2034. [CrossRef]
79. Vu, T.H.N.; Morozkina, S.N.; Uspenskaya, M.V. Study of the Nanofibers Fabrication Conditions from the Mixture of Poly(vinyl alcohol) and Chitosan by Electrospinning Method. *Polymers* **2022**, *14*, 811. [CrossRef] [PubMed]
80. Abdul Hameed, M.M.; Mohamed Khan, S.A.P.; Thamer, B.M.; Rajkumar, N.; El-Hamshary, H.; El-Newehy, M. Electrospun nanofibers for drug delivery applications: Methods and mechanism. *Polym. Adv. Technol.* **2023**, *34*, 6–23. [CrossRef]
81. Abdulhussain, R.; Adebisi, A.; Conway, B.R.; Asare-Addo, K. Electrospun nanofibers: Exploring process parameters, polymer selection, and recent applications in pharmaceuticals and drug delivery. *J. Drug Deliv. Sci. Technol.* **2023**, *90*, 105156. [CrossRef]
82. Hashem, H.M.; Motawea, A.; Kamel, A.H.; Bary, E.M.A.; Hassan, S.S.M. Fabrication and characterization of electrospun nanofibers using biocompatible polymers for the sustained release of venlafaxine. *Sci. Rep.* **2022**, *12*, 18037. [CrossRef]
83. Chou, S.F.; Carson, D.; Woodrow, K.A. Current strategies for sustaining drug release from electrospun nanofibers. *J. Control Release* **2015**, *220*, 584–591. [CrossRef]
84. CeCe, R.; Jining, L.; Islam, M.; Korvink, J.G.; Sharma, B. An Overview of the Electrospinning of Polymeric Nanofibers for Biomedical Applications Related to Drug Delivery. *Adv. Eng. Mater.* **2024**, *26*, 2301297. [CrossRef]
85. Chen, C.; Tang, Y.; Vlahovic, B.; Yan, F. Electrospun Polymer Nanofibers Decorated with Noble Metal Nanoparticles for Chemical Sensing. *Nanoscale Res. Lett.* **2017**, *12*, 451. [CrossRef] [PubMed]
86. Babayevska, N.; Przysiecka, Ł.; Nowaczyk, G.; Jarek, M.; Järvekülg, M.; Kangur, T.; Janiszewska, E.; Jurga, S.; Iatsunskyi, I. Fabrication of Gelatin-ZnO Nanofibers for Antibacterial Applications. *Materials* **2021**, *14*, 103. [CrossRef] [PubMed]

87. Sanchaniya, J.V.; Lasenko, I.; Gobins, V.; Kobeissi, A.; Goljandin, D. A Finite Element Method for Determining the Mechanical Properties of Electrospun Nanofibrous Mats. *Polymers* **2024**, *16*, 852. [CrossRef]
88. Riaz, U.; Ashraf, S. *Characterization of Polymer Blends with FTIR Spectroscopy*; Wiley: Hoboken, NJ, USA, 2014; pp. 625–678.
89. Drobota, M.; Gradinaru, L.M.; Vlad, S.; Bargan, A.; Butnaru, M.; Angheloiu, M.; Aflori, M. Preparation and Characterization of Electrospun Collagen Based Composites for Biomedical Applications. *Materials* **2020**, *13*, 3961. [CrossRef]
90. Bunaciu, A.A.; UdriŞTioiu, E.; Aboul-Enein, H. X-Ray Diffraction: Instrumentation and Applications. *Crit. Rev. Anal. Chem./CRC* **2015**, *45*, 289–299. [CrossRef] [PubMed]
91. Leyva-Porras, C.; Cruz-Alcantar, P.; Espinosa-Solís, V.; Martínez-Guerra, E.; Piñón-Balderrama, C.I.; Compean Martínez, I.; Saavedra-Leos, M.Z. Application of Differential Scanning Calorimetry (DSC) and Modulated Differential Scanning Calorimetry (MDSC) in Food and Drug Industries. *Polymers* **2020**, *12*, 5. [CrossRef] [PubMed]
92. Alkilani, A.Z.; McCrudden, M.T.; Donnelly, R.F. Transdermal Drug Delivery: Innovative Pharmaceutical Developments Based on Disruption of the Barrier Properties of the stratum corneum. *Pharmaceutics* **2015**, *7*, 438–470. [CrossRef] [PubMed]
93. Du, P.; Chen, X.; Chen, Y.; Li, J.; Lu, Y.; Li, X.; Hu, K.; Chen, J.; Lv, G. In vivo and in vitro studies of a propolis-enriched silk fibroin-gelatin composite nanofiber wound dressing. *Heliyon* **2023**, *9*, e13506. [CrossRef]
94. Hadipour-Goudarzi, E.; Hemmatinejad, N.; Shokrgozar, M.A. Fabrication and DOE Optimization of Electrospun Chitosan/Gelatin/PVA Nanofibers for Skin Tissue Engineering. *Macromol. Mater. Eng.* **2023**, *308*, 2200562. [CrossRef]
95. Ghazal, H.; Waqar, A.; Yaseen, F.; Shahid, M.; Sultana, M.; Tariq, M.; Bashir, M.K.; Tahseen, H.; Raza, T.; Ahmad, F. Role of nanoparticles in enhancing chemotherapy efficacy for cancer treatment. *Next Mater.* **2024**, *2*, 100128. [CrossRef]
96. Liu, M.; Duan, X.-P.; Li, Y.-M.; Yang, D.-P.; Long, Y.-Z. Electrospun nanofibers for wound healing. *Mater. Sci. Eng. C* **2017**, *76*, 1413–1423. [CrossRef] [PubMed]
97. Gupta, A.K.; Shemer, A.; Economopoulos, V.; Talukder, M. Diabetic Foot and Fungal Infections: Etiology and Management from a Dermatologic Perspective. *J. Fungi* **2024**, *10*, 577. [CrossRef] [PubMed]
98. Pradeepa, R.; Mohan, V. Chapter 2—Epidemiology of chronic complications of diabetes: A global perspective. In *Chronic Complications of Diabetes Mellitus*; Mauricio, D., Alonso, N., Eds.; Academic Press: Cambridge, MA, USA, 2024; pp. 11–23.
99. Gao, C.; Zhang, L.; Wang, J.; Jin, M.; Tang, Q.; Chen, Z.; Cheng, Y.; Yang, R.; Zhao, G. Electrospun nanofibers promote wound healing: Theories, techniques, and perspectives. *J. Mater. Chem. B* **2021**, *9*, 3106–3130. [CrossRef]
100. Venugopal, J.; Ramakrishna, S. Applications of polymer nanofibers in biomedicine and biotechnology. *Appl. Biochem. Biotechnol.* **2005**, *125*, 147–157. [CrossRef]
101. Akombaetwa, N.; Bwanga, A.; Makoni, P.A.; Witika, B.A. Applications of Electrospun Drug-Eluting Nanofibers in Wound Healing: Current and Future Perspectives. *Polymers* **2022**, *14*, 2931. [CrossRef] [PubMed]
102. Lu, X.; Zhou, L.; Song, W. Recent Progress of Electrospun Nanofiber Dressing in the Promotion of Wound Healing. *Polymers* **2024**, *16*, 2596. [CrossRef] [PubMed]
103. Amiri, N.; Ajami, S.; Shahroodi, A.; Jannatabadi, N.; Amiri Darban, S.; Fazly Bazzaz, B.S.; Pishavar, E.; Kalalinia, F.; Movaffagh, J. Teicoplanin-loaded chitosan-PEO nanofibers for local antibiotic delivery and wound healing. *Int. J. Biol. Macromol.* **2020**, *162*, 645–656. [CrossRef] [PubMed]
104. Nair, L.S.; Laurencin, C.T. Nanofibers and Nanoparticles for Orthopaedic Surgery Applications. *J. Bone Jt. Surg.* **2008**, *90*, 128–131. [CrossRef]
105. Baek, D.H.; Ki, C.S.; Um, I.C.; Park, Y.H. Metal ion adsorbability of electrospun wool keratose/silk fibroin blend nanofiber mats. *Fibers Polym.* **2007**, *8*, 271–277. [CrossRef]
106. Linh, N.T.B.; Min, Y.K.; Song, H.-Y.; Lee, B.-T. Fabrication of polyvinyl alcohol/gelatin nanofiber composites and evaluation of their material properties. *J. Biomed. Mater. Res. Part B Appl. Biomater.* **2010**, *95B*, 184–191. [CrossRef]
107. Frohbergh, M.E.; Katsman, A.; Botta, G.P.; Lazarovici, P.; Schauer, C.L.; Wegst, U.G.K.; Lelkes, P.I. Electrospun hydroxyapatite-containing chitosan nanofibers crosslinked with genipin for bone tissue engineering. *Biomaterials* **2012**, *33*, 9167–9178. [CrossRef] [PubMed]
108. Wang, J.; Planz, V.; Vukosavljevic, B.; Windbergs, M. Multifunctional electrospun nanofibers for wound application—Novel insights into the control of drug release and antimicrobial activity. *Eur. J. Pharm. Biopharm.* **2018**, *129*, 175–183. [CrossRef]
109. Eskinazi-Budge, A.; Manickavasagam, D.; Czech, T.; Novak, K.; Kunzler, J.; Oyewumi, M.O. Preparation of emulsifying wax/glyceryl monooleate nanoparticles and evaluation as a delivery system for repurposing simvastatin in bone regeneration. *Drug Dev. Ind. Pharm.* **2018**, *44*, 1583–1590. [CrossRef]
110. Ozkizilcik, A.; Davidson, P.; Turgut, H.; Sharma, H.S.; Sharma, A.; Tian, Z.R. Nanocarriers as CNS Drug Delivery Systems for Enhanced Neuroprotection. In *Drug and Gene Delivery to the Central Nervous System for Neuroprotection: Nanotechnological Advances*; Sharma, H.S., Muresanu, D.F., Sharma, A., Eds.; Springer International Publishing: Cham, Switzerland, 2017; pp. 33–55.
111. D'Amato, A.R.; Ziemba, A.M.; Johnson, C.D.L.; Gilbert, R.J. 18—Advances in the use of electrospun fibers for the central nervous system. In *Electrofluidodynamic Technologies (EFDTs) for Biomaterials and Medical Devices*; Guarino, V., Ambrosio, L., Eds.; Woodhead Publishing: Sawston, UK, 2018; pp. 377–398.

112. Haidar, M.K.; Timur, S.S.; Demirbolat, G.M.; Nemutlu, E.; Gürsoy, R.N.; Ulubayram, K.; Öner, L.; Eroğlu, H. Electrospun Nanofibers for Dual and Local Delivery of Neuroprotective Drugs. *Fibers Polym.* **2021**, *22*, 334–344. [CrossRef]
113. Kellaway, S.C.; Ullrich, M.M.; Dziemidowicz, K. Electrospun drug-loaded scaffolds for nervous system repair. *WIREs Nanomed. Nanobiotechnology* **2024**, *16*, e1965. [CrossRef] [PubMed]
114. Norouzi, M. Recent advances in brain tumor therapy: Application of electrospun nanofibers. *Drug Discov. Today* **2018**, *23*, 912–919. [CrossRef]
115. Bini, T.B.; Gao, S.; Wang, S.; Ramakrishna, S. Poly(l-lactide-co-glycolide) biodegradable microfibers and electrospun nanofibers for nerve tissue engineering: An in vitro study. *J. Mater. Sci.* **2006**, *41*, 6453–6459. [CrossRef]
116. Yang, Y. Nanofibrous scaffolds supporting optimal central nervous system regeneration: An evidence-based review. *J. Neurorestoratology* **2015**, *3*, 123–131. [CrossRef]
117. Dolci, L.S.; Perone, R.C.; Di Gesù, R.; Kurakula, M.; Gualandi, C.; Zironi, E.; Gazzotti, T.; Tondo, M.T.; Pagliuca, G.; Gostynska, N.; et al. Design and In Vitro Study of a Dual Drug-Loaded Delivery System Produced by Electrospinning for the Treatment of Acute Injuries of the Central Nervous System. *Pharmaceutics* **2021**, *13*, 848. [CrossRef] [PubMed]
118. Kamsani, N.H.; Haris, M.S.; Pandey, M.; Taher, M.; Rullah, K. Biomedical application of responsive 'smart' electrospun nanofibers in drug delivery system: A minireview. *Arab. J. Chem.* **2021**, *14*, 103199. [CrossRef]
119. Pérez-González, G.L.; Villarreal-Gómez, L.J.; Serrano-Medina, A.; Torres-Martínez, E.J.; Cornejo-Bravo, J.M. Mucoadhesive electrospun nanofibers for drug delivery systems: Applications of polymers and the parameters' roles. *Int. J. Nanomed.* **2019**, *ume 14*, 5271–5285. [CrossRef]
120. Vidyadhari, A.; Singh, N.; Singh, A.K.; Ralli, T.; Solanki, P.; Mirza, M.A.; Parvez, S.; Kohli, K. Investigation of Luliconazole-Loaded Mucoadhesive Electrospun Nanofibers for Anticandidal Activity in the Management of Vaginal Candidiasis. *ACS Omega* **2023**, *8*, 42102–42113. [CrossRef] [PubMed]
121. Nematpour, N.; Moradipour, P.; Zangeneh, M.M.; Arkan, E.; Abdoli, M.; Behbood, L. The application of nanomaterial science in the formulation a novel antibiotic: Assessment of the antifungal properties of mucoadhesive clotrimazole loaded nanofiber versus vaginal films. *Mater. Sci. Eng. C* **2020**, *110*, 110635. [CrossRef] [PubMed]
122. Ball, C.; Krogstad, E.; Chaowanachan, T.; Woodrow, K.A. Drug-eluting fibers for HIV-1 inhibition and contraception. *PLoS ONE* **2012**, *7*, e49792. [CrossRef] [PubMed]
123. Bellu, E.; Diaz, N.; Kralovič, M.; Divin, R.; Sarais, G.; Fadda, A.; Satta, R.; Montesu, M.A.; Medici, S.; Brunetti, A.; et al. Myrtle-Functionalized Nanofibers Modulate Vaginal Cell Population Behavior While Counteracting Microbial Proliferation. *Plants* **2022**, *11*, 1577. [CrossRef] [PubMed]
124. Huang, C.; Soenen, S.J.; van Gulck, E.; Vanham, G.; Rejman, J.; Van Calenbergh, S.; Vervaet, C.; Coenye, T.; Verstraelen, H.; Temmerman, M.; et al. Electrospun cellulose acetate phthalate fibers for semen induced anti-HIV vaginal drug delivery. *Biomaterials* **2012**, *33*, 962–969. [CrossRef] [PubMed]
125. Aggarwal, U.; Goyal, A.K.; Rath, G. Development and characterization of the cisplatin loaded nanofibers for the treatment of cervical cancer. *Mater. Sci. Eng. C* **2017**, *75*, 125–132. [CrossRef]
126. Ahani, E.; Mianehro, A. Electrospun scaffold based on poly(ethylene oxide)/poly(ε-caprolactone) for prolonged intravaginal antiviral drug release. *J. Drug Deliv. Sci. Technol.* **2023**, *88*, 104856. [CrossRef]
127. Tuğcu-Demiröz, F.; Saar, S.; Tort, S.; Acartürk, F. Electrospun metronidazole-loaded nanofibers for vaginal drug delivery. *Drug Dev. Ind. Pharm.* **2020**, *46*, 1015–1025. [CrossRef]
128. Cejkova, J.; Cejka, C.; Trosan, P.; Zajicova, A.; Sykova, E.; Holan, V. Treatment of alkali-injured cornea by cyclosporine A-loaded electrospun nanofibers—An alternative mode of therapy. *Exp. Eye Res.* **2016**, *147*, 128–137. [CrossRef] [PubMed]
129. Thakkar, S.; Misra, M. Electrospun polymeric nanofibers: New horizons in drug delivery. *Eur. J. Pharm. Sci.* **2017**, *107*, 148–167. [CrossRef] [PubMed]
130. Hsu, K.-H.; Fang, S.-P.; Lin, C.-L.; Liao, Y.-S.; Yoon, Y.-K.; Chauhan, A. Hybrid Electrospun Polycaprolactone Mats Consisting of Nanofibers and Microbeads for Extended Release of Dexamethasone. *Pharm. Res.* **2016**, *33*, 1509–1516. [CrossRef]
131. Mirzaeei, S.; Taghe, S.; Asare-Addo, K.; Nokhodchi, A. Polyvinyl Alcohol/Chitosan Single-Layered and Polyvinyl Alcohol/Chitosan/Eudragit RL100 Multi-layered Electrospun Nanofibers as an Ocular Matrix for the Controlled Release of Ofloxacin: An In Vitro and In Vivo Evaluation. *AAPS PharmSciTech* **2021**, *22*, 170. [CrossRef]
132. da Silva, G.R.; Song, E.; Chen, K.M.; Chen, F.; Jiang, L.; Kim, H.; Kang, N.-W.; Koh, W.-G.; Myung, D. Gelatin nanofibers coated with hyaluronic acid as a mesenchymal stromal cell scaffold for corneal regeneration. *Int. J. Pharm.* **2025**, *669*, 125009. [CrossRef] [PubMed]
133. Persano, L.; Camposeo, A.; Tekmen, C.; Pisignano, D. Industrial Upscaling of Electrospinning and Applications of Polymer Nanofibers: A Review. *Macromol. Mater. Eng.* **2013**, *298*, 504–520. [CrossRef]
134. Kailasa, S.; Reddy, M.S.B.; Maurya, M.R.; Rani, B.G.; Rao, K.V.; Sadasivuni, K.K. Electrospun Nanofibers: Materials, Synthesis Parameters, and Their Role in Sensing Applications. *Macromol. Mater. Eng.* **2021**, *306*, 2100410. [CrossRef]

135. De Vrieze, S.; Camp, T.; Nelvig, A.; Hagström, B.; Westbroek, P.; Clerck, K. The effect of temperature and humidity on electrospinning. *J. Mater. Sci.* **2009**, *44*, 1357–1362. [CrossRef]
136. Hojabri, M.; Tayebi, T.; Kasravi, M.; Aghdaee, A.; Ahmadi, A.; Mazloomnejad, R.; Tarasi, R.; Shaabani, A.; Bahrami, S.; Niknejad, H. Wet-spinnability and crosslinked Fiber properties of alginate/hydroxyethyl cellulose with varied proportion for potential use in tendon tissue engineering. *Int. J. Biol. Macromol.* **2023**, *240*, 124492. [CrossRef] [PubMed]
137. Bhardwaj, N.; Kundu, S.C. Electrospinning: A fascinating fiber fabrication technique. *Biotechnol. Adv.* **2010**, *28*, 325–347. [CrossRef]
138. Mehta, P.P.; Pawar, V.S. 22—Electrospun nanofiber scaffolds: Technology and applications. In *Applications of Nanocomposite Materials in Drug Delivery*; Inamuddin, Asiri, A.M., Mohammad, A., Eds.; Woodhead Publishing: Sawston, UK, 2018; pp. 509–573.
139. Jurić, M.; Goksen, G.; Donsì, F.; Jurić, S. Innovative Applications of Electrospun Nanofibers Loaded with Bacterial Cells Towards Sustainable Agri-Food Systems and Regulatory Compliance. *Food Eng. Rev.* **2024**, *16*, 270–303. [CrossRef]
140. Campbell, B.; Anderson, Z.; Han, D.; Nebor, I.; Forbes, J.; Steckl, A.J. Electrospinning of cyanoacrylate tissue adhesives for human dural repair in endonasal surgery. *J. Biomed. Mater. Res. Part B Appl. Biomater.* **2022**, *110*, 660–667. [CrossRef]
141. Uhljar, L.É.; Ambrus, R. Electrospinning of Potential Medical Devices (Wound Dressings, Tissue Engineering Scaffolds, Face Masks) and Their Regulatory Approach. *Pharmaceutics* **2023**, *15*, 417. [CrossRef]

polymers

MDPI

Review

Advances in the Fabrication, Properties, and Applications of Electrospun PEDOT-Based Conductive Nanofibers

Emanuele Alberto Slejko [1,*], Giovanni Carraro [1], Xiongchuan Huang [2] and Marco Smerieri [1]

[1] IMEM-CNR, Institute of Materials for Electronics and Magnetism of the National Research Council of Italy, Via Dodecaneso 33, 16146 Genova, Italy; giovanni.carraro@cnr.it (G.C.); marco.smerieri@cnr.it (M.S.)

[2] School of Information Science and Technology, Fudan University, Handan Rd. 220, Shanghai 200433, China; huangxiongchuan@fudan.edu.cn

* Correspondence: emanuelealberto.slejko@cnr.it

Abstract: The production of nanofibers has become a significant area of research due to their unique properties and diverse applications in various fields, such as biomedicine, textiles, energy, and environmental science. Electrospinning, a versatile and scalable technique, has gained considerable attention for its ability to fabricate nanofibers with tailored properties. Among the wide array of conductive polymers, poly(3,4-ethylenedioxythiophene) (PEDOT) has emerged as a promising material due to its exceptional conductivity, environmental stability, and ease of synthesis. The electrospinning of PEDOT-based nanofibers offers tunable electrical and optical properties, making them suitable for applications in organic electronics, energy storage, biomedicine, and wearable technology. This review, with its comprehensive exploration of the fabrication, properties, and applications of PEDOT nanofibers produced via electrospinning, provides a wealth of knowledge and insights into leveraging the full potential of PEDOT nanofibers in next-generation electronic and functional devices by examining recent advancements in the synthesis, functionalization, and post-treatment methods of PEDOT nanofibers. Furthermore, the review identifies current challenges, future directions, and potential strategies to address scalability, reproducibility, stability, and integration into practical devices, offering a comprehensive resource on conductive nanofibers.

Keywords: nonwovens; conductive polymer; electrospinning; PEDOT:PSS; electrospun nanofibers

1. Introduction

The production of nanofibers has emerged as a pivotal area of research due to their unique properties and wide-ranging applications in fields such as biomedicine, textiles, energy, and environmental science [1,2]. Among the various methods available for nanofiber fabrication, electrospinning has garnered significant attention for its versatility, scalability, and ability to produce nanofibers with tailored properties [3]. Electrospinning is a technique that involves applying a high-voltage electrical field to a polymer solution or melt. This process induces the formation of a thin jet, which elongates and solidifies into continuous nanofibers as the solvent evaporates or the polymer solidifies. The resulting nanofibers exhibit diameters ranging from a few nanometers to several micrometers, with exceptional uniformity and purity.

One advantage of electrospun nanofibers is their high surface area-to-volume ratio, which enhances their interactions with surrounding environments [4]. This property makes electrospun nanofibers suitable for filtration, tissue engineering, drug delivery, and sensing applications. Moreover, the properties of electrospun nanofibers can be easily tailored by adjusting various parameters, including polymer concentration, solvent type, applied voltage, and collection distance, among many others [5]. This tunability enables the customization of nanofibers to meet specific application requirements, further enhancing their utility across diverse fields. In addition to their tunable properties, electrospun nanofibers offer scalability and cost-effectiveness, making them suitable for industrial production. The simplicity of the electrospinning process, coupled with its ability to produce uniform and pure nanofibers, makes it an attractive option for large-scale manufacturing.

Conductive polymers have garnered significant attention in recent years for their unique combination of electrical conductivity, mechanical flexibility, and processability, opening new avenues for developing advanced electronic devices and energy storage systems [6]. Several experiments have been conducted on polyaniline (PANI), polypyrrole (PPy), polythiophenes, and many others [7]. Among the various conductive polymers, poly(3,4-ethylenedioxythiophene) (PEDOT) is a promising material due to its exceptional conductivity, environmental stability, and ease of synthesis. PEDOT belongs to the family of polythiophenes and is characterized by its π-conjugated structure, consisting of alternating double and single bonds along the polymer backbone [5]. This π-conjugation imparts high electrical conductivity to PEDOT, making it one of the most conductive polymers known to date. Furthermore, ethylenedioxy (EDOT) functional groups enhance PEDOT's solubility and processability, enabling facile fabrication of thin films, coatings, and composites. The first scientific contribution investigating the fabrication and properties of PEDOT-based nanofibers is dated 2008 [8]. The highest PEDOT conductivity reported, approximately 4380 S/cm and almost comparable to that of ITO, was achieved through a controlled sulfuric acid post-treatment, which induced the formation of crystallized PEDOT nanofibrils by removing PSS via charge separation [9].

One of the key advantages of PEDOT is its environmental stability, attributed to the presence of ether linkages in the polymer backbone. PEDOT exhibits remarkable stability under ambient conditions, unlike many other conductive polymers, which are prone to degradation upon air, moisture, or UV radiation exposure. This property makes PEDOT particularly well suited for applications requiring long-term reliability and durability, such as organic electronic devices and biosensors. Moreover, PEDOT offers tunable electrical and optical properties through chemical doping and structural modifications. By doping PEDOT with various dopant molecules, such as polyanions or conducting salts, its conductivity can be tailored over a wide range to meet specific application requirements. Additionally, the optical transparency of PEDOT can be adjusted by controlling the polymer chain morphology, making it suitable for optoelectronic applications, including transparent electrodes and displays. In recent years, PEDOT has found widespread use in diverse fields, including organic electronics, energy storage, biomedicine, and wearable technology [1,2]. Its compatibility with flexible substrates and biocompatibility has made PEDOT an attractive candidate for flexible electronics, implantable medical devices, and bioelectronic interfaces [10,11]. Furthermore, its high charge storage capacity and rapid redox kinetics have enabled the development of high-performance supercapacitors and batteries with enhanced cycling stability and power density.

The most common commercially available product of PEDOT is its dispersion in water. PEDOT is stabilized by polystyrene sulfonate (PSS), which balances the positively charged PEDOT thanks to its negative charges (Figure 1). Several investigations have been conducted on the morphology and chemical configuration of PEDOT:PSS water dispersions: studies have validated the prevailing model of PEDOT films [12], confirming that they are composed of spherical grains with diameters ranging from 50 to 80 nm. Energy-dispersive X-ray analysis indicates that these grains likely have a 5–10 nm thick PSS-rich shell surrounding a PEDOT-rich core. High-magnification High-Angle Annular Dark-Field Scanning Transmission Electron Microscopy images suggest that the grains are made up of individual tangles approximately 5 nm or smaller in diameter. The cohesion between grains is primarily achieved through hydrogen bonds between PSS groups in the shells. In the context of PEDOT:PSS, electrospinning offers several advantages. First, it enables the fabrication of nanofibers with high surface area-to-volume ratios [4], which is beneficial for applications such as sensors, actuators, and tissue engineering [13]. Second, it allows for precise control over fiber diameter and alignment, which can influence properties like conductivity and mechanical strength. Third, electrospinning can be used to create complex three-dimensional structures, providing versatility in design. The significance of electrospinning for PEDOT:PSS nanofibers lies in its potential to enhance the material's performance and expand its applications in bioelectronics. By producing PEDOT:PSS nanofibers through electrospinning, researchers can create bioactive interfaces with enhanced conductivity, mechanical properties, and biocompatibility. These interfaces are crucial for developing advanced biomedical devices such as biosensors, neural implants, and drug delivery systems. The evolution of PEDOT nanofibers produced by electrospinning has seen significant advancements, particularly in their conductivity and mechanical properties. Initially, conductive PEDOT nanofibers were fabricated through electrospinning of insulating polymers followed by vapor-phase polymerization of PEDOT, achieving high conductivity and resistive heating capabilities [14,15]. It was only during the 2010s that studies introduced techniques like oxidative polymerization and the blending of PEDOT with other polymers for solution-based electrospinning, further enhancing the structural stability, conductivity, and flexibility of the nanofibers [16–18]. In 2017, Yu et al. were able to propose a one-step fabrication of core–shell nanofibers based on PEDOT:PSS and poly(ethylene oxide), while three years later, Du et al. developed a core–shell structure with chitosan nanofibers with improved biocompatibility for biomedical research and regenerative medicine [19,20]. These innovations have positioned PEDOT nanofibers as promising materials for applications in flexible electronics, sensors, and biomedical devices.

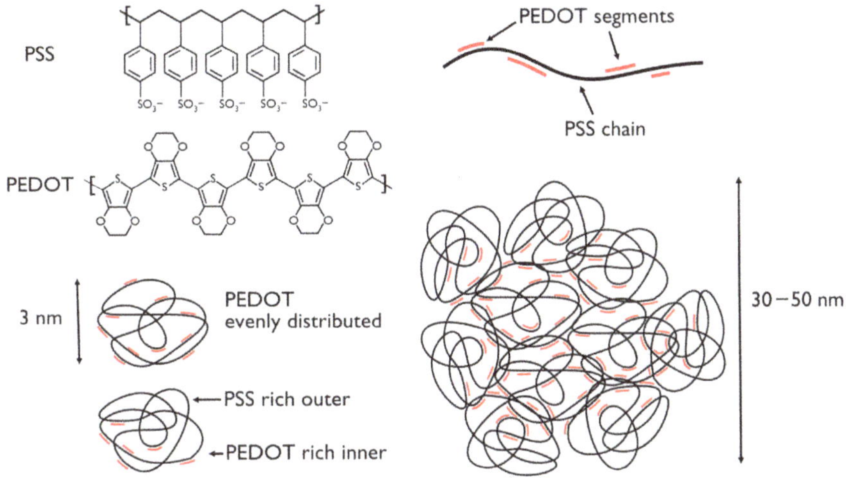

Figure 1. It is proposed that PEDOT:PSS grain consists of numerous tangles, where a single PSS chain forms each tangle with several PEDOT segments attached. These tangles are organized inside the grain so that the PEDOT segments are evenly distributed. On the outer layer of the grain, the tangles are arranged such that the hydrophilic PSS parts are oriented outward, allowing the grains to be dispersed in an aqueous solution. This arrangement facilitates the even distribution and dispersion of PEDOT grains in water.

This review systematically examines the state-of-the-art techniques for producing PEDOT nanofibers via electrospinning, enhancing their electrical conductivity, and exploring their diverse applications. Firstly, we provide an overview of the electrospinning process, discussing key parameters such as solution properties, processing conditions, and collector configuration that influence the morphology and properties of the resulting nanofibers. Next, we delve into recent advancements in the synthesis and functionalization of PEDOT-based precursors for electrospinning, including the incorporation of dopants, additives, and nanomaterials to tailor the electrical conductivity and mechanical properties of the nanofibers. Furthermore, we explore various post-treatment methods employed to enhance further the conductivity of PEDOT nanofibers, such as chemical doping, thermal annealing, and surface modification, evaluating their effectiveness in improving electrical performance while preserving nanofiber morphology. Subsequently, we investigate the emerging applications of PEDOT nanofibers in electronic devices, energy storage systems, sensors, actuators, and biomedical devices, highlighting their unique advantages, such as high surface-to-volume ratio, flexibility, and biocompatibility [2]. By consolidating recent research findings and insights from the literature, this work provides researchers and engineers with a comprehensive understanding of the opportunities and challenges associated with PEDOT nanofiber production via electrospinning. Finally, we identify current challenges and future directions in the field, including scalability, reproducibility, stability, and integration into practical devices, and propose potential strategies to address these issues. Overall, it serves as a valuable resource for researchers seeking to harness the full potential of PEDOT nanofibers in next-generation electronic and functional devices.

2. Fundamentals of Electrospinning

Electrospinning is a versatile and widely used technique for producing nanofibers with controllable morphology, high surface area, and diverse properties. The process involves the application of an electric field to a polymer solution or melt, causing the formation of an electrified jet that undergoes stretching and solidification to form nanofibers. The physical mechanisms underlying electrospinning are complex and involve several key phenomena, including electrostatic forces, surface tension, fluid dynamics, and polymer chain entanglement. Understanding these mechanisms is crucial for optimizing the electrospinning process and tailoring the properties of the resulting nanofibers.

At the heart of electrospinning is the interaction between electric fields and polymer solutions or melts [21,22]. When a high electric field is applied to a polymer solution or melt, charges accumulate on the surface of the droplet or meniscus due to the electrostatic induction. The repulsion between like charges leads to the formation of a Taylor cone at the tip of the droplet or meniscus, where the electric field strength is the highest. The curvature of the Taylor cone increases the electric field strength, promoting the ejection of a fine jet of polymer solution or melt from the cone apex. The stretching and solidification of the ejected jet are governed by a balance of forces, including electrostatic forces, surface tension, and viscous forces. As the jet travels towards the collector under the influence of the electric field, it undergoes elongation and thinning due to the electrostatic repulsion between charged polymer chains. The stretching of the jet promotes molecular alignment and orientation along the axis of the fiber, resulting in nanofibers with high aspect ratios and aligned structures. Surface tension also plays a crucial role in electrospinning by resisting the elongation of the jet and promoting the formation of a stable Taylor cone. The surface tension of the polymer solution or melt determines the size and shape of the Taylor cone, as well as the stability of the jet during flight. Low-surface-tension solutions tend to form finer and more uniform nanofibers due to reduced instability and jet whipping. Fluid dynamics govern the motion of the polymer solution or melt during electrospinning and influence the morphology of the resulting nanofibers. The viscosity and rheological properties of the polymer solution or melt dictate the jet behavior, including its stretching, bending, and whipping. High-viscosity solutions tend to produce thicker and less uniform nanofibers due to increased resistance to stretching and bending. Polymer chain entanglement also plays a critical role in electrospinning by promoting the formation of stable jets and preventing premature jet breakup. Entangled polymer chains exhibit viscoelastic behavior, allowing them to deform and stretch under the influence of the electric field without undergoing catastrophic rupture. Polymer solutions with higher molecular weights and concentrations typically exhibit greater chain entanglement, forming finer and more uniform nanofibers.

A conventional solution-electrospinning setup, as illustrated in Figure 2, comprises several essential components, including a high-voltage power source, a syringe pump, a syringe loaded with the polymeric solution, a metal spinneret, usually a blunt-cut needle, connected to the syringe, and a fiber collector, typically a metal plate connected to a grounded electrode. The syringe pump regulates the flow rate of the polymer solution. The high-voltage power supply, in conjunction with the spinneret that directs the charged polymer solution into a high electric field and ultimately on the collector, provides the force necessary to elongate the charged polymer solution into fibrous structures.

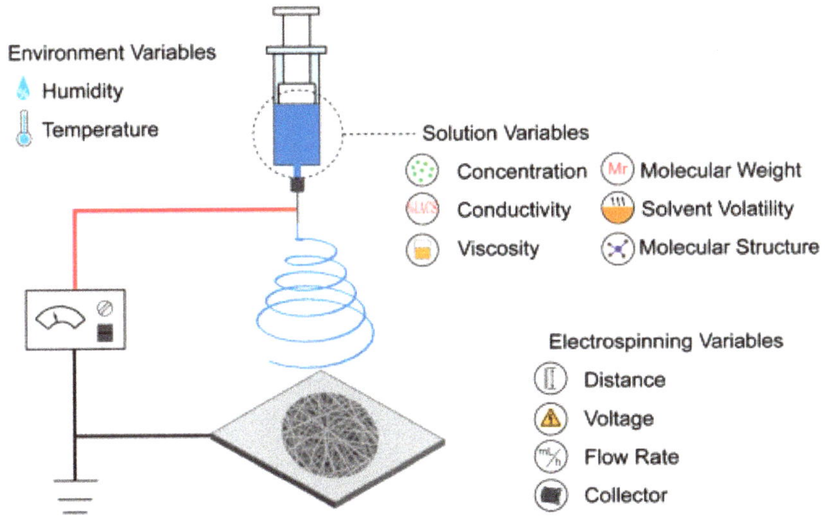

Figure 2. Schematic representation of a conventional electrospinning setup, alongside the environmental conditions, the characteristics of the solution utilized, and the various electrospinning parameters that can significantly influence the resultant fiber morphology and structure. An electrospinning system is made up of four main parts: a syringe pump, a voltage power supply, a needle, and a collector. Reproduced with permission from Long et al. in Electrospinning: Nanofabrication and Applications, published by Elsevier: Amsterdam, 2019 [23].

Understanding the equipment setup and operational parameters is crucial for controlling the electrospinning process and tailoring the properties of the resulting nanofibers. The most important components are detailed in the following.

- High-Voltage Power Supply: The high-voltage power supply is a critical component of the electrospinning setup, providing the electric field necessary to generate the electrospun jet. Voltages typically range from a few kilovolts to tens of kilovolts, depending on the specific polymer solution and desired fiber properties. The power supply must deliver a stable and controllable voltage to ensure consistent electrospinning performance.
- Syringe Pump or Spinneret: The polymer solution or molten polymer is commonly inserted into a syringe attached to either a syringe pump or spinneret. The syringe pump regulates the polymer solution flow, enabling accurate management of the deposition rate and fiber diameter. In contrast, a spinneret can extrude the polymer solution or molten polymer at a controlled rate.
- Collector: The collector is a conductive substrate placed at a fixed distance from the spinneret to collect the electrospun fibers. Depending on the desired fiber alignment and morphology, the collector can take various forms, including flat plates, rotating drums, or mandrels. The collector can also be grounded to provide an electrical path for the charged fibers, preventing charge accumulation and promoting uniform deposition.
- Grounded Electrode: A grounded electrode is typically placed close to the collector to provide an electrical path for the charged polymer solution or melt. The grounded electrode helps to neutralize the charges on the collected fibers, preventing them

from repelling each other and forming aggregates. The grounded electrode can be positioned below or surrounding the collector, depending on the specific electrospinning setup.

Operational parameters for the electrospinning process must be set up depending on the fabrication of single polymeric nanofibers. The most important ones include the following:

- Flow Rate: The flow rate of the polymer solution or melt is a critical parameter that determines the deposition rate and fiber diameter during electrospinning. Higher flow rates generally result in thicker fibers due to increased polymer deposition, while lower flow rates produce finer fibers. The flow rate can be adjusted using the syringe pump or spinneret to optimize the electrospinning process for a given polymer system.
- Voltage: The applied voltage plays a crucial role in controlling the electrospinning process by governing the formation and behavior of the electrified jet. Higher voltages increase electrostatic repulsion between the charged polymer chains, promoting jet elongation and thinning. However, excessively high voltages can cause jet instability and lead to beads or non-uniform fibers forming. The voltage should be optimized based on the polymer solution properties and desired fiber morphology.
- Distance: The distance between the spinneret and the collector is another important parameter influencing the electrospinning process. The distance affects the trajectory and stretching of the ejected jet and the deposition pattern and fiber alignment on the collector. Closer distances generally result in finer fibers due to increased stretching, while larger distances produce thicker fibers. The distance should be carefully adjusted to achieve the desired fiber morphology and alignment.
- Solution Properties: The properties of the polymer solution or melt, including viscosity, conductivity, surface tension, and concentration, influence the electrospinning process and fiber morphology. Higher-viscosity solutions tend to produce thicker fibers due to increased resistance to jet elongation, while lower-viscosity solutions produce finer fibers. Similarly, higher-conductivity solutions promote more efficient charge dissipation and jet stabilization, leading to finer and more uniform fibers. The concentration of the polymer solution also affects the fiber morphology, with higher concentrations generally resulting in thicker fibers.

3. Synthesis of PEDOT-Based Nanofibers via Electrospinning

Electrospinning is a highly effective technique for fabricating nanofibers, offering advantages such as simplicity, versatility, and cost-effectiveness. PEDOT stands out among the various materials suitable for electrospinning due to its excellent electrical conductivity, environmental stability, and biocompatibility [10]. However, as PEDOT is a conductive material, other non-conducting polymers or chemicals, like poly(ethylene oxide)—PEO—are typically added to help with fiber production [5,24].

Electrospinning offers a unique advantage for fabricating PEDOT-based conductive nanofibers compared to other fabrication methods and fiber materials, particularly in terms of applications (see Table 1). This technique allows for the creation of nanofibers with high surface area, porosity, and flexibility, which are crucial for various advanced applications such as wearable electronics, energy storage, and biomedical devices. The ability to control fiber morphology and composition through electrospinning enhances the performance of PEDOT-based nanofibers, making them superior to other methods and materials. The specific advantages of electrospinning for PEDOT-based conductive nanofibers are the following.

Table 1. In the table are the main advantages of electrospinning for PEDOT-based nanofibers over other additive manufacturing methods such as 3D printing and bioprinting.

Feature/Method	Electrospinning (PEDOT)	Other A.M. Methods
Surface area	High	Moderate to low
Porosity	High	Variable
Flexibility	High	Low to moderate
Conductivity	Enhanced with additives	Lower without additives
Material versatility	High	Limited
Mechanical properties	Improved with additives	Variable
Morphological control	Precise	Less precise
Applications	Broad (wearables, energy, biomedical)	Limited to specific fields

- High surface area and porosity: Electrospinning produces nanofibers with a high surface area-to-volume ratio, which is valuable for applications requiring high reactivity and interaction with the environment, such as sensors and catalysts [25].
- Flexibility and comfort: The nanofibers mats produced are highly flexible, making them suitable for wearable electronics where comfort and adaptability to body movements are essential [26].
- Enhanced conductivity: By incorporating materials like reduced graphene oxide (rGO), the conductivity of PEDOT-based nanofibers can be significantly enhanced, which is crucial for electronic applications [27].
- Versatility in material composition: Electrospinning allows for the incorporation of various materials, such as nanocarbons, to enhance the mechanical and electrical properties of the fibers, making them suitable for a wide range of applications from photovoltaics to biomedical devices [28].
- Improved mechanical properties: The process can improve the mechanical properties of the fibers, such as tensile strength, which is important for structural applications [27].
- Controlled morphology: Electrospinning allows for precise control over fiber diameter and morphology, which can be tailored to specific application needs, such as creating coaxial structures for electronic devices [29,30].

PEDOT nanofibers have been synthesized using several electrospinning methods, including solution-based and melt-electrospinning. Each method has distinct advantages and limitations, which will be discussed in the following sections.

3.1. Solution-Based Electrospinning

Solution-based electrospinning is the most employed method for synthesizing PEDOT nanofibers. In this approach, PEDOT or its precursors are dissolved in a suitable solvent to form a homogeneous solution that can be electrospun into nanofibers. The first step in solution-based electrospinning is to prepare a polymer solution with appropriate viscosity, conductivity, and surface tension. For PEDOT nanofibers, PEDOT:PSS (poly(3,4-ethylenedioxythiophene):poly(styrenesulfonate)) is often used due to its solubility in water and various organic solvents. PEDOT:PSS solutions can be mixed with co-solvents such as dimethyl sulfoxide (DMSO) or ethylene glycol to improve the conductivity and film-forming properties of the resulting nanofibers. The electrospinning process begins by loading the polymer solution into a syringe with a needle or spinneret. A high-voltage power supply is connected to the needle, creating an electric field between the needle and a grounded collector. As the voltage increases, the electrostatic forces overcome the surface tension of the solution, forming a Taylor cone at the tip of the needle. A jet of polymer solution is then ejected from the Taylor cone and undergoes stretching and thinning as it travels toward the collector. Operational parameters such as flow rate, applied voltage, and the distance between the needle and collector (working distance) significantly influence the morphology and properties of the resulting nanofibers. For instance, a higher voltage can increase the stretching of the jet, leading to finer fibers, while the flow rate must be optimized to ensure a continuous and stable jet without bead formation. The electrified jet solidifies upon solvent evaporation, and the fibers are collected on a grounded substrate, which can be a flat plate, rotating drum, or wire mesh. The choice of collector affects the alignment and orientation of the nanofibers. For example, a rotating drum collector can produce aligned nanofibers, which is desirable for applications requiring anisotropic properties.

Solution-based electrospinning is straightforward and adaptable to a wide range of polymers, including PEDOT:PSS [31]. It allows for easy incorporation of various solvents and additives to modify the properties of the nanofibers. By adjusting the concentration of the polymer solution, applied voltage, flow rate, and working distance, the morphology of the nanofibers can be precisely controlled. This flexibility is beneficial for tailoring the fibers for specific applications. The equipment and materials required for solution-based electrospinning are relatively inexpensive and widely available, making it a cost-effective method for producing nanofibers. The main limitations of solution-based electrospinning are related to using solvents, which can introduce several challenges, including toxicity, volatility, and environmental concerns. Handling and disposing of solvents safely can also add to the overall cost and complexity of the process. Traces of solvent may remain in the nanofibers after electrospinning, potentially affecting their properties and performance, particularly in applications where purity is critical. Not all polymers are soluble in suitable solvents, which can limit the choice of materials that can be electrospun using this method.

One of the first routes to produce PEDOT-based nanofibers using solution-electrospinning relied on fabricating a carrier polymer nonwoven structure followed by the vapor deposition of PEDOT on the samples in a controlled environment [15]. This way, a PEDOT coating was obtained over the carrier polymer. More recently, the process has been optimized to electrospin PEDOT-based nanofibers directly from the starting solution without any subsequent vapor deposition [32].

3.2. *Melt-Electrospinning*

Melt-electrospinning, an advanced method for producing PEDOT nanofibers without solvents, requires high precision and control. This technique involves melting the polymer and electrospinning it in the molten state. Melt-electrospinning offers several advantages, including avoiding solvent-related issues such as toxicity, volatility, and environmental impact. However, it also demands higher temperatures and specialized equipment to handle the molten polymer. The polymer is heated above its melting point to obtain a molten state suitable for electrospinning, requiring precise temperature control to ensure a stable and homogeneous melt. Additives or plasticizers may be incorporated to lower the melting temperature and improve the spinnability of the polymer. Like solution-based electrospinning, melt-electrospinning involves applying a high voltage to the molten polymer, creating an electrified jet. However, due to the higher viscosity of the molten polymer compared to a solution, the applied voltage and working distance must be carefully controlled to achieve continuous and uniform fiber formation. Using a heated spinneret and a controlled temperature environment around the electrospinning setup is crucial to maintain the polymer in a molten state throughout the process. The molten polymer jet solidifies upon cooling, and the fibers are collected on a grounded substrate. The solidification process in melt-electrospinning is typically faster than in solution-based electrospinning due to the rapid cooling of the molten polymer, resulting in the formation of fibers with different morphological characteristics, such as increased crystallinity and mechanical strength.

Melt-electrospinning, while offering significant advantages, also has some limitations. The method eliminates the need for solvents, avoiding the associated health, safety, and environmental issues. It is an attractive method for producing nanofibers for applications requiring high purity and environmental safety. The fibers produced by melt-electrospinning often exhibit increased crystallinity and mechanical strength due to the rapid cooling and solidification of the molten polymer. However, the need to maintain the polymer in a molten state requires high temperatures, which can limit the range of polymers that can be processed and may degrade temperature-sensitive materials. The equipment needed for melt-electrospinning, including heated spinnerets and controlled-temperature environments, is more complex and expensive than solution-based electrospinning. The high viscosity of molten polymers can make achieving continuous and uniform fiber formation challenging, requiring precise control of operational parameters. Melt-electrospinning is particularly suitable for thermoplastic polymers, expanding the range of materials that can be processed into nanofibers.

To date, this process for producing PEDOT nanofibers has yet to be reported in the literature. The main reason is that PEDOT is usually available as a dispersion in water. Therefore, the most common way to fabricate nanofibers is using solution-electrospinning.

3.3. *Coaxial Electrospinning*

Coaxial electrospinning is a variation of the traditional electrospinning process, where two or three polymer solutions are simultaneously electrospun through a coaxial needle [33,34]. This method enables the fabrication of core–shell nanofibers with PEDOT as either the core or the shell material. Coaxial electrospinning is particularly useful for creating nanofibers with enhanced functionalities, such as improved electrical conductivity, mechanical properties, or biocompatibility. Preparing coaxial solutions involves selecting compatible core and shell materials with appropriate viscosities and conductivities. PEDOT:PSS can be combined with various polymers, such as poly(vinyl alcohol) (PVA) or polycaprolactone (PCL), to form core–shell structures. The core and shell solutions are loaded into separate syringes and fed through a coaxial needle, where they are electrospun together to form composite nanofibers. The electrospinning process for coaxial electrospinning is like traditional electrospinning but requires precise control of the flow rates for both the core and shell solutions. The applied voltage and working distance must also be optimized to ensure the formation of continuous and uniform core–shell fibers. The coaxial structure provides additional functionality to the nanofibers, such as enhanced conductivity from the PEDOT core or improved mechanical properties from the polymer shell. The core–shell nanofibers are collected on a grounded substrate, and the morphology and properties of the fibers can be tailored by adjusting the electrospinning parameters. Coaxial electrospinning offers a versatile approach to creating multifunctional nanofibers with potential applications in electronics, sensors, and biomedical devices.

Coaxial electrospinning allows the production of core–shell nanofibers, combining the properties of different polymers in a single fiber. This can enhance the functionality of the nanofibers, such as improving electrical conductivity, mechanical strength, or biocompatibility. By selecting different materials for the core and shell, the properties of the nanofibers can be finely tuned for specific applications. This method enables the creation of nanofibers with unique and enhanced characteristics. For example, coaxial electrospinning was employed by Olive et al. as a one-step alternative to fabricating nano-reinforced polyvinylidene fluoride (PVDF) fibers. This method enabled the dispersion of iron oxide (Fe_3O_4) nanoparticles with different size distributions (monomodal or bimodal) within the polymeric fibers (Figure 3). Coaxial electrospinning can encapsulate active ingredients, such as drugs or bioactive molecules, within the core of the fibers, protecting them from the environment and controlling their release. Like the previously discussed processes, coaxial electrospinning presents some limitations: the instrument setup is more complex than traditional electrospinning, requiring precise alignment and control of multiple polymer solutions. This increases the complexity and cost of the process. Achieving the desired core–shell structure requires careful optimization of the flow rates, viscosities, and conductivities of both solutions. Inconsistent parameters can lead to irregularities in the fiber structure. Finding solvents mutually compatible with the core and shell materials can be challenging, potentially limiting the range of materials used in coaxial electrospinning.

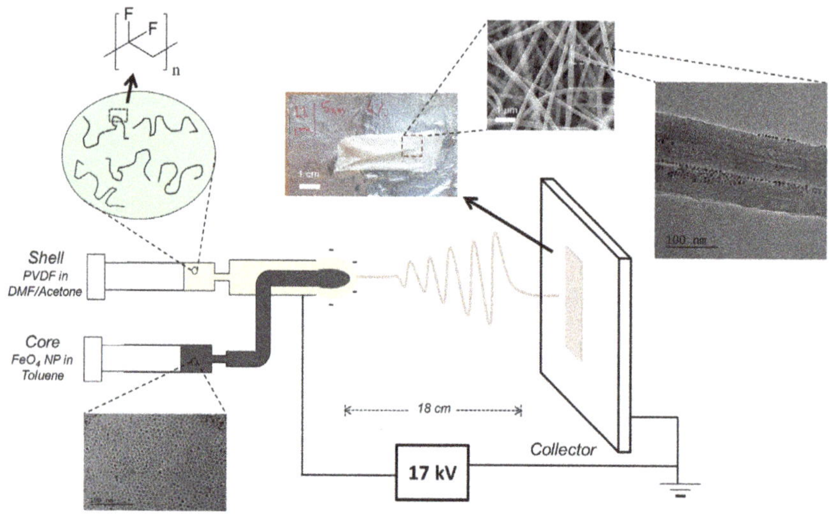

Figure 3. Schematic representation of the coaxial electrospinning process. Reproduced with permission from Oliva et al. in Polymer Composites, published by Wiley, 2021 [33].

4. Main Characterization Techniques for PEDOT Nanofibers

Characterizing PEDOT nanofibers is essential for understanding their morphology, structure, and properties, directly influencing their performance in various applications. This section outlines the common techniques used for characterizing PEDOT nanofibers, including scanning electron microscopy (SEM), Transmission Electron Microscopy (TEM), X-ray Diffraction (XRD), Fourier Transform Infrared Spectroscopy (FTIR), and conductivity measurements. Nevertheless, all the characterization techniques used in materials science can be applied to investigating nonwoven materials. Each method provides unique insights into the physical and chemical attributes of the nanofibers, enabling comprehensive analysis and optimization for specific applications.

4.1. Scanning Electron Microscopy (SEM)

Scanning electron microscopy (SEM) is a powerful technique used to analyze the surface morphology and topography of PEDOT nanofibers. SEM operates by scanning a focused beam of electrons across the sample surface, which generates secondary electrons that are detected to produce high-resolution images. SEM provides detailed images of the nanofiber diameter, distribution, and surface texture. It allows for the observation of fiber alignment, uniformity, and the presence of any defects or bead formation. SEM images help us understand the effect of electrospinning parameters on fiber morphology, which is crucial for optimizing the process.

For instance, Zarrin et al. (2018) used SEM to analyze the morphology and diameter distribution of electrospun PEDOT nanofibers, while Costa et al. (2023) used SEM to obtain information about the thickness of nanofibers mats before and after vapor treatment with DMSO [31,35].

4.2. Transmission Electron Microscopy (TEM)

Transmission Electron Microscopy (TEM) involves transmitting a beam of electrons through a thin sample. The interaction of electrons with the sample produces an image that reveals internal structures at the nanoscale. TEM offers insights into the internal structure and crystallinity of PEDOT nanofibers. It can reveal the presence of core–shell structures, crystallites, and any phase separation within the fibers. TEM is particularly useful for understanding the distribution of conductive domains within the fibers, which impacts their electrical properties. Zarrin et al. (2018) used TEM to examine the protrusions of PEDOT nanofibers and related them to the surface roughness of nanofibers observed with SEM [31]. Examples of SEM and TEM images are reported in Figure 4.

Figure 4. (Top) SEM micrograph and corresponding nanofiber diameter distribution are shown for (**A**) electrospun EDOT-PAA (67:33) and (**B**) electrospun PEDOT-PAA (67:33) directly oxidized in an oxidation bath containing $FeCl_3{\cdot}6H_2O$. (mid and bottom). Typical TEM images of PEDOT-PAA (50:50) nanofibers are also presented. Reproduced with permission from Zarrin et al. in Synthetic Mets, published by Elsevier, 2018 [31].

4.3. *X-ray Diffraction (XRD)*

X-ray Diffraction (XRD) is a crucial tool for understanding the crystalline structure of PEDOT nanofibers. When X-rays are directed at a sample, they are diffracted by the crystal lattice, and the resulting diffraction pattern provides information about the lattice parameters and crystallinity. This analysis reveals the degree of crystallinity and the presence of any crystalline phases in the nanofibers. Such information is critical for understanding how the electrospinning process and subsequent treatments (e.g., annealing) affect the crystalline structure and, consequently, the electrical and mechanical properties of the nanofibers. Lim et al. (2023) employed XRD to study the crystallinity of PVDF-PEDOT:PSS nanofibers and found that the amount of PEDOT influences the relative quantity of the beta phase in PVDF [36].

4.4. *Fourier Transform Infrared Spectroscopy (FTIR)*

Fourier Transform Infrared Spectroscopy (FTIR) is a key technique for identifying chemical modifications in PEDOT nanofibers. It measures the absorption of infrared radiation by the sample as a function of wavelength, providing information about the chemical bonds and molecular structure of the material. FTIR analysis of PEDOT nanofibers identifies functional groups and confirms the chemical structure of the polymer. It can detect any chemical modifications, such as doping or the presence of additives, and monitor the interaction between PEDOT and other components in composite nanofibers. Huang et al. (2013) and later Verpoorten et al. (2020) used Differential Scanning Calorimetry and FTIR, respectively, to investigate the chemical modifications in the PEDOT:PSS-PEO system upon thermal treatment and associated it with the cross-linking between PSS and PEO chains [6,37].

4.5. Conductivity Measurements

Conductivity measurements assess the electrical properties of PEDOT nanofibers. Techniques such as four-point probes and impedance spectroscopy are commonly used to measure the electrical conductivity and impedance of the nanofiber mats or individual fibers. Conductivity measurements provide quantitative data on the electrical performance of PEDOT nanofibers. For example, Babaie et al. (2020) measured the electrical conductivity of cross-linked scaffolds [10]: three samples, each measuring 10×1 mm^2, were cut from each electrospun sheet, and the thickness d of each sheet was recorded. The resistance (R) of nanofibers was determined by measuring the current while applying a voltage range from -10 V to 10 V in 0.2 V steps with a distance l between the electrodes (width w was 1 mm). The electrical conductivity σ of the fibers was calculated using Equation (1).

$$\sigma = \frac{l}{wdR} \quad (1)$$

These measurements are essential for evaluating the effect of electrospinning parameters, post-treatment processes (e.g., solvent annealing), and the incorporation of conductive additives on the overall conductivity of the nanofibers. Park et al. (2016) conducted conductivity measurements on PEDOT:PSS nanofibers with PEO or PVA as the carrier polymer treated with DMSO and EG and observed a significant increase in conductivity when using PEO and EG thanks to the extended quinoid structure of the nanofibers [18]. Another common characterization of conductive nanofiber-based sensors is cyclic voltammetry, during which the electrochemical reaction of the system can be investigated as a function of the applied potential. For example, Barroso et al. (2023) reported that a foaming process has beneficial effects in terms of active surface area and porosity of sensors, improving their electrochemical response [13].

5. Properties of PEDOT Nanofibers

5.1. Electrical Properties

PEDOT nanofibers are known for their excellent electrical conductivity, which stems from the conjugated backbone of the PEDOT polymer. The conductivity of PEDOT nanofibers can range from 1 to 1000 S/cm, depending on factors such as the dopant used, the degree of crystallinity, and post-treatment processes like solvent annealing or thermal treatment [18,38]. Including dopants such as PSS (poly(styrene sulfonate)) or small molecules like ethylene glycol significantly enhances conductivity by increasing the carrier density and mobility [24]. Treatments such as solvent annealing with DMSO or ethylene glycol can remove excess insulating components (e.g., PSS) and improve the ordering of the PEDOT chains, leading to higher conductivity. For example, Costa et al. evaluated electrospun nanofibers of PEDOT:PSS and Ag nanowires chemically treated with DMSO vapor [35]. After the treatment, they observed an improvement of two orders of magnitude in the electrical conductivity of the film. The applied voltage, solution concentration, and flow rate during electrospinning affect the fiber diameter and uniformity, which in turn influence the overall conductivity of the nanofiber mat. High conductivity makes PEDOT nanofibers suitable for applications in flexible electronics, sensors, and energy storage devices like supercapacitors and batteries [5,11]. Their use in transparent conductive films and organic photovoltaics is also significant due to their excellent electrical properties and optical transparency [38]. Cyclic voltammetry (CV) and electrochemical impedance spectroscopy (EIS) are common electrical characterizations that can reveal the redox behavior of materials in nonwoven films by showing the current response as a function of the applied potential [39]. Peaks in the CV curves correspond to oxidation and reduction processes, providing insights into the electroactive species present in the film (Figure 5). The shape and symmetry of the CV curves indicate the reversibility of redox reactions. Sharp, symmetric peaks suggest reversible reactions, while broad or asymmetric peaks indicate quasi-reversible or irreversible processes. By cycling the potential multiple times, CV can assess the stability of nonwoven films under electrochemical conditions. Stable films will show consistent peak currents and shapes over successive cycles. The area under the CV curve can be used to estimate the capacitance of the film, which is crucial for applications in energy storage devices like supercapacitors. The current response at different scan rates can help distinguish between capacitive and faradaic (battery-like) behaviors. EIS measures the impedance of a system over a range of frequencies, providing detailed information about the charge transfer resistance at the interface between the nonwoven film and the electrolyte. This resistance is a critical parameter for evaluating the efficiency of electron transfer processes in devices like sensors and batteries. The impedance data can be used to determine the double-layer capacitance at the film–electrolyte interface. This capacitance is indicative of the surface area and the ability of the film to store charge. EIS can reveal information about ion diffusion and mass transport within the nonwoven film. Features such as the Warburg impedance, which appears at low frequencies, are associated with diffusion-controlled processes. The intrinsic electrical conductivity of the nonwoven film can be extracted by fitting the EIS data to equivalent circuit models. This includes separating contributions from bulk resistance, grain boundary resistance, and other interfacial resistances. When used together, CV and EIS can provide a comprehensive understanding of the electrochemical behavior of nonwoven films. CV can identify active redox processes, while EIS can quantify the resistive and capacitive components associated with these processes.

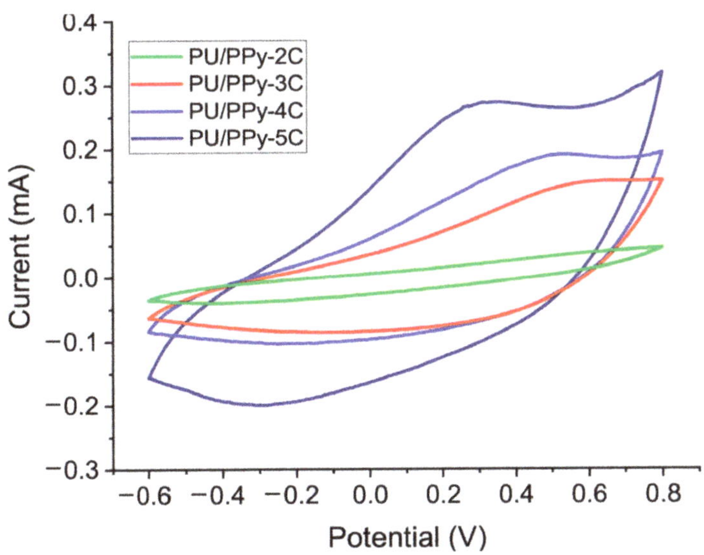

Figure 5. Cyclic voltammetry of PU/PPy nanofibrous actuators in the 0.1 M $LiClO_4$ electrolyte solution between potentials of 0.6 to 0.8 V with a scan rate of 5 mV s^{-1}. Reproduced with permission from Ebadi et al. in Smart Materials and Structures, published by IOP Publishing, 2020 [40].

5.2. Mechanical Properties

PEDOT nanofibers typically exhibit moderate mechanical strength and flexibility. Their tensile strength ranges from 10 to 100 MPa, while the elastic modulus can vary widely based on the fabrication method and the presence of reinforcing agents [41]. Thinner fibers tend to have higher tensile strength due to the reduction in defects. Uniform fiber morphology also contributes to enhanced mechanical properties. Blending PEDOT with other polymers (e.g., PVDF, PVA, PLA) or incorporating nanofillers (e.g., carbon nanotubes, graphene) can significantly improve the mechanical strength and elasticity of the nanofibers. The solution viscosity, polymer concentration, and solvent type used during electrospinning affect the fiber formation process and, thus, the mechanical properties of the final nanofibers. For example, adding various concentrations of PEDOT:PSS to PVDF solutions has a direct impact on the deposition area, the diameter of the fiber, and the crystallinity of PVDF, as reported by Lim et al. The mechanical properties of PEDOT nanofibers are crucial for their use in wearable electronics, where flexibility and durability are essential [6,36]. They are also important in biomedical applications such as tissue engineering and drug delivery, where the mechanical strength of the scaffold influences cell attachment and growth [10].

5.3. Thermal Properties

PEDOT nanofibers exhibit good thermal stability, typically decomposing above 200 °C. The molecular structure and the presence of dopants or additives influence the thermal properties. Higher crystallinity and molecular weight generally enhance the thermal stability of PEDOT nanofibers. Certain dopants can either stabilize or destabilize the polymer matrix. For instance, PSS tends to increase the thermal stability of PEDOT due to strong ionic interactions. Incorporating thermally stable materials like inorganic nanoparticles can further improve the thermal stability of PEDOT nanofibers. The thermal stability of PEDOT nanofibers is critical for applications in high-temperature environments such as in sensors and actuators. It also plays a role in the long-term reliability of electronic devices, where thermal degradation can affect performance.

6. Applications of PEDOT Nanofibers

PEDOT (poly(3,4-ethylenedioxythiophene)) nanofibers have garnered significant attention across various fields due to their exceptional electrical, mechanical, and thermal properties. These properties make PEDOT nanofibers versatile materials suitable for energy storage, sensors, actuators, and tissue engineering. Recent research and developments in these application areas highlight the innovative use of PEDOT nanofibers and their potential to revolutionize technology.

6.1. Sensors

6.1.1. Strain and Pressure Sensors

The applications of conventional sensors are limited by their boundedness, which is dependent on hard materials, uncontrolled self-multilayer structure building, and dense substrates, which are not ideal in terms of stretchability and permeability. The exceptional stretchability and air permeability of elastomeric nanofibers are attributed to their network structure. Stretchable conductors may be efficiently produced by elastomeric nanofibers by the introduction of metal nanofillers, intrinsic conductive polymers, carbon materials, and other techniques. This shows significant potential in the realm of flexible sensors [42]. Recently, the characteristics of composite solutions with poly(vinylidene fluoride) added to PEDOT:PSS were investigated, and the changes in electrospinning behavior and morphology were analyzed accordingly [36]. The research paper discusses the application of PVDF, a piezoelectric polymer, in various fields due to its flexibility and high piezoelectric constant. It also explores the potential application of PEDOT:PSS, a conductive polymer, in improving piezoelectric performance when combined with a piezoelectric polymer. The study suggests that the findings could lead to replacing ceramic-based materials with enhanced polymer-based piezoelectric elements, thereby accelerating the adoption of wearable piezoelectric devices. The methods used in this paper involved investigating the characteristics of composite solutions with PEDOT:PSS added to PVDF, analyzing changes in electrospinning behavior and morphology, and evaluating the crystallinity of PVDF in deposited fibers using Fourier Transform Infrared Spectroscopy and X-ray Diffraction analysis. Additionally, the study involved analyzing the changes in the radius of the jet, deposition area, fiber diameter, and crystallinity of the β phase as the PEDOT:PSS content increased, along with investigating the piezoelectric performance of the PVDF/PEDOT composite webs with various PEDOT:PSS contents [43]. Ding et al. showcase a method for producing mechanically stretchable and electrically conductive nonwoven textiles made of polyurethane (PU) through electrospinning and dip-coating with the conducting polymer PEDOT:PSS [44]. By adjusting the number of dip-coating times, nonwovens with initial sheet resistance ranging from 35 to 240 Ω/sq (electrical conductivity of 30–200 S m^{-1}) can be conveniently prepared. The resistance of the nonwoven PEDOT:PSS@PU stabilizes after the first 10 stretch–release cycles, enabling its use as a stretchable conductor within a certain strain range for applications such as electric circuits.

Traditionally, integrating different sensors into one platform involves complex fabrication processes. Verpoorten et al. explored a simplified solution using a single piezoresistive and electrochemically active electrospun nanofiber mat [45]. This material serves as the sensitive element in a wearable physiological flex-sensing platform, reducing the process flow to just two steps. The resulting NFs pH-Flex Sensor can monitor both human joint deformation and skin pH. The pH sensing utilizes EIS, showing linear relationships between pH value and both double-layer capacitance and charge transfer resistance. Additionally, the sensor's electrical resistance varies linearly with bending deformation. The gauge factors, 45.84 in traction and 208.55 in compression, indicate the nanostructured NFs' remarkable piezoresistive behavior.

Recent studies have focused on thermoplastic polyurethane (TPU) as a primary material for flexible strain sensors [46–48]. TPU is highlighted for its mechanical flexibility and compatibility with various conductive materials. Different conductive fillers, such as carbon black (CB), carbon nanotubes (CNTs), graphene flakes (GRs), MXene, metallic materials, and conductive polymers like PEDOT have been incorporated inside TPU. These fillers enhance the electrical conductivity and mechanical properties of the composite materials. The use of electrospinning is emphasized for fabricating nanofiber composites. This method enables the creation of highly stretchable and conductive fibers, essential for developing strain sensors with high sensitivity, stretchability, and a wide sensing range. The effectiveness of these sensors is discussed in various applications, including motion detection and biomedical uses. Each study meticulously analyzes the conductivity, mechanical properties, and strain-sensing performance of the composites (see Figure 6), exploring the effects of filler dimensionality and synergy, as well as the structural evolution and re-arrangement of filler networks under mechanical stress.

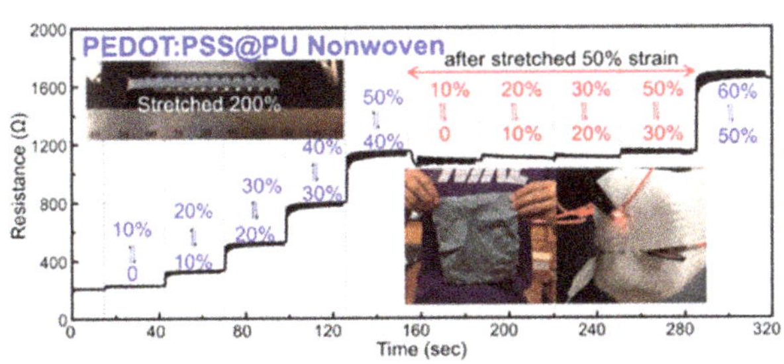

Figure 6. Resistance of the PEDOT:PSS/PU composite nonwoven as a function of stretching time. Reproduced with permission from Ding et al. in ACS Appl. Mater. Interfaces, published by American Chemical Society, 2017 [44].

6.1.2. Chemical Sensors

The high surface area and conductivity of PEDOT nanofibers make them ideal for chemical sensing applications, where they can detect gases, ions, and biomolecules [13]. Huang et al. (2015) achieved the controlled and continuous electrospinning of nearly pure PEDOT nanofibers by using magnesium nitrate as a cross-linker to enhance polymer entanglement [49]. Compared to bulky PEDOT films, these electrospun nanofibers demonstrated significantly higher sensitivities and faster response times in detecting gas molecules of dimethylformamide, dimethyl sulfoxide, and propylene carbonate. Shiu et al. (2022) recently developed a gas sensor for ethanol monitoring [50]. In their experiment, a gas sensitivity measurement system (WS-30B) was utilized to evaluate the gas sensitivity of PVP/PEDOT/TiO_2 nanofiber membranes. The sensor operated with a voltage of 5 V, and a steady baseline was established between 0 and 1. After two minutes, the specified gas or liquid was introduced, and the detector measured the change in electrical resistivity. Ethanol was efficiently evaporated using a heater, and the resulting gas was evenly diffused throughout the test case with a blowing air set. Tests were conducted at room temperature, with air as the carrier gas, to simulate a typical sensor environment. The materials' gas sensitivity and its relationship to relative humidity were assessed by comparing the electrical resistivity of the sensor before and after exposure to the test conditions (Figure 7). They observed a higher sensitivity for electrospun nanofiber-based sensors than other fabrication techniques (like PEDOT in situ polymerization). Regmi et al. investigated the hydrogen gas sensing mechanism of suspended graphene/PEDOT:PSS/PEO composite nanoscale channels created through near-field electrospinning [51]. Due to effective charge transfer at the graphene–PEDOT:PSS-PEO interface, these channels offer a higher surface area-to-volume ratio for efficient gas diffusion and enhanced response. The sensor achieves a 2% response to 1 ppm H_2 concentration with good linearity across a wide dynamic range. Additionally, arrays of nanoscale channels for improved sensitivity and integrated microheaters for rapid device recovery are implemented.

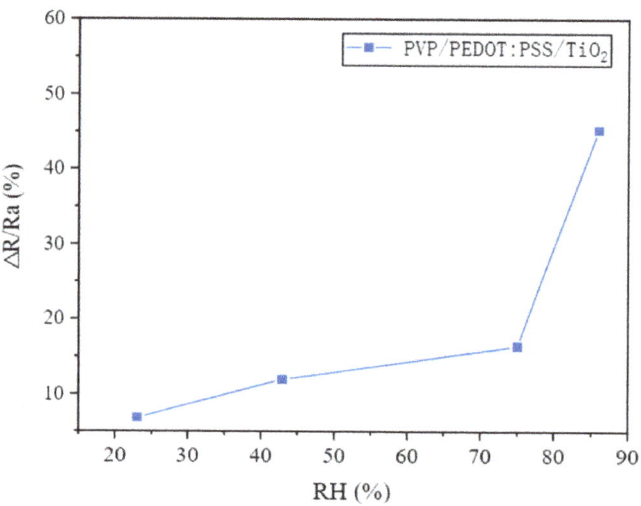

Figure 7. The influence of relative humidity, in the range of 11–86%, on the gas sensitivity response of the PVP/PEDOT:PSS/TiO_2 sensors [50].

Electrospun PEDOT:PSS/PVP composite nanofibers have been successfully used for sensing CO [52].

Numerous studies have explored the use of conducting polymers like polypyrrole, polythiophene, and polyaniline for gas sensor applications. In recent years, there has been a growing interest in poly(3,4-ethylenedioxythiophene)-poly(styrenesulfonate) (PE-

DOT:PSS) as a sensing material due to its exceptional electrical conductivity and environmental stability. Additionally, PEDOT:PSS/insulating polymer composites exhibit responsiveness to chemical vapors. For instance, electrospun polyvinyl alcohol/PEDOT:PSS (PVA/PEDOT:PSS) fibers have been developed for the detection of formaldehyde, ethanol, and other organic gases using the resistance measuring method. These fibers possess high porosity and specific surface area, offering improved sensitivity and faster response time compared to thin-film sensors.

Zhang et al. conducted a study involving creating conductive fibers by adding PEDOT:PSS conductive water solution into a 10% PVA water solution [53]. Electrospinning experiments were conducted under high-pressure airflow assistance, with the maximum spinning voltage set at 70 kV and a receiving distance of 120 cm. As a comparison, PVA/PEDOT:PSS composite fibers were also prepared using a traditional electrospinning device with a spinning voltage of 20 kV and a receiving distance of 15 cm. The primary objective was to investigate the impact of the specific surface area (size effect) of electrospun fibers on their response to external stimuli. Comparative tests were carried out on ultrafine nanofiber membranes and traditional electrospun fiber membranes to evaluate the response characteristics of PVA/PEDOT:PSS fibers, focusing on changes in electrical resistance in an ammonia environment. The study observed and discussed the size effect of electrospun fibers on gas-sensing properties such as sensitivity and response time.

J. Choi et al. fabricated PEDOT:PSS/PVP nanofibers via electrospinning, and the verification of the presence of PEDOT:PSS in the nanofibers was achieved through FT–Raman spectroscopy analysis [54]. Furthermore, the detection capabilities of the electrospun PEDOT:PSS/PVP nanofibers were investigated by assessing their reaction to repeated exposure to organic vapors like ethanol, methanol, THF, and acetone at ambient temperature. The exposure of PEDOT:PSS/PVP nanofibers to different solvents demonstrated contrasting electrical behaviors between protic and aprotic solvents. These outcomes suggest that the electrospun PEDOT:PSS/PVP nanofibers hold substantial promise as a material for sensing organic vapors.

6.1.3. Biosensors

PEDOT nanofibers are employed in biosensors to detect biological molecules such as glucose, DNA, and proteins, owing to their biocompatibility and conductive nature. For example, combining PEDOT with carbon nanofibers (CNFs) resulted in a novel composite material exhibiting low impedance, high surface area, high charge injection capability, and efficient neurotransmitter monitoring, making it highly suitable for neuronal therapies [55]. The PEDOT-modified microelectrodes showed superior electrochemical performance, with a low specific impedance of 1.28 MΩ μm^2 at 1 kHz and a high charge injection limit of 10.03 mC/cm^2, while demonstrating excellent sensitivity and low detection limits for dopamine and serotonin, and no cytotoxicity, highlighting PEDOT's potential for next-generation neural microelectrodes. Meng et al. (2020) present a method for creating a bi-functional PEDOT interface with a 3D nanofibrous network and carboxylic acid groups, enabling the development of all-polymer-based biosensors [56]. The Nano-PEDOT-COOH, with tunable fiber diameters and carboxylic acid group densities, showed excellent electrochemical properties and the capability for stable enzyme immobilization, resulting in a highly sensitive and responsive biosensor for lactate detection.

Electrospun PEDOT:PSS-based nanofibers have been developed for the fabrication of wearable biosensors for dopamine and ascorbic acid [45,57]. The nanofibers showed excellent biocompatibility and electrical conductivity, making them suitable for biosensing applications.

6.2. Biomedical

PEDOT exhibits several advantageous properties that make it highly beneficial for biomedical applications. Its excellent electrical conductivity and biocompatibility make PEDOT an ideal material for neural interfaces [55], where stable and reliable electrical performance is critical for stimulating and recording neuronal activity. The biocompatibility of PEDOT helps minimize inflammatory responses and tissue damage, which is crucial for long-term implantation [58]. Additionally, PEDOT's flexibility allows it to conform to the complex geometries of biological tissues, enhancing the interface between the device and the tissue [59]. This flexibility also makes PEDOT suitable for wearable biomedical devices, where it can maintain its conductive properties under mechanical deformation (Figure 8). Furthermore, PEDOT can be fabricated into various forms, such as thin films and electrospun nanofibers, enabling its use in several applications, from biosensors to biofuel cells. These properties collectively make PEDOT versatile and valuable in advancing biomedical technologies.

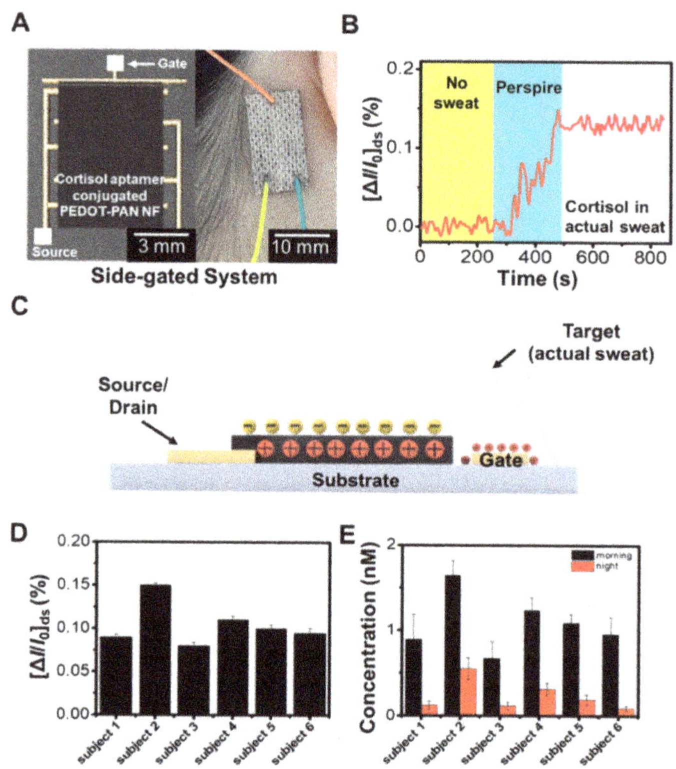

Figure 8. Application of the wearable cortisol aptasensor for detecting cortisol in actual sweat. (**A**) Illustration and optical image of the side-gated system cortisol aptasensor detecting cortisol in actual sweat. (**B**) Real-time monitoring using an actual sweat sample from subject 2. (**C**) Schematic illustration of the side-gated system mechanism. (**D**) Bar graph comparing the intensity change in subjects' actual sweat samples. (**E**) ELISA results of the subject in the morning (black) and at night (red) after cycling. Reproduced with permission from An et al. in ACS Sensors, published by American Chemical Society, 2022 [60].

For example, electrospun PEDOT:PSS-based nanofibers showed good biocompatibility and electrical conductivity, making them suitable for biomedical devices like biosensors and neural implants [61]. In another study, electrical conductive composite nanofibers were fabricated using PEDOT:PSS with cellulose nanofibrils (CNFs) [62]. The fabrication process involved electrospinning the composite nanofibers to create a water-resistant and cytocompatible material. Recently, highly electrically conductive nanofibers made of poly [2,2′-m-(phenylene)-5,5′-bibenzimidazole] (PBI) have been manufactured by electrospinning and then coated with cross-linked poly (3,4-ethylenedioxythiophene) doped with poly (styrene sulfonic acid) by spin-coating or dip-coating [63]. The methods used in the paper include the manufacture of highly electrically conductive nanofibers made of poly [2,2′-m-(phenylene)-5,5′-bibenzimidazole] (PBI) by electrospinning. These nanofibers were then coated with cross-linked poly (3,4-ethylenedioxythiophene) doped with poly (styrene sulfonic acid) (PEDOT:PSS) using spin-coating or dip-coating techniques. Characterization of the scaffolds was carried out through scanning electron microscopy (SEM) imaging and attenuated total reflectance Fourier Transform Infrared (ATR-FTIR) Spectroscopy. The electrical conductivity of the fibers was measured using the four-probe method, showing significant increases in conductivity for both spin-coated and dip-coated samples. Experimental parameters were carefully explored to achieve reproducible conductive nanofibers synthesis in large quantities, focusing on the critical importance of relative humidity during the electrospinning process and doping post-treatment involving glycols and alcohols. The synthesized fibers were assembled as a mat on glass substrates to form a conductive and transparent electrode, which was then fully characterized for its optoelectronic properties.

6.2.1. Tissue Engineering

The biocompatibility and conductive properties of PEDOT nanofibers make them suitable for scaffolds in tissue engineering, where they can support cell growth and differentiation [10]. For example, Abedi et al. (2019) investigated chitosan scaffolds with varying concentrations of PEDOT for cardiac tissue engineering [64]. The findings indicate that the addition of PEDOT enhances the scaffolds' mechanical properties, electrical conductivity, biocompatibility, and cell viability, with the 1 wt% PEDOT scaffold showing a significant 30–40% reduction in fiber diameter, a 100-fold increase in electrical conductivity, and a tensile strength increase of about 9 MPa, making it a promising candidate for cardiac tissue engineering applications. In another study, Jin et al. (2013) developed PEDOT nanofiber mats with superior mechanical properties (tensile strength: 8.7 ± 0.4 Mpa; Young's modulus: 28.4 ± 3.3 Mpa) and flexibility, alongside high biocompatibility comparable to tissue culture plates when tested with human cancer stem cells [17]. With outstanding electrical conductivity (7.8 ± 0.4 S cm^{-1}), these PEDOT nanofiber mats are promising for various biotechnology applications, including tissue engineering, drug delivery, cell culture, and implanted electrodes.

Electroconductive biomaterials have emerged to facilitate the recovery of degenerated electrically conductive tissues, particularly cardiac tissues following myocardial infarction. F. Furlani et al. investigated the fabrication of electroconductive scaffolds for the regeneration of cardiac tissue utilizing the biocompatible and conductive polymer PEDOT:PSS, in conjunction with a biomimetic gelatin polymer network [65]. The utilization of dehydrothermal (DHT) treatment under vacuum conditions produced structurally stable scaffolds without the need for additional cross-linking agents. These resultant scaffolds emulate the Young's modulus of native cardiac tissue and possess a well-interconnected porosity, along with excellent swelling capacity and stability under physiological conditions. Moreover, the incorporation of PEDOT:PSS serves to augment the electroconductivity of the resulting materials. All scaffolds exhibit non-cytotoxicity towards H9c2 cardiomyoblasts, with PEDOT:PSS promoting cell adhesion, particularly in the initial stages, essential for favorable outcomes post-implantation, as well as cell proliferation and spreading on the scaffolds. Given the favorable interaction of the scaffolds with cardiomyoblasts, these biomimetic and electroconductive scaffolds demonstrate potential as implantable biomaterials for regenerating electroconductive tissues, especially cardiac tissue, and as a promising 3D tissue model for in vitro screening of biomolecules.

6.2.2. Drug Delivery Systems

PEDOT nanofibers can be employed in drug delivery systems due to their ability to load and release therapeutic agents in a controlled manner. Chen et al. (2017) introduce multifunctional core–shell hybrid microfibers for biomedical applications, synthesized using coaxial spinning and dip-coating [66]. These microfibers are composed of a core made of bacterial cellulose and a shell layer of PEDOT, offering improved mechanical characteristics, controlled drug release, and responsiveness to electrical stimulation, demonstrating potential for applications in tissue engineering and drug delivery.

6.3. *Wearable Electronics*

The absence of mechanically strong fibers and yarns with good electrical conductivity impedes the development of electronic textiles. Conjugated polymer fibers show great promise in fulfilling those demands [67]. Electrospun composite nanofibers of PEDOT:PSS/cellulose acetate and PEDOT:PSS/PEO have been studied for electromagnetic interference (EMI) shielding [68]. The nanofibers showed excellent EMI shielding effectiveness and mechanical properties suitable for electronic devices. The EMI shielding is based on reflection and absorption loss. The former is the weakening of the electromagnetic wave when passing through materials with mismatching dielectric constants. The latter is the weakening when propagating inside a material. The reflection loss is enhanced in nanofiber-based materials due to the large surface to volume ratio and can be further improved with coaxial or multiwire fiber geometries. Absorption loss can be achieved if the thickness of the fabric is sufficient and can be improved by adding nanofillers with magnetic properties to the fibers.

Stretchable conductors and organic electrochemical transistors (OECTs) were fabricated from PEDOT:Tos (poly (3,4-ethylenedioxythiophene):iron tosylate) nanofibers [69]. The devices were prepared by a combination of electrospinning and electrode printing followed by vapor-phase polymerization (VPP). The impact of both the processing time and the composition of three electrospinning mixtures on the electrospun fiber mats was evaluated by scanning electron microscopy and cyclic voltammetry. Fibrillar mats prepared from the different mixtures maintained their electrical properties and could be stretched up to 140% of their original length. Stretchable OECTs were fabricated by printing silver drain and source electrodes directly on the conductive spun fibers. The fabricated devices showed transistor behavior up to ~50% strain.

Novel evidence is shown on the capability of tuning the conductance and photoresponse of composite core–shell nanofibers, based on the doping of the PEDOT:PSS phase with different solvents and PbS nanoparticles and the arrangement of the core–shell phases [70]. The study employed coaxial electrospinning to synthesize core–shell PEDOT:PSS–polyvinylpirrolidone nanofibers. The fibers were doped with different solvents such as dimethyl sulfoxide (DMSO), isopropyl alcohol (IPA), ethylene glycol, and PbS nanoparticles at varying concentrations.

Flexible Soft Electronics

Recent advancements in flexible and conductive nanofiber materials highlight their potential in various applications, including wearable electronics, and in flexible and conductive materials [71]. With the expansion of the Internet of Things (IoT), a demand arises for portable and wearable devices that exceed the capacities of conventional silicon-based electronics, thereby leading to the emergence of novel nano-systems integrating conductive electrospun fibers. Furthermore, flexible and transparent materials represent alternatives to expensive and environmentally unsustainable electrodes based on indium and tin used in various fields, such as smart displays, energy, and wearables. However, few alternatives have all the desired properties. Due to the limitations of indium tin oxide, alternative electrodes like metal films and nanowires have emerged. Nanofibers are seen as an ideal material for high-performance transparent electrodes [72]. Electrodes of silver nanoparticles dispersed in electrospun PEDOT nanofibers have been studied for their high conductivity and mechanical flexibility, with a transparency of 77% [73]. A highly conductive nanocomposite created by grafting polyvinyl alcohol onto surface-modified carbon nanofibers demonstrated significant improvements in flexibility, conductivity, and biodegradability, making it suitable for biofuel cell electrodes and wearable electronics [60].

A recent work by Gramont et al. develops highly stretchable poly-(3,4-ethylenedioxythiphene) doped with tosylate (PEDOT:Tos) nanofibers that can be stretched up to 140% of the initial length, maintaining high conductivity [74]. The methods used in the paper involved a two-step process which included electrospinning of a carrier polymer with an oxidant and vapor-phase polymerization to produce highly stretchable poly-(3,4-ethylenedioxythiphene) (PEDOT) doped with tosylate (PEDOT:Tos) nanofibers on a polydimethylsiloxane substrate.

A process to synthesize continuous conducting nanofibers was developed using PEDOT:PSS as a conducting polymer and an electrospinning method, producing a conformable conductive and transparent coating that is well adapted to nonplanar surfaces, with very large aspect ratio features [75].

The rise of lightweight, flexible, and portable electronics and wearable devices has led to concerns over increased electromagnetic interference (EMI) levels. To mitigate this issue, the urgent development of lightweight, hydrophobic, and flexible EMI shielding materials is necessary.

Lee et al. developed core–shell nanofibers of PVDF with a highly conductive PEDOT shell for EMI shielding, showcasing the core–shell formation through vapor-phase polymerization [76]. The resulting nanofibers, known for flexibility and mechanical strength, exhibited an enhanced EMI SE of 40 dB in the X-band range due to increased internal reflections from porosity, with specific shielding effectiveness reaching 16,230 dB cm^2 g^{-1} at a 14 μm thickness, attributed to the PEDOT shell's high conductivity, durability, and hydrophobic properties.

Through the fabrication of poly(3,4-ethylenedioxythiophene): polystyrene sulfonate (PEDOT:PSS) and polyvinylpyrrolidone (PVP)-coated, carbonized electrospun polyacrylonitrile (PAN) nanofibers (CNFs) and their flexible polydimethylsiloxane (PDMS) composites, a highly effective EMI shielding material with a shielding effectiveness of 44 dB in the frequency range of 8–26.5 GHz and an absolute EMI SE of 5678 dB cm^2 g^{-1} has been created [77]. The shielding mechanism of PEDOT:PSS-PVP/CNF composites is primarily absorption-based, highlighting their potential as practical applications for thin, flexible, hydrophobic, and lightweight EMI shielding materials.

6.4. Energy Storage

PEDOT nanofibers are widely used in supercapacitors due to their high electrical conductivity and large surface area [24]. Their fibrous morphology provides a porous structure, enhancing ion diffusion and storage capacity [78], as represented in Figure 9.

This enhancement can be explained in terms of the hierarchical substructure of the yarns and mats, which can be easily tuned to allow optimized electronic transportation and ion diffusion from the interior of the material to the surface by shortening the ion diffusion distance. The large surface area-to-volume ratio and the easy tunability of the fiber structure are indeed intrinsic advantages of the electrospinning technique. On the other hand, efficient electron charge transfer is enabled by the high conductivity of the PEDOT polymer.

PEDOT nanofibers produced through electrospinning and vapor-phase polymerization had a 350 nm diameter and high conductivity (60 ± 10 S cm^{-1}), the highest reported for polymer nanofibers [79]. The mats, with soldered intersections, exhibited strong dimensional stability. These ultra-porous mats enhanced electrochemical properties. All-textile flexible supercapacitors integrated the mats with carbon cloths and PAN nanofibrous membranes. The supercapacitors were flexible, utilizing a solid electrolyte with an ionic liquid and PVDF-co-HFP. These devices displayed stable performance in ambient conditions.

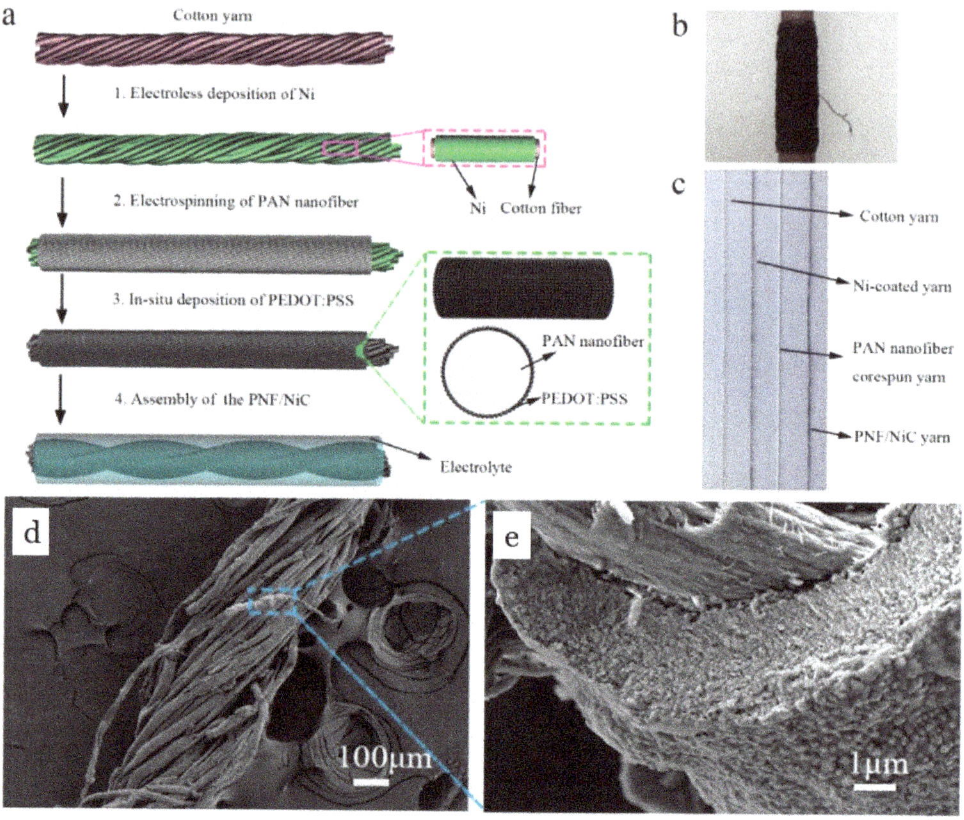

Figure 9. (**a**) Scheme of the fabrication of the PNF/NiC capacitor yarn. (**b**) Photograph of the Ni-coated cotton yarn wound on the spinning cone. (**c**) Photograph of various yarns during processing. (**d**) High-magnification image of polyacrylonitrile nanofiber core-spun yarn and (**e**) its cross-section [78].

A recent study by Du et al. (2022) presents a low-cost method for creating mechanically strong and conductive PEDOT bulk films using cellulose nanofibrils (CNFs) as building blocks [80]. The resulting PEDOT/CNF nano-paper exhibits excellent flexibility, high tensile strength (72 MPa), and high electrical conductivity (66.67 S/cm), making it a promising electrode material for flexible supercapacitors with outstanding cycling stability and high performance.

Finally, PEDOT:PSS electrodes were used in supercapacitors with polyvinyl alcohol/phosphoric acid (PVA/H_3PO_4) gel polyelectrolyte, forming an all-solid-state SC [17]. The methods employed in this paper involved fabricating flexible and transparent supercapacitors using electrospun PEDOT:PSS electrodes. These electrodes were prepared through electrospinning techniques to enhance their flexibility and transparency.

PEDOT nanofibers serve also as conductive additives in lithium-ion and sodium-ion batteries, improving charge transport and overall battery performance [81]. For example, in a recent work by Zeng et al. (2021), a network of carbon nanofibers (NCNF) was prepared to form a composite with sulfur for lithium–sulfur batteries but initially showed low specific capacity and poor cycle stability due to poor conductivity and large pore size [82]. Coating the NCNF with PEDOT significantly improved conductivity and addressed lithium polysulfide dissolution/loss, resulting in a composite electrode with a high sulfur loading that achieved a specific capacity of 793 mAh/g at 0.1 C, maintaining 623 mAh/g after 200 cycles, thus enhancing low current rate cyclability.

A. Laforgue created Poly(3,4-ethylenedioxythiophene)/manganese oxide coated on porous carbon nanofibers (P-CNFs/PEDOT/MnO_2) through a combination of electrospinning, carbonization, and electrodeposition, forming an advanced anode material [83]. The structural and morphological analysis shows that the rough surface of the electrode provides active sites for Li+ storage due to the presence of PEDOT nanoparticles and irregular block-shaped MnO_2. The electrode demonstrates superior electrochemical performance with a discharge capacity of 1477 mAh/g at 2 mA/g, outperforming other materials like P-CNFs/PEDOT (1191 mAh/g) and P-CNFs/MnO_2 (763 mAh/g). After 20 cycles, the P-CNFs/PEDOT/MnO_2 electrode maintains satisfactory performance with a Coulombic efficiency above 90% and a lower charge transfer resistance compared to other electrodes. Overall, the results suggest that this electrode has the potential to replace commercial graphite in lithium-ion batteries.

7. Challenges and Future Perspectives

Electrospun PEDOT-based nanofibers hold significant promise for various applications, yet they face several challenges related to scalability, reproducibility, stability, and integration into practical devices.

Scalability is one of the primary challenges of the electrospinning process, which is often limited by the need for precise control over fiber morphology and uniformity. This issue can be addressed by optimizing the electrospinning parameters and developing advanced fabrication techniques that ensure consistent fiber production at a larger scale. Recent strategies include the use of multi-needle systems, auxiliary electrodes, and controlled environmental conditions, each contributing to overcoming specific limitations of traditional electrospinning, focusing on enhancing production capacity, and ensuring consistent fiber quality. The use of multiple electrospinning needles operating in parallel is a prominent strategy to increase production capacity. This approach, however, introduces challenges such as uneven electric field distribution due to Coulombic repulsion between charged jets and needles. To mitigate this, auxiliary electrodes have been employed to uniformize the electric field, thereby improving the deposition area and fiber morphology. Implementing parallel lateral plates as auxiliary electrodes has been shown to effectively address the uneven electric field distribution in multi-needle systems. This setup helps achieve a uniform electric field, which is crucial for consistent fiber production. The use of finite element simulations has further validated the impact of these electrodes on the electric field intensity, leading to high-quality polymer nanofibers [84].

Reproducibility is another critical concern, as variations in fiber properties can lead to inconsistent device performance. Enhancing reproducibility can be achieved by implementing standardized protocols and integrating real-time monitoring systems in the electrospinning process to guarantee consistency in the characteristics of the fibers. Reproducibility in electrospinning is significantly influenced by environmental conditions, particularly relative humidity (RH). A miniaturized electrospinning setup, designed to fit within a biological safety cabinet, allows precise control of RH, enhancing the reproducibility of fiber production. This setup demonstrated that controlling RH can significantly affect fiber diameter, thus ensuring consistent quality across production batches [85]. While these optimization strategies have advanced the scalability and reproducibility of electrospinning, challenges remain, particularly in balancing high throughput with fiber quality. Future research may focus on integrating these advanced electrospinning configurations with automated systems to further enhance production efficiency and consistency.

Real-time monitoring of nanofiber quality is conducted by means of interferometry: in industrial roll-to-roll setups, fiber thickness can be estimated with an optical interferometer, which provides feedback to the flow rate controller of the electrospinning instrument. Furthermore, a novel optical system integrated through a coaxial needle has recently been proposed to monitor the process, aiming to enhance the reproducibility and quality of nanofibers [86]. The system uses an optical fiber to pass light through the polymeric solution and couple it to the electrospun fiber. Experimental tests confirmed that the system effectively tracks changes in the electrospun fiber's waveguide properties and nanofiber diameter, offering a promising solution for industrial-scale electrospinning using multi- or moving needles.

Stability, particularly in harsh environmental conditions, is essential for the practical application of these nanofibers. To the best of our knowledge, no specific study has been published on the stability of PEDOT-based nanofibers yet. Nevertheless, a few recently published studies have addressed the stability of PEDOT-based thin films investigating the aging/degradation mechanism and possible enhancement methods [87–89]. Such improvements could be obtained by surface modifications [90], stabilizing agents [88], or by coatings that protect the fibers from degradation due to moisture, temperature, and UV exposure [87,89].

Integrating practical devices, such as wearable sensors and electronic skins, requires the nanofibers to exhibit excellent mechanical properties, biocompatibility, and electrical performance. This can be facilitated by developing composite materials that combine PEDOT with other polymers or nanomaterials to enhance their mechanical strength and flexibility while maintaining their electrical properties. For conducting polymers like PEDOT:PSS, solution formulation improvements have been crucial. By optimizing the polymer concentration and solution properties, researchers have achieved significant enhancements in fiber conductivity and power factor, which are essential for applications in wearable electronics [26]. For example, stretchable conducting polymeric films were fabricated by blending PEDOT:PSS with poly(vinyl alcohol) in polar organic solvents like DMSO and EG [91], resulting in enhanced electrical and mechanical properties. The study found that DMSO and EG induce crystalline domains and structural changes in PEDOT, significantly boosting electrical performance and stretchability. It is reasonable to expect a similar behavior in nonwoven materials, too.

Future research should focus on developing multifunctional nanofibers that simultaneously address multiple performance criteria, such as conductivity, flexibility, and durability. Additionally, exploring advanced characterization techniques to understand the structure–property relationships in PEDOT-based nanofibers will be crucial for optimizing their performance in various applications. Adopting green and sustainable electrospinning processes will also be vital to minimize the environmental impact of large-scale production. In fact, this fabrication technique, which is a form of additive manufacturing, offers several environmental benefits. Firstly, electrospinning is highly efficient, allowing for precise control over material usage, thereby minimizing waste. The process typically operates at ambient temperatures, especially when carried out in a solution-based configuration, reducing energy consumption compared to traditional manufacturing methods. Additionally, the ability to utilize biodegradable or recyclable polymers in electrospinning further enhances its environmental credentials. Furthermore, traditional fiber production methods like Chemical Vapor Deposition (CVD), melt blowing, or solution blow spinning often require large amounts of solvents or other chemicals, which can lead to hazardous waste and environmental pollution. Electrospinning can use greener solvents or even solvent-free approaches, thereby minimizing chemical waste and the environmental impact. However, while the potential for sustainability is clear, comprehensive Life Cycle Assessment (LCA) studies are still needed to fully quantify the environmental impact of large-scale electrospinning and to identify areas for further improvement. By addressing these challenges through targeted research and development, PEDOT-based nanofibers can be effectively scaled up and integrated into a wide range of practical devices, facilitating widespread adoption in different fields, such as healthcare, environmental monitoring, and flexible electronics. Current wearable sensors often lack comfort and face issues with signal interference from different stimuli. Functional sensors that address these challenges are essential for providing reliable interaction between humans and their environment. PEDOT-based nonwovens are solid candidates for the development, for example, of flexible, stretchable electronic skins that mimic natural skin for integration in smart clothing [92,93].

8. Conclusions

The production of nanofibers has become a crucial area of study because of their distinct characteristics and diverse applications in various fields like biomedicine, textiles, energy, and environmental science. Electrospinning has gained notable attention as a method for creating nanofibers due to its adaptability, scalability, and capability to produce nanofibers with specific properties. It involves applying a high-voltage electrical field to a polymer solution or melt, forming a thin jet that elongates and solidifies into continuous nanofibers as the solvent evaporates or the polymer solidifies. These nanofibers have diameters ranging from a few nanometers to several micrometers, exhibiting exceptional uniformity and purity. The high surface area-to-volume ratio of electrospun nanofibers enhances their interactions with the surroundings, making them ideal for filtration, tissue engineering, drug delivery, and sensing applications. Additionally, the properties of these nanofibers can be easily adjusted by modifying parameters such as polymer concentration, solvent type, applied voltage, and collection distance, offering customization to meet specific needs across various fields. Besides their customizable properties, electrospun nanofibers are also scalable and cost-effective, suitable for industrial-scale production. The simplicity of the electrospinning process, along with its ability to produce uniform and pure nanofibers, presents an appealing option for large-scale manufacturing. Conductive polymers have gained attention for their unique combination of electrical conductivity, mechanical flexibility, and processability, opening new possibilities for advanced electronic devices and energy storage systems. Various experiments have been carried out on polymers like polyaniline (PANI), polypyrrole (PPy), polythiophenes, etc. Among these, poly(3,4-ethylenedioxythiophene) (PEDOT) is notable for its exceptional conductivity, environmental stability, and ease of synthesis. Belonging to the polythiophene family, PEDOT is characterized by its π-conjugated structure with alternating double and single bonds along the polymer backbone, providing high electrical conductivity. The presence of ethylenedioxy (EDOT) functional groups enhances the solubility and processability of PEDOT, facilitating the production of thin films, coatings, and composites. Electrospun PEDOT-based nanofibers have adjustable electrical and optical properties, suitable for applications in organic electronics, energy storage, biomedicine, and wearable technology. While electrospinning offers significant advantages for PEDOT-based conductive nanofibers, it also presents challenges such as the need for precise control over processing parameters to achieve desired fiber characteristics. Additionally, the inherent low molecular weight and rigidity of conductive polymers like PEDOT can complicate the electrospinning process, often requiring the use of carrier polymers or coaxial spinning techniques to achieve stable fiber formation. Despite these challenges, the versatility and adaptability of electrospinning make it a promising method for advancing the development of high-performance conductive nanofibers across various applications.

Author Contributions: Conceptualization, E.A.S. and M.S.; writing, original draft preparation, E.A.S.; writing, review, and editing, E.A.S., G.C., X.H. and M.S.; funding acquisition, M.S. All authors have read and agreed to the published version of the manuscript.

Funding: This work was supported in part by the Italian Ministry of Foreign Affairs and International Cooperation, grant number CN24GR09 (CUP B53C24001100001). Funded by the European Union—NextGenerationEU and by the Ministry of University and Research (MUR), National Recovery and Resilience Plan (NRRP), Mission 4, Component 2, Investment 1.5, project "RAISE—Robotics and AI for Socio-economic Empowerment" (ECS00000035). E.A. Slejko, G. Carraro, and M. Smerieri are part of RAISE Innovation Ecosystem. We also acknowledge financial support under the National Recovery and Resilience Plan (NRRP), Mission 4, Component 2, Investment 1.1, Call for tender No. 104 published on 2.2.2022 by the Italian Ministry of University and Research (MUR), funded by the European Union—NextGenerationEU—Project Title "SMART non-woven fabrics for multifUnctional wearable haPtic devices (SMART_UP)"—CUP B53D23008810006—Grant Assignment Decree No. 966 adopted on 30.06.2023 by the Italian Ministry of Ministry of University and Research (MUR).

Institutional Review Board Statement: Not applicable.

Data Availability Statement: No new data were created or analyzed in this study. Data sharing is not applicable to this article.

Conflicts of Interest: The authors declare no conflicts of interest.

References

1. Yi, H.; Patel, R.; Patel, K.D.; Bouchard, L.-S.; Jha, A.; Perriman, A.W.; Patel, M. Conducting Polymer-Based Scaffolds for Neuronal Tissue Engineering. *J. Mater. Chem. B* **2023**, *11*, 11006–11023. [CrossRef]
2. Seiti, M.; Giuri, A.; Corcione, C.E.; Ferraris, E. Advancements in Tailoring PEDOT: PSS Properties for Bioelectronic Applications: A Comprehensive Review. *Biomater. Adv.* **2023**, *154*, 213655. [CrossRef] [PubMed]
3. Keirouz, A.; Wang, Z.; Reddy, V.S.; Nagy, Z.K.; Vass, P.; Buzgo, M.; Ramakrishna, S.; Radacsi, N. The History of Electrospinning: Past, Present, and Future Developments. *Adv. Mater. Technol.* **2023**, *8*, 2201723. [CrossRef]
4. Asai, H. Actuators with Nanofiber Mat Electrodes: Effect of Electrode Preparation Method on Actuator Performance. *Polym. J.* **2021**, *53*, 1083–1091. [CrossRef]
5. Long, Y.-Z.; Li, M.-M.; Gu, C.; Wan, M.; Duvail, J.-L.; Liu, Z.; Fan, Z. Recent Advances in Synthesis, Physical Properties and Applications of Conducting Polymer Nanotubes and Nanofibers. *Prog. Polym. Sci.* **2011**, *36*, 1415–1442. [CrossRef]
6. Verpoorten, E.; Massaglia, G.; Ciardelli, G.; Pirri, C.F.; Quaglio, M. Design and Optimization of Piezoresistive PEO/PEDOT:PSS Electrospun Nanofibers for Wearable Flex Sensors. *Nanomaterials* **2020**, *10*, 2166. [CrossRef] [PubMed]
7. Gawali, C.R.; Daweshar, E.; Kolhe, A.; Kumar, S. Recent Advances in Nanostructured Conducting Polymer Electrospun for Application in Electrochemical Biosensors. *Microchem. J.* **2024**, *200*, 110326. [CrossRef]
8. Nguyen, H.D.; Ko, J.M.; Kim, H.J.; Kim, S.K.; Cho, S.H.; Nam, J.D.; Lee, J.Y. Electrochemical Properties of Poly(3,4-Ethylenedioxythiophene) Nanofiber Non-Woven Web Formed by Electrospinning. *J. Nanosci. Nanotechnol.* **2008**, *8*, 4718–4721. [CrossRef]
9. Kim, N.; Kee, S.; Lee, S.H.; Lee, B.H.; Kahng, Y.H.; Jo, Y.; Kim, B.; Lee, K. Transparent Electrodes: Highly Conductive PEDOT:PSS Nanofibrils Induced by Solution-Processed Crystallization (Adv. Mater. 14/2014). *Adv. Mater.* **2014**, *26*, 2109. [CrossRef]
10. Babaie, A.; Bakhshandeh, B.; Abedi, A.; Mohammadnejad, J.; Shabani, I.; Ardeshirylajimi, A.; Moosavi, S.R.; Amini, J.; Tayebi, L. Synergistic Effects of Conductive PVA/PEDOT Electrospun Scaffolds and Electrical Stimulation for More Effective Neural Tissue Engineering. *Eur. Polym. J.* **2020**, *140*, 110051. [CrossRef]
11. Luo, Y.; Zhao, L.; Luo, G.; Dong, L.; Xia, Y.; Li, M.; Li, Z.; Wang, K.; Maeda, R.; Jiang, Z. Highly Sensitive Piezoresistive and Thermally Responsive Fibrous Networks from the In Situ Growth of PEDOT on MWCNT-Decorated Electrospun PU Fibers for Pressure and Temperature Sensing. *Microsyst. Nanoeng.* **2023**, *9*, 113. [CrossRef]
12. Lang, U.; Müller, E.; Naujoks, N.; Dual, J. Microscopical Investigations of PEDOT:PSS Thin Films. *Adv. Funct. Mater.* **2009**, *19*, 1215–1220. [CrossRef]
13. Barroso-Solares, S.; Pinto, J.; Salvo-Comino, C.; Cuadra-Rodríguez, D.; García-Cabezón, C.; Rodríguez-Pérez, M.A.; Rodríguez-Méndez, M.L. Tuning the Electrochemical Response of PCL-PEDOT:PSS Fibers-Based Sensors by Gas Dissolution Foaming. *Appl. Surf. Sci.* **2023**, *638*, 158062. [CrossRef]
14. Laforgue, A. Electrically Controlled Colour-Changing Textiles Using the Resistive Heating Properties of PEDOT Nanofibers. *J. Mater. Chem.* **2010**, *20*, 8233–8235. [CrossRef]
15. Laforgue, A.; Robitaille, L. Production of Conductive PEDOT Nanofibers by the Combination of Electrospinning and Vapor-Phase Polymerization. *Macromolecules* **2010**, *43*, 4194–4200. [CrossRef]
16. Park, H.; Lee, S.J.; Kim, S.; Ryu, H.W.; Lee, S.H.; Choi, H.H.; Cheong, I.W.; Kim, J.-H. Conducting Polymer Nanofiber Mats via Combination of Electrospinning and Oxidative Polymerization. *Polymer* **2013**, *54*, 4155–4160. [CrossRef]
17. Jin, L.; Wang, T.; Feng, Z.-Q.; Leach, M.K.; Wu, J.; Mo, S.; Jiang, Q. A Facile Approach for the Fabrication of Core–Shell PEDOT Nanofiber Mats with Superior Mechanical Properties and Biocompatibility. *J. Mater. Chem. B* **2013**, *1*, 1818–1825. [CrossRef]
18. Park, S.W.; Oh, T.H.; Hwang, J.S.; Lee, Y.J. Effect of Solvent and Blended Polymer on Electrical Conductivity of PEDOT:PSS/Polymer Blended Nanofibers. *Fibers Polym.* **2016**, *17*, 1171–1174. [CrossRef]
19. Du, L.; Li, T.; Jin, F.; Wang, Y.; Li, R.; Zheng, J.; Wang, T.; Feng, Z.-Q. Design of High Conductive and Piezoelectric Poly (3,4-Ethylenedioxythiophene)/Chitosan Nanofibers for Enhancing Cellular Electrical Stimulation. *J. Colloid Interface Sci.* **2020**, *559*, 65–75. [CrossRef]
20. Yu, C.-C.; Ho, B.-C.; Juang, R.-S.; Hsiao, Y.-S.; Naidu, R.V.R.; Kuo, C.-W.; You, Y.-W.; Shyue, J.-J.; Fang, J.-T.; Chen, P. Poly(3,4-Ethylenedioxythiophene)-Based Nanofiber Mats as an Organic Bioelectronic Platform for Programming Multiple Capture/Release Cycles of Circulating Tumor Cells. *ACS Appl. Mater. Interfaces* **2017**, *9*, 30329–30342. [CrossRef]
21. Angkawinitwong, U.; Williams, G.R. *Electrospun Polymers and Composites*; Elsevier: Amsterdam, The Netherlands, 2021; pp. 405–432. [CrossRef]
22. Blachowicz, T.; Ehrmann, A. Conductive Electrospun Nanofiber Mats. *Materials* **2019**, *13*, 152. [CrossRef] [PubMed]

23. Long, Y.-Z.; Yan, X.; Wang, X.-X.; Zhang, J.; Yu, M. *Electrospinning: Nanofabrication and Applications*; Elsevier: Amsterdam, The Netherlands, 2019; pp. 21–52. [CrossRef]
24. Cárdenas-Martínez, J.; España-Sánchez, B.L.; Esparza, R.; Ávila-Niño, J.A. Flexibleand Transparent Supercapacitors Using Electrospun PEDOT:PSS Electrodes. *Synth. Met.* **2020**, *267*, 116436. [CrossRef]
25. Sahar, H.; Naseer, S.; Gilani, M.R.H.S.; Naqvi, S.A.R.; Aamir, M.; Akhtar, J. *Sustainable Materials for Electrochemical Capacitors*; John Wiley & Sons: Hoboken, NJ, USA, 2023; pp. 361–390. [CrossRef]
26. Ewaldz, E.; Rinehart, J.M.; Miller, M.; Brettmann, B. Processability of Thermoelectric Ultrafine Fibers via Electrospinning for Wearable Electronics. *ACS Omega* **2023**, *8*, 30239–30246. [CrossRef] [PubMed]
27. Liu, J.; Zhuo, Y.; Lu, J.; Wang, L.; Ren, X. Novel Conductive RGO/PEDOT:PSS/PVA Nanofibers by Electrospinning. *Integr. Ferroelectr.* **2022**, *229*, 305–312. [CrossRef]
28. Kausar, A.; Ahmad, I. Electrospinning Processing of Polymer/Nanocarbon Nanocomposite Nanofibers—Design, Features, and Technical Compliances. *J. Compos. Sci.* **2023**, *7*, 290. [CrossRef]
29. Serrano-Garcia, W.; Ramakrishna, S.; Thomas, S.W. Electrospinning Technique for Fabrication of Coaxial Nanofibers of Semiconductive Polymers. *Polymers* **2022**, *14*, 5073. [CrossRef]
30. Acosta, M.; Santiago, M.D.; Irvin, J.A. Electrospun Conducting Polymers: Approaches and Applications. *Materials* **2022**, *15*, 8820. [CrossRef]
31. Zarrin, N.; Tavanai, H.; Abdolmaleki, A.; Bazarganipour, M.; Alihosseini, F. An Investigation on the Fabrication of Conductive Polyethylene Dioxythiophene (PEDOT) Nanofibers through Electrospinning. *Synth. Met.* **2018**, *244*, 143–149. [CrossRef]
32. Pinto, N.J.; Rivera, D.; Melendez, A.; Ramos, I.; Lim, J.H.; Johnson, A.T.C. Electrical Response of Electrospun PEDOT-PSSA Nanofibers to Organic and Inorganic Gases. *Sens. Actuators B Chem.* **2011**, *156*, 849–853. [CrossRef]
33. Oliva, F.S.N.; Picart, L.; Leon-Valdivieso, C.Y.; Benalla, A.; Lenglet, L.; Ospina, A.; Jestin, J.; Bedoui, F. Coaxial Electrospinning Process toward Optimal Nanoparticle Dispersion in Polymeric Matrix. *Polym. Compos.* **2021**, *42*, 1565–1573. [CrossRef]
34. Han, D.; Steckl, A.J. Coaxial Electrospinning Formation of Complex Polymer Fibers and Their Applications. *ChemPlusChem* **2019**, *84*, 1453–1497. [CrossRef] [PubMed]
35. Costa, E.L.; Muniz, E.C.; Cava, C.E. PEDOT:PSS/AgNWs Nanofibers Obtained by Electrospun and the Post-Treatment via DMSO Vapor Exposure. *Synth. Met.* **2023**, *298*, 117442. [CrossRef]
36. Lim, J.; Choi, S.; Kim, H.S. Electrospinning Behavior and Piezoelectric Property of PVDF/PEDOT:PSS Composite Web by PEDOT:PSS Addition. *Polym. Adv. Technol.* **2023**, *34*, 1959–1966. [CrossRef]
37. Huang, T.-M.; Batra, S.; Hu, J.; Miyoshi, T.; Cakmak, M. Chemical Cross-Linking of Conducting Poly(3,4-Ethylenedioxythiophene): Poly(Styrenesulfonate) (PEDOT:PSS) Using Poly(Ethylene Oxide) (PEO). *Polymer* **2013**, *54*, 6455–6462. [CrossRef]
38. Chotimah; Winandianto, B.; Munir, M.; Kartini, I.; Kusumaatmaja, A.; Triyana, K. The Electrical Properties of PEDOT:PSS Nanofibers. *AIP Conf. Proc.* **2016**, *1755*, 150013. [CrossRef]
39. Basri, N.A.F.; Mustafa, M.N.; Sulaiman, Y. Facile Fabrication of PVA Nanofiber Coated with PEDOT as a Counter Electrode for Dye-Sensitized Solar Cell. *J. Mater. Sci. Mater. Electron.* **2019**, *30*, 8705–8711. [CrossRef]
40. Ebadi, S.V.; Fashandi, H.; Semnani, D.; Rezaei, B.; Fakhrali, A. Electroactive Actuator Based on Polyurethane Nanofibers Coated with Polypyrrole through Electrochemical Polymerization: A Competent Method for Developing Artificial Muscles. *Smart Mater. Struct.* **2020**, *29*, 045008. [CrossRef]
41. Shaker, A.; Hassanin, A.H.; Shaalan, N.M.; Hassan, M.A.; El-Moneim, A.A. Micropatterned Flexible Strain Gauge Sensor Based on Wet Electrospun Polyurethane/PEDOT: PSS Nanofibers. *Smart Mater. Struct.* **2019**, *28*, 075029. [CrossRef]
42. Zhu, C.; Zheng, J.; Fu, J. Electrospinning Nanofibers as Stretchable Sensors for Wearable Devices. *Macromol. Biosci.* **2024**, *24*, e2300274. [CrossRef]
43. Chen, G.; Chen, G.; Pan, L.; Chen, D. Electrospun Flexible PVDF/GO Piezoelectric Pressure Sensor for Human Joint Monitoring. *Diam. Relat. Mater.* **2022**, *129*, 109358. [CrossRef]
44. Ding, Y.; Xu, W.; Wang, W.; Fong, H.; Zhu, Z. Scalable and Facile Preparation of Highly Stretchable Electrospun PEDOT:PSS@PU Fibrous Nonwovens toward Wearable Conductive Textile Applications. *ACS Appl. Mater. Interfaces* **2017**, *9*, 30014–30023. [CrossRef] [PubMed]
45. Verpoorten, E.; Massaglia, G.; Pirri, C.F.; Quaglio, M. Electrospun PEO/PEDOT:PSS Nanofibers for Wearable Physiological Flex Sensors. *Sensors* **2021**, *21*, 4110. [CrossRef]
46. Tang, J.; Wu, Y.; Ma, S.; Yan, T.; Pan, Z. Sensing Mechanism of a Flexible Strain Sensor Developed Directly Using Electrospun Composite Nanofiber Yarn with Ternary Carbon Nanomaterials. *iScience* **2022**, *25*, 105162. [CrossRef]
47. Zhou, Z.; Tang, W.; Xu, T.; Zhao, W.; Zhang, J.; Bai, C. Flexible Strain Sensors Based on Thermoplastic Polyurethane Fabricated by Electrospinning: A Review. *Sensors* **2024**, *24*, 4793. [CrossRef] [PubMed]
48. Seyedin, M.Z.; Razal, J.M.; Innis, P.C.; Wallace, G.G. Strain-Responsive Polyurethane/PEDOT:PSS Elastomeric Composite Fibers with High Electrical Conductivity. *Adv. Funct. Mater.* **2014**, *24*, 2957–2966. [CrossRef]
49. Huang, Y.-C.; Lo, T.-Y.; Chen, C.-H.; Wu, K.-H.; Lin, C.-M.; Whang, W.-T. Electrospinning of Magnesium-Ion Linked Binder-Less PEDOT:PSS Nanofibers for Sensing Organic Gases. *Sens. Actuators B Chem.* **2015**, *216*, 603–607. [CrossRef]
50. Shiu, B.-C.; Liu, Y.-L.; Yuan, Q.-Y.; Lou, C.-W.; Lin, J.-H. Preparation and Characterization of PEDOT:PSS/TiO_2 Micro/Nanofiber-Based Gas Sensors. *Polymers* **2022**, *14*, 1780. [CrossRef]

51. Regmi, A.; Shin, D.; Na, N.; Chang, J. Suspended Graphene/PEDOT: PSS-PEO Channel for H_2 Gas Sensing Fabricated Using Direct-Write Functional Fibers. *Adv. Mater. Technol.* **2023**, *8*, 2201505. [CrossRef]
52. Zhang, H.-D.; Yan, X.; Zhang, Z.-H.; Yu, G.-F.; Han, W.-P.; Zhang, J.-C.; Long, Y.-Z. Electrospun PEDOT:PSS/PVP Nanofibers for CO Gas Sensing with Quartz Crystal Microbalance Technique. *Int. J. Polym. Sci.* **2016**, *2016*, 3021353. [CrossRef]
53. Zhang, Q.; Wang, X.; Fu, J.; Liu, R.; He, H.; Ma, J.; Yu, M.; Ramakrishna, S.; Long, Y. Electrospinning of Ultrafine Conducting Polymer Composite Nanofibers with Diameter Less than 70 Nm as High Sensitive Gas Sensor. *Materials* **2018**, *11*, 1744. [CrossRef]
54. Choi, J.; Lee, J.; Choi, J.; Jung, D.; Shim, S.E. Electrospun PEDOT:PSS/PVP Nanofibers as the Chemiresistor in Chemical Vapour Sensing. *Synth. Met.* **2010**, *160*, 1415–1421. [CrossRef]
55. Saunier, V.; Flahaut, E.; Blatché, M.-C.; Bergaud, C.; Maziz, A. Carbon Nanofiber-PEDOT Composite Films as Novel Microelectrode for Neural Interfaces and Biosensing. *Biosens. Bioelectron.* **2020**, *165*, 112413. [CrossRef]
56. Meng, L.; Turner, A.P.F.; Mak, W.C. Tunable 3D Nanofibrous and Bio-Functionalised PEDOT Network Explored as a Conducting Polymer-Based Biosensor. *Biosens. Bioelectron.* **2020**, *159*, 112181. [CrossRef] [PubMed]
57. Shi, J.; Wang, J.; Zhao, Z.; Dong, F. Fabrication of Reduced Graphene Oxide Doped Poly(3,4-Ethylenedioxythiophene): Polystyrene (Sulfonate-Co-Acrylic Acid) Conductive Films with High Conductivity and Stability. *Thin Solid Films* **2022**, *762*, 139525. [CrossRef]
58. Hu, Z.; Niu, Q.; Hsiao, B.S.; Yao, X.; Zhang, Y. Bioactive Polymer-Enabled Conformal Neural Interface and Its Application Strategies. *Mater. Horiz.* **2022**, *10*, 808–828. [CrossRef]
59. Sharma, V.K.; Chakraborty, G.; Narendren, S.; Katiyar, V. Fabrication and Characterization of Conductive Electrospun Nanofiber Mats of Carbon Nanofiber/Poly(Vinyl Alcohol)/Poly(Lactic Acid) Ternary Nanocomposites for Flexible Electronics Applications. *Mater. Adv.* **2023**, *4*, 6294–6303. [CrossRef]
60. An, J.E.; Kim, K.H.; Park, S.J.; Seo, S.E.; Kim, J.; Ha, S.; Bae, J.; Kwon, O.S. Wearable Cortisol Aptasensor for Simple and Rapid Real-Time Monitoring. *ACS Sens.* **2022**, *7*, 99–108. [CrossRef]
61. Ferraris, S.; Spriano, S.; Scalia, A.C.; Cochis, A.; Rimondini, L.; Cruz-Maya, I.; Guarino, V.; Varesano, A.; Vineis, C. Topographical and Biomechanical Guidance of Electrospun Fibers for Biomedical Applications. *Polymers* **2020**, *12*, 2896. [CrossRef]
62. Latonen, R.-M.; Cabrera, J.A.W.; Lund, S.; Kosourov, S.; Vajravel, S.; Boeva, Z.; Wang, X.; Xu, C.; Allahverdiyeva, Y. Electrospinning of Electroconductive Water-Resistant Nanofibers of PEDOT–PSS, Cellulose Nanofibrils and PEO: Fabrication, Characterization, and Cytocompatibility. *ACS Appl. Bio Mater.* **2021**, *4*, 483–493. [CrossRef]
63. Sordini, L.; Silva, J.C.; Garrudo, F.F.F.; Rodrigues, C.A.V.; Marques, A.C.; Linhardt, R.J.; Cabral, J.M.S.; Morgado, J.; Ferreira, F.C. PEDOT:PSS-Coated Polybenzimidazole Electroconductive Nanofibers for Biomedical Applications. *Polymers* **2021**, *13*, 2786. [CrossRef]
64. Abedi, A.; Hasanzadeh, M.; Tayebi, L. Conductive Nanofibrous Chitosan/PEDOT:PSS Tissue Engineering Scaffolds. *Mater. Chem. Phys.* **2019**, *237*, 121882. [CrossRef]
65. Furlani, F.; Campodoni, E.; Sangiorgi, N.; Montesi, M.; Sanson, A.; Sandri, M.; Panseri, S. Electroconductive Scaffolds Based on Gelatin and PEDOT:PSS for Cardiac Regeneration. *Int. J. Biol. Macromol.* **2023**, *224*, 266–280. [CrossRef]
66. Chen, C.; Chen, X.; Zhang, H.; Zhang, Q.; Wang, L.; Li, C.; Dai, B.; Yang, J.; Liu, J.; Sun, D. Electrically-Responsive Core-Shell Hybrid Microfibers for Controlled Drug Release and Cell Culture. *Acta Biomater.* **2017**, *55*, 434–442. [CrossRef]
67. Sarabia-Riquelme, R.; Andrews, R.; Anthony, J.E.; Weisenberger, M.C. Highly Conductive Wet-Spun PEDOT:PSS Fibers for Applications in Electronic Textiles. *J. Mater. Chem. C* **2020**, *8*, 11618–11630. [CrossRef]
68. Yan, K.; Xu, L. Electrospun Composite Nanofiber Membranes for Electromagnetic Protection. *ACS Appl. Nano Mater.* **2024**, *7*, 4615–4629. [CrossRef]
69. Lerond, M.; Subramanian, A.; Skene, W.G.; Cicoira, F. Combining Electrospinning and Electrode Printing for the Fabrication of Stretchable Organic Electrochemical Transistors. *Front. Phys.* **2021**, *9*, 708914. [CrossRef]
70. Mendoza-Diaz, M.I.; Garcia-Gutierrez, D.F.; Sepulveda-Guzman, S.; Moreno-Cortez, I.E.; Garcia-Gutierrez, D.I. Tuning the Optoelectronic Properties of PEDOT:PSS-PVP Core–Shell Electrospun Nanofibers by Solvent-Quantum Dot Doping and Phase Inversion. *Nanotechnology* **2019**, *30*, 395601. [CrossRef] [PubMed]
71. Bagherzadeh, R.; Gorji, M.; Bafgi, M.S.S.; Saveh-Shemshaki, N. Section Three: Electrospinning for Particular Fiber Types and Applications. In *Electrospun Nanofibers*; Elsevier: Amsterdam, The Netherlands, 2017; pp. 467–519. [CrossRef]
72. Zhang, H.; Zhu, X.; Tai, Y.; Zhou, J.; Li, H.; Li, Z.; Wang, R.; Zhang, J.; Zhang, Y.; Ge, W.; et al. Recent Advances in Nanofiber-Based Flexible Transparent Electrodes. *Int. J. Extreme Manuf.* **2023**, *5*, 032005. [CrossRef]
73. Karagiorgis, X.; Shakthivel, D.; Khandelwal, G.; Ginesi, R.; Skabara, P.J.; Dahiya, R. Highly Conductive PEDOT:PSS: Ag Nanowire-Based Nanofibers for Transparent Flexible Electronics. *ACS Appl. Mater. Interfaces* **2024**, *16*, 19551–19562. [CrossRef]
74. de Gramont, F.B.; Zhang, S.; Tomasello, G.; Kumar, P.; Sarkissian, A.; Cicoira, F. Highly Stretchable Electrospun Conducting Polymer Nanofibers. *Appl. Phys. Lett.* **2017**, *111*, 093701. [CrossRef]
75. Bessaire, B.; Mathieu, M.; Salles, V.; Yeghoyan, T.; Celle, C.; Simonato, J.-P.; Brioude, A. Synthesis of Continuous Conductive PEDOT:PSS Nanofibers by Electrospinning: A Conformal Coating for Optoelectronics. *ACS Appl. Mater. Interfaces* **2017**, *9*, 950–957. [CrossRef] [PubMed]
76. Lee, S.; Park, J.; Kim, M.C.; Kim, M.; Park, P.; Yoon, I.-J.; Nah, J. Polyvinylidene Fluoride Core–Shell Nanofiber Membranes with Highly Conductive Shells for Electromagnetic Interference Shielding. *ACS Appl. Mater. Interfaces* **2021**, *13*, 25428–25437. [CrossRef]

77. Sharma, G.K.; James, N.R. Highly Flexible, PEDOT:PSS-Polyvinylpyrrolidone Coated Carbon Nanofiber-Polydimethylsiloxane Composite for Electromagnetic Interference Shielding. *Synth. Met.* **2023**, *296*, 117376. [CrossRef]
78. Sun, X.; He, J.; Qiang, R.; Nan, N.; You, X.; Zhou, Y.; Shao, W.; Liu, F.; Liu, R. Electrospun Conductive Nanofiber Yarn for a Wearable Yarn Supercapacitor with High Volumetric Energy Density. *Materials* **2019**, *12*, 273. [CrossRef]
79. Laforgue, A. All-Textile Flexible Supercapacitors Using Electrospun Poly(3,4-Ethylenedioxythiophene) Nanofibers. *J. Power Sources* **2011**, *196*, 559–564. [CrossRef]
80. Du, H.; Zhang, M.; Liu, K.; Parit, M.; Jiang, Z.; Zhang, X.; Li, B.; Si, C. Conductive PEDOT:PSS/Cellulose Nanofibril Paper Electrodes for Flexible Supercapacitors with Superior Areal Capacitance and Cycling Stability. *Chem. Eng. J.* **2022**, *428*, 131994. [CrossRef]
81. Chen, X.; Zhao, C.; Yang, K.; Sun, S.; Bi, J.; Zhu, N.; Cai, Q.; Wang, J.; Yan, W. Conducting Polymers Meet Lithium–Sulfur Batteries: Progress, Challenges, and Perspectives. *Energy Environ. Mater.* **2023**, *6*, e12483. [CrossRef]
82. Zeng, S.-Z.; Yao, Y.; Tu, J.; Tian, Z.; Zou, J.; Zeng, X.; Zhu, H.; Kong, L.B.; Han, P. Improving the Low-Rate Stability of Lithium-Sulfur Battery through the Coating of Conductive Polymer. *Ionics* **2021**, *27*, 3887–3893. [CrossRef]
83. Abdah, M.A.A.M.; Mokhtar, M.; Khoon, L.T.; Sopian, K.; Dzulkurnain, N.A.; Ahmad, A.; Sulaiman, Y.; Bella, F.; Su'ait, M.S. Synthesis and Electrochemical Characterizations of Poly(3,4-Ethylenedioxythiophene/Manganese Oxide Coated on Porous Carbon Nanofibers as a Potential Anode for Lithium-Ion Batteries. *Energy Rep.* **2021**, *7*, 8677–8687. [CrossRef]
84. Beaudoin, É.J.; Kubaski, M.M.; Samara, M.; Zednik, R.J.; Demarquette, N.R. Scaled-Up Multi-Needle Electrospinning Process Using Parallel Plate Auxiliary Electrodes. *Nanomaterials* **2022**, *12*, 1356. [CrossRef]
85. Morkus, P.; Sibbald, S.; Choi, L.; Rassenberg, S.; Filipe, C.D.M.; Latulippe, D.R. Miniaturization of an Enclosed Electrospinning Process to Enhance Reproducibility in the Fabrication of Rapidly Dissolving Cell-based Biosensors. *Biotechnol. J.* **2024**, *19*, e2300306. [CrossRef]
86. Lunni, D.; Giordano, G.; Pignatelli, F.; Filippeschi, C.; Linari, S.; Sinibaldi, E.; Mazzolai, B. Light-Assisted Electrospinning Monitoring for Soft Polymeric Nanofibers. *Sci. Rep.* **2020**, *10*, 16341. [CrossRef] [PubMed]
87. Schultheiss, A.; Gueye, M.; Carella, A.; Benayad, A.; Pouget, S.; Faure-Vincent, J.; Demadrille, R.; Revaux, A.; Simonato, J.-P. Insight into the Degradation Mechanisms of Highly Conductive Poly(3,4-Ethylenedioxythiophene) Thin Films. *ACS Appl. Polym. Mater.* **2020**, *2*, 2686–2695. [CrossRef]
88. Bontapalle, S.; Varughese, S. Understanding the Mechanism of Ageing and a Method to Improve the Ageing Resistance of Conducting PEDOT:PSS Films. *Polym. Degrad. Stab.* **2020**, *171*, 109025. [CrossRef]
89. Duzcukoglu, H.; Kaybal, H.B.; Asmatulu, R. Enhancing the Coating Durability and Electrical Stability of Fiber Composites with SPEEK/PEDOT:PSS Permanent Coatings: A Novel Approach. *Polym. Degrad. Stab.* **2024**, *228*, 110908. [CrossRef]
90. Yemata, T.A.; Zheng, Y.; Kyaw, A.K.K.; Wang, X.; Song, J.; Chin, W.S.; Xu, J. Improved Thermoelectric Properties and Environmental Stability of Conducting PEDOT:PSS Films Post-Treated With Imidazolium Ionic Liquids. *Front. Chem.* **2020**, *7*, 870. [CrossRef]
91. Sau, S.; Kundu, S. Stretchable Polymeric Film with Enhanced Electrical and Mechanical Properties. *Colloids Surf. A Physicochem. Eng. Asp.* **2022**, *636*, 128130. [CrossRef]
92. Qi, K.; He, J.; Wang, H.; Zhou, Y.; You, X.; Nan, N.; Shao, W.; Wang, L.; Ding, B.; Cui, S. A Highly Stretchable Nanofiber-Based Electronic Skin with Pressure-, Strain-, and Flexion-Sensitive Properties for Health and Motion Monitoring. *ACS Appl. Mater. Interfaces* **2017**, *9*, 42951–42960. [CrossRef]
93. Huang, C.-Y.; Chiu, C.-W. Facile Fabrication of a Stretchable and Flexible Nanofiber Carbon Film-Sensing Electrode by Electrospinning and Its Application in Smart Clothing for ECG and EMG Monitoring. *ACS Appl. Electron. Mater.* **2021**, *3*, 676–686. [CrossRef]

polymers

MDPI

Review

Application of Electrospun Drug-Loaded Nanofibers in Cancer Therapy

Yaoyao Yang *, Rui Zhang , Zhiyuan Liang, Junli Guo, Bingying Chen, Shengwei Zhou and Dengguang Yu *

School of Materials and Chemistry, University of Shanghai for Science and Technology, 516 Jungong Road, Shanghai 200093, China; 212203081@st.usst.edu.cn (R.Z.); 212203084@st.usst.edu.cn (Z.L.); 223353223@st.usst.edu.cn (J.G.); 223353224@st.usst.edu.cn (B.C.); 213353182@st.usst.edu.cn (S.Z.)

* Correspondence: yyyang@usst.edu.cn (Y.Y.); ydg017@usst.edu.cn (D.Y.)

Abstract: In the 21st century, chemotherapy stands as a primary treatment method for prevalent diseases, yet drug resistance remains a pressing challenge. Utilizing electrospinning to support chemotherapy drugs offers sustained and controlled release methods in contrast to oral and implantable drug delivery modes, which enable localized treatment of distinct tumor types. Moreover, the core–sheath structure in electrospinning bears advantages in dual-drug loading: the core and sheath layers can carry different drugs, facilitating collaborative treatment to counter chemotherapy drug resistance. This approach minimizes patient discomfort associated with multiple-drug administration. Electrospun fibers not only transport drugs but can also integrate metal particles and targeted compounds, enabling combinations of chemotherapy with magnetic and heat therapies for comprehensive cancer treatment. This review delves into electrospinning preparation techniques and drug delivery methods tailored to various cancers, foreseeing their promising roles in cancer treatment.

Keywords: electrospinning; nanofibers; nanostructure; drug release; cancer therapy

1. Introduction

The shape and location of tumors vary, typically categorized as benign or malignant. Benign tumors usually cause minimal harm, often resulting in localized obstructions and pressure. Malignant tumors, known as "cancer", consist of cells differing from normal ones in the body. While normal cell growth is limited, cancer cells proliferate without restraint. As they multiply, cancer cells consume essential nutrients meant for normal cells, leading to the decline and demise of healthy cells. Additionally, these cancer cells produce toxins that harm the body. Chemotherapy is a vital aspect of cancer treatment, aiming to impede the division and growth of cancer cells. However, chemotherapy also brings about systemic side effects due to its high toxicity, drug resistance, and limited selectivity [1]. Challenges in tumor therapy include rapid drug elimination, insufficient delivery to tumor tissues, and the risk of metastasis to other organs. Consequently, cancer chemotherapy often results in dose-related toxicity and side effects such as immunosuppression, neuropathy, and hair loss [2].

Electrospinning can be used to incorporate drugs into nanofibers, resulting in a decelerated drug release rate, extended release duration, minimized drug dispersal, and effective mitigation of potential adverse effects from multiple-drug delivery [3]. Moreover, electrospun drug-loaded nanofibers have the capacity to concurrently accommodate various drugs, fostering a synergistic effect among them. This synergy revitalizes initially drug-resistant compounds, generating novel therapeutic effects [4,5]. Figure 1 depicts the upward trajectory of articles published in the past 20 years on the "Web of Science" database, focusing on "electrospinning & drug delivery" and "electrospinning & cancer", respectively. As shown in Figure 1a,b, the data reveal that the applications of electrospinning in drug delivery continues to increase, with the primary research areas including sustained release, drug combination, and targeted release. The data in Figure 1c,d demonstrate the growing application of electrospinning in cancer treatment research. Most of the studies are about lung cancer and breast cancer, and research on other types of cancer is also increasing year by year.

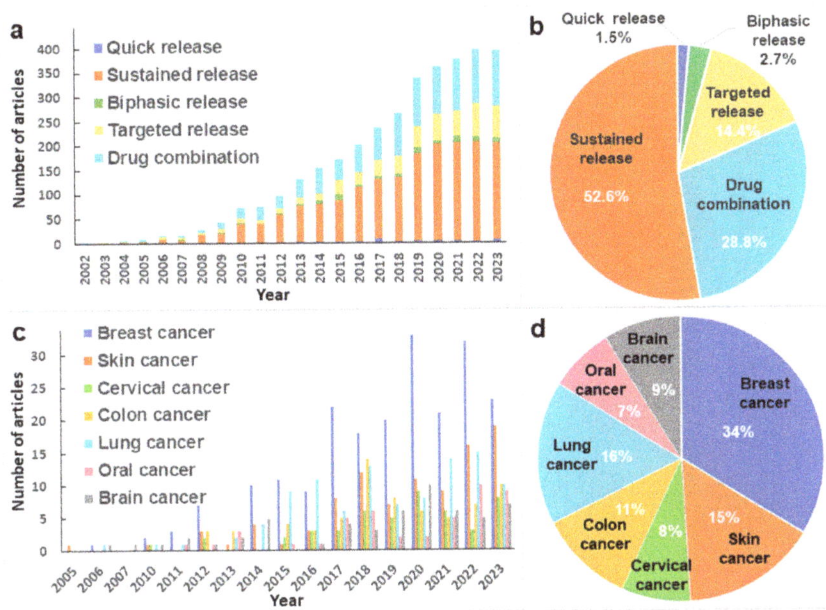

Figure 1. Statistics of studies retrieved from the literature on the "Web of Science" platform with the subjects of (**a**,**b**) "electrospinning & drug delivery" and (**c**,**d**) "electrospinning & cancer", respectively.

The primary aim of this review is to assess the efficacy and role of electrospun nanofibers in addressing the formidable challenges of chemotherapy in cancer treatment, emphasizing their potential in mitigating drug-related toxicity and enhancing drug delivery efficiency.

2. Preparation of Drug-loaded Nanofibers via Electrospinning

Electrospinning is an electrohydrodynamic technique used to create nanofibers. This process involves energizing polymer droplets to generate jets. The electrostatic repulsion between the charges generated on the droplets causes them to form a Taylor cone, which elongates due to stretching induced by the electric current, resulting in the production of fine fibers. Substances that are suitable for electrospinning are primarily organic polymers present in a solution or in a molten state. These include small-molecule polymers that dissolve in conductive polymers at optimal dissolution concentrations or those that can self-assemble to achieve sufficient chain entanglement [6]. Currently, as shown in Figure 2, advancements in multi-fluid electrospinning technology have enriched the structures of electrospinning fibers. Varieties such as bead-on-string structures [7], smooth fibers [8,9], core–sheath structures [10–16], Janus structures [17], "pig nose" structures [18], and others are continually emerging.

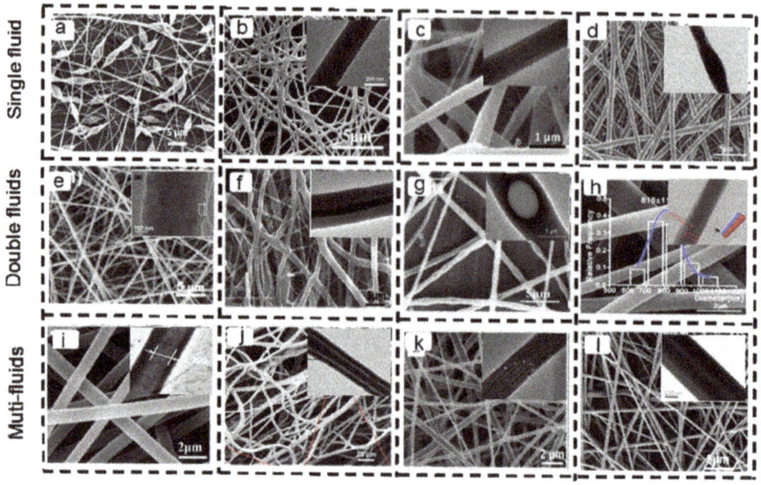

Figure 2. SEM and TEM images of electrospun drug-loaded nanofibers with various structures. Single-fluid electrospinning: (**a**) Bead-on-string fibers [7]. Copyright 2018, reproduced with permission from *Nature*. (**b**) Smooth fibers [8]. Copyright 2022, reproduced with permission from MDPI. (**c**,**d**) Core–sheath fibers [10,11]. Copyright 2023, 2019, reproduced with permission from Elsevier. Double-fluid electrospinning: (**e**) Smooth fibers [9]. Copyright 2021, reproduced with permission from MDPI. (**f**) Core–sheath fibers [12]. Copyright 2020, reproduced with permission from Elsrvier. (**g**) Beaded core–sheath fibers [13]. Copyright 2019, reproduced with permission from WILEY. (**h**) Janus fibers [17]. Copyright 2021, reproduced with permission from Elsevier. Multi-fluid electrospinning: (**i**) Two-layer core–sheath fibers [14]. Copyright 2016, reproduced with permission from Elsevier. (**j**) Three-layer core–sheath fibers [15]. Copyright 2020, reproduced with permission from MDPI. (**k**) Four-layer core–sheath fibers [16]. Copyright 2021, reproduced with permission from Elsevier. (**l**) Two core fibers and one sheath fiber [18]. Copyright 2020, reproduced with permission from Elsevier.

2.1. Single-Fluid Electrospinning

Single-fluid electrospinning stands as a commonly used technique to produce nanofibers. The morphology of these electrospun fibers is influenced by system parameters (polymer molecular weight, solvent properties, spinning liquid concentration, etc.), process factors (voltage, flow rate, and acceptance distance), and environmental parameters (temperature and humidity).

Typically, smooth and continuous nanofibers are the most prevalent morphologies. Noteworthy patterns include a decreasing solution viscosity, leading to a smaller fiber diameter or increased fusiform defects, and a higher average polymer molecular weight favoring easier bonding and network formation [19]. A disruption in the balance between the surface tension and electric field force can result in jet splitting and fiber branching [20]. Common substrates for drug-carrying nanofibers include polyvinylpyrrolidone (PVP) [7], polycaprolactone (PCL) [21,22], and cellulose acetate (CA) [23]. Loading drugs onto these nanofibers enables the rapid, controlled, and targeted release of the drug. Using various polymers to create composite fibers for drug loading not only ensures effective drug release properties but also enhances the overall fiber properties [24,25].

When preparing drug-loaded nanofibers using electrospinning, it is crucial to consider the compatibility between the drugs and polymer substrates. For example, if hydrophilic drugs are paired with hydrophobic substrates, this can cause drug aggregation within the fibers, resulting in uneven drug dispersion [26].

In another study, Bukhary et al. utilized polyvinylpyrrolidone as a polymer matrix, employing electrospinning to create fast-dissolving oral Fixed-Dose Combinations (FDCs) containing amlodipine besylate and valsartan, which are commonly used drugs for hypertension. These FDCs exhibited drug encapsulation rates exceeding 85% of the theoretical dose. Achieving a rapid release within 360 s, the resulting fiber membrane demonstrated both folding durability and adequate thickness, rendering it suitable for use as an oral patch [67].

3.2. Sustained Release

Sustained release stands out as a protracted mode of drug delivery that is extensively employed in treating chronic conditions like cancer. For instance, biodegradable electrospun scaffolds, composed of chitosan (CS), polyvinyl alcohol, ibuprofen, and silver nanoparticles, exhibit both analgesic and antibacterial effects, and they are applied in tissue engineering [74].

In another application, electrospun nanofibers, supported by the hydrophobic polymer poly(lactic-co-glycolic acid) copolymer (PLGA) with pirfenidone as the outer layer and the hydrophilic polymer PVP with moxifloxacin as the inner layer, extend the release duration of both drugs. This extended release significantly prolongs drug residence on corneal abrasions, reducing the frequency of doses [75].

Resveratrol (RES) combined with other natural drugs exhibits inhibitory effects on multiple pathways, influencing cancer cell growth and cancer-causing signaling. In RES- and xanthohumol (XAN)-loaded fiber membranes, produced via coaxial electrospinning, drug release accelerates with an increased drug content. RES and XAN are released progressively, and their releases are sustained for up to 350 h [76].

Effective local chemotherapy drug delivery to tumors necessitates the development of Drug Delivery Systems (DDSs), ensuring long-term drug release. Several DDSs for local drug delivery, including polymer nanoparticles, hydrogels, nanoemulsions, and nanofibers, have been developed [77]. Electrospun drug-loaded nanofibers, functioning as carriers for amorphous drugs, enhance drug bioavailability compared to conventional drugs. This characteristic minimizes the total dosage and administration frequency, offering an advantage in cancer therapy.

3.3. Biphasic Release

Biphasic release offers the combined advantages of rapid and sustained release, holding promising applications in various fields such as cancer treatment [78–80], bone tissue regeneration [81–84], wound healing [85–88], rotator cuff tendon tears [89], and eye diseases [90].

When preparing electrospun drug-loaded nanofibers, due to the rapid evaporation of solvents and high ionic strength in the solution, most drug compounds tend to accumulate on or near the fiber surface, leading to an initial burst release. This initial burst release is unavoidable [91]. Consequently, the biphasic release rate of drug-loaded nanofibers is influenced by factors such as the polymer molecular weight [92], concentration [93], sheath thickness [94], additives, and more. For instance, introducing graphene oxide can enhance the mechanical properties of electrospun membranes and slow the release of model drugs by reducing the matrix swelling [95].

Apart from the nanofiber composition, the internal structures and external shapes of the nanofibers play significant roles in determining the drug release rate [17].

3.4. Targeted Release

Variations in pH across different parts of the human body inspire the use of pH-sensitive materials, like the Eudragit polymer family, for targeted drug delivery [96,97]. For instance, the gastric juice pH ranges from 0.9 to 1.8, contrasting with the intestinal tract's pH of around 7. Electrospun core–sheath nanofibers, composed of a lipid drug core and an

Eudragit S100 shell, exhibit promise in providing targeted drug release, specifically to the colon [14].

The slightly elevated acidity of the tumor microenvironment (TME) compared to normal physiological conditions presents an opportunity for pH-responsive DDSs in tumor therapy [98]. An example is γ-cyclodextrin (γ-CD) linked with 3-(diethylamino)propylamine (DEAP), electrospun to create microparticles with a cellular porous structure for encapsulating paclitaxel (PTX). Under acidic conditions, the protonation of pH-responsive DEAP disrupts the particle structure via charge repulsion, triggering the release of PTX from the collapsed particle [99]. In another approach, core–sheath nanofibers, prepared via coaxial electrospinning, feature chitosan and polyethylene oxide as the sheath, and polycaprolactone with sodium bicarbonate as the core. The sheath and core layers are loaded with lidocaine hydrochloride (Lid) and curcumin (CUR), respectively. The acidic environment protonates the $-NH_2$ group of chitosan, forming $-NH_3^+$, whereas sodium bicarbonate reacts with hydrogen ions to generate CO_2, creating pores on the fiber surface. Consequently, both Lid and CUR display release curves that are responsive to an acidic pH (~5.4). Due to chitosan's sensitivity to acidity, the total release of Lid was higher under acidic conditions than under alkaline conditions [66].

3.5. Drug Combination

Drug resistance easily develops during tumor treatment, often necessitating the use of two or more antitumor drugs in combination. This approach is applicable to cancers that have failed chemotherapy and targeted therapy or exhibit frequent drug resistance. Such scenarios are commonly observed in the treatment of glioblastoma [100], triple-negative breast cancer [101], chronic myeloid leukemia [102], and other cancers [103].

The natural drug PTX stands as one of the most commonly used drugs in combination therapies. It demonstrates synergic effects with evolimus [100], tetrandrine [104], CUR [105], silybin (SB) [106], melittin [107], honokiol (HNK) [108], fatty acid amide hydrolase (FAAH) [109], and epirubicin (EPI) [110]. The combination of PTX and HNK enhances the cytotoxicity of PTX in drug-resistant MCF-7/ADR tumor cells by increasing paclitaxel's cellular uptake and preventing the invasion and migration of MDA-MB-231 cells, both in vitro and in vivo, thereby reducing metastasis in nude mice [108]. Electrospun fibers loaded with paclitaxel and tetrandrine inhibit the expressions of p-STAT3 and p-JAK, initiate the release of cytochrome c, and promote the expression of caspase protein [104]. Moreover, the combined treatment of PTX and silybin (SB) exhibits a synergistic effect by sensitizing chemotherapy and regulating the tumor microenvironment, effectively inhibiting tumor growth through enhanced intratumoral penetration [106]. Moreover, doxorubicin (DOX) exhibits synergistic effects with berberine (BBR) [111], CUR combined with resveratrol [112], and apatinib (AP) [13]. Y. He et al. demonstrated the simultaneous loading of DOX and AP onto nanofibers using microfluidic electrospinning, as depicted in Figure 4. Achieving an encapsulation efficiency of 99%, the two drugs followed a preset release model: the rapid release of DOX and the gradual release of AP. Sustained AP release effectively inhibits drug resistance in tumor cells, while the increased intracellular accumulation of DOX leads to outstanding antitumor results with minimal systemic toxicity [17].

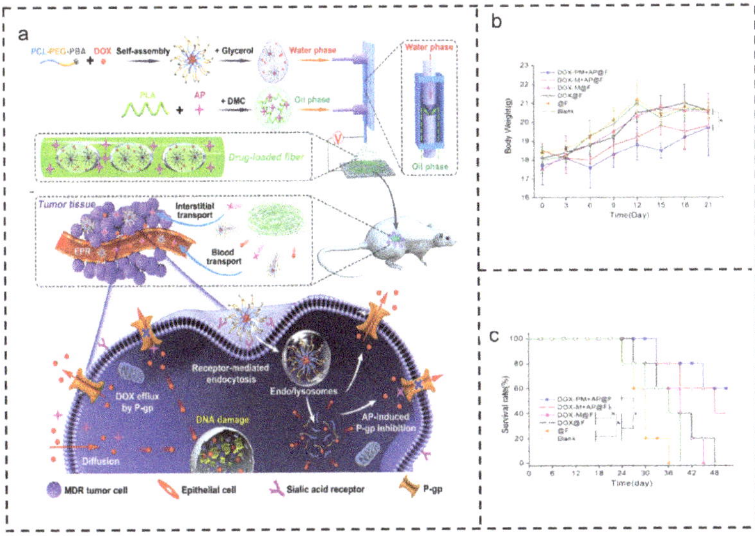

Figure 4. Application of doxorubicin and apatinib in combination to overcome multidrug resistance in cancer [13]. Copyright 2019, reproduced with permission from WILEY. (**a**) Fabrication and application diagram of drug-loaded fiber device. (**b**) Weight change. (**c**) Survival curves of tumor-bearing nude mice.

4. Application of Drug-loaded Nanofibers in Cancer Therapy

As previously highlighted, electrospun drug-loaded nanofibers offer high drug loading, exceptional encapsulation efficiency, and specific drug release profiles, including sustained, multiphase, targeted, and combined release. These capabilities are achieved through meticulous material selection and design, presenting specific and extensive applications in cancer treatment [113–115]. This section delineates their applications across various tumor therapies.

4.1. Breast Cancer

Breast cancer stands as one of the most prevalent cancers in today's society, and it is commonly treated with surgical resection or post-tumor-resection chemotherapy. Electrospun fibers serve as vehicles for chemotherapy drug delivery localized to the site [116], enabling dosage adjustment based on patch thickness and total area [117]. Abasalta et al. utilized coaxial electrospinning to load DOX, an anticarcinostatic drug, onto nanofibers. The presence of DOX-loaded coaxial nanofibers resulted in a mortality rate of approximately 57% for MCF-7 breast cancer cells after 7 days [118].

Another development by Li et al. introduced an implantable hierarchical microfiber device using coaxial electrospinning, forming periodically arranged chambers within a fiber matrix. Doxorubicin hydrochloride within these chambers exhibited rapid release, effectively eliminating the remaining tumor cells and curbing tumor recurrence. Additionally, the continuous release of the matrix metalloproteinases-2 (MMP-2) inhibitor disulfide from the fiber matrix inhibited tumor invasion and prevented metastasis. This fiber-based device showed promising results as a postoperative local adjuvant chemotherapy, demonstrating excellent efficacy in preventing recurrence and metastasis [119]. Studies have demonstrated that the addition of surfactants to electrospun fibers improves their therapeutic effects [120,121].

Enhancing chemotherapy's efficacy often involves combining it with complementary cancer treatment methods like magnetic therapy [122,123], radiation therapy [124], and hyperthermia [125], among others. Hence, the integration methods of metal or magnetic nanoparticles with electrospinning are common approaches. For instance, spherical nickel ferrite nanoparticles synthesized via the sol-gel method were employed in electrospinning to produce magnetic core–sheath nanofibers loaded with DOX, achieving an encapsulation efficiency exceeding 90%. These drug-loaded nanofibers exhibited sustained and controlled DOX release for 25 days under physiological pH conditions and within 7 days under acidic pH conditions without an initial burst release. In the presence of an external magnetic field, these nanofibers demonstrated up to 83% maximum cytotoxicity toward MCF-7 breast cancer cells, showcasing their potential to be used for breast cancer treatment [126].

Another example involved incorporating the anticancer drug CUR into mesoporous silica nanoparticles (MSNs) embedded in PLGA via electrospinning. This approach examined CUR's in vitro release mode from the fibers and its anticancer efficacy against MCF-7 breast cancer cells. The findings revealed sustained and prolonged drug release properties for CUR, with the drug-loaded fibers exhibiting higher cytotoxicity and reduced mobility, thereby increasing apoptosis induction in cancer cells [127].

Moreover, in a setup where the anticancer drugs paclitaxel and 5-Fluorouracil (5-FU) were incorporated into the core layer of chitosan-containing core–sheath nanofibers, magnetic gold nanoparticles coated the nanofibers' surfaces. The high swelling of the chitosan copolymer and nanofibers led to an initial rapid release of paclitaxel and 5-FU within 48 h, followed by a sustained release for up to 25 days. This super-strong and sustained release profile of paclitaxel and 5-FU, facilitated by the gold nanoparticle layer, resulted in a 78% maximum killing rate of magnetic gold-coated nanofibers on 4T1 breast cancer cells when an applied magnetic field was present [128] (Figure 5).

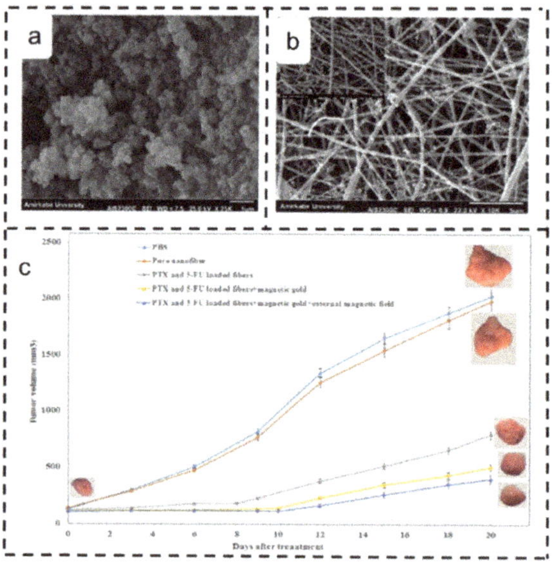

Figure 5. Application of electrospun magnetic core–sheath drug-carrying nanofibers in breast cancer treatment [128]. Copyright 2020, reproduced with permission from Elsevier. SEM images of (**a**) magnetic gold nanoparticles and (**b**) nanofibers coated with magnetic gold nanoparticles. (**c**) Tumor growth after fibroblast treatment.

4.2. Skin Cancer

Skin cancer manifests primarily in three forms: basal cell carcinoma, squamous cell carcinoma, and the highly malignant melanoma, which is particularly prone to local invasion and distant metastasis. Surgery serves as the primary treatment for most melanomas. However, owing to their invasive nature, residual tumor cells post-surgery significantly elevate the risk of local recurrence and micrometastase. Consequently, drug-loaded electrospun fiber membranes are frequently applied in postoperative chemotherapy for melanoma, commonly utilized as patches [129,130] or implantable devices [131,132].

S. Samadzadeh et al. developed temperature-responsive single-layer electrospun nanofibers with dual capabilities of simultaneous heating and two-stage drug release, triggered by an alternating magnetic field switch. Their smart hyperthermia nanofiber scaffolds exhibited a combination of rapid initial drug release followed by prolonged release. Through the dual effects of heat and the two-stage drug release, they effectively induced cytotoxicity, harnessing synergistic effects over an extended period. This two-stage drug release method, compared to standard chemotherapy, offers preferable cytotoxicity induction through prolonged synergistic effects. These temperature-responsive drug-loaded electrospun nanofibers could serve as effective implantable devices to prevent local cancer recurrence [132].

Muthulakshmi et al. extracted bioactive compounds from the medicinal plant Terminalia catappa (TC), loading them into a sodium alginate polymer. Electrospun fiber membranes prepared from this combination exhibited the ability to heighten oxidative stress and induce apoptosis by regulating caspase protein in skin melanoma cells, showcasing remarkable efficacy. The presence of anticancer elements within TC contributes to its enhanced anticancer activity, while the electrospinning technique plays a crucial role in amplifying this effect [133].

In another approach, Xue et al. engineered a therapeutic amorphous calcium carbonate (ACC) nano-formulation within a single layer of gelatin/polycaprolactone (GP) nanofibers using electrospinning. Combining ACC nano-formulation with Fe^{2+}-pre-activated bleomycin results in a biocatalytically enhanced therapeutic effect. Hydrolyzed ACC functions as a proton scavenger, effectively adjusting the acidity of tumor tissues in situ, thereby continuously suppressing tumor recurrence and metastasis. The acid-triggered decomposition of ACC releases Ca^{2+}, activating the downstream Wnt/β-catenin signaling pathway. This process, combined with the wound regeneration effect of GP substrates, provides adjunct therapy for the postoperative management of melanoma [134] (Figure 6).

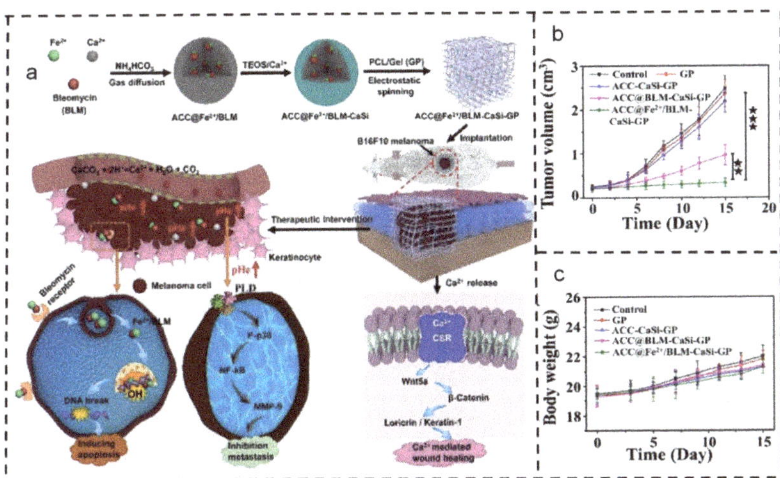

Figure 6. Application of gelatin/polycaprolactone (GP) nanofiber scaffolds in inhibiting proliferation of skin cancer cells [134]. Copyright 2021, reproduced with permission from WILEY. (**a**) The synthesis process of the ACC@Fe^{2+}/BLM-CaSi-GP stent and its complementary ability to simultaneously treat melanoma and wound healing. (**b**) Weight change (n = 5) and (**c**) survival curves of nude mice bearing tumors under different treatments with MCF-7/ADR at 21 days.

Combining chemotherapy, magnetic therapy, and hyperthermia represents a significant trend in cancer treatment. Cimen et al. utilized electrospinning to fabricate spherical silver nanoparticles (AgNPs) coated with thyme extract using the bioreduction method, embedding them in a polycaprolactone/polylactic acid (PCL/PLA) core–sheath nanofiber scaffold. SiO_2 and Ag nanoparticles were integrated into the polymer during preparation. This method facilitates the actual treatment combination of hyperthermia and magnetic therapy while exhibiting substantial antibacterial effects against Staphylococcus aureus and Escherichia coli. Demonstrating a higher efficiency compared to other nanofiber scaffolds, it notably inhibits the proliferation of skin cancer cells (SK-MEL-30 cells). Additionally, the fiber membrane exhibits good biodegradability, mitigating harm to patients after repeated removal [135].

4.3. Cervical Cancer

Vaginal drug administration offers several advantages, including ease of application, prolonged drug retention, and extensive coverage, and it is commonly employed in cervical cancer treatment. However, traditional carriers like gels and tablets face limitations such as instability, leakage, mucus adhesion, and an inadequate residence time. Consequently, drug-carrying fiber membranes for cervical cancer treatment often require enhanced adhesion. Aggarwal et al. investigated polycaprolactin/CS loaded with cisplatin composite electrospun nanofibers, demonstrating superior mucosal adhesion and exceptional efficacy in local chemotherapy for cervical cancer in mice [136]. Additionally, Yi et al. and B. Chen et al. developed biodegradable drug-loaded nanofibers capable of sustained drug release over four weeks, aiding in cervical cancer treatment and preventing its recurrence [137]. Given the unique nature of the cervix, these fibrous membranes typically possess antibacterial effects [138] and promote wound healing [139] alongside their therapeutic roles in treating cervical cancer.

Combining chemotherapy with phototherapy holds significant promise in cervical cancer treatment. In their study, illustrated in Figure 7, Wang et al. developed MSNs carrying both DOX and indocyanine green (ICG) (named DIMSNs). These nanoparticles were then incorporated with CS/PVA, leading to the creation of multifunctional composite nanofibers (DIMSN/F) through electrospinning. These implantable nanofibers exhibited optimal drug accumulation and specific drug release within the vaginal region. When subjected to simulated vaginal secretion erosion, DIMSN/F displayed site-specific drug release, achieving a remarkable tumor inhibition rate of 72.5% for primary cervical/vaginal cancer. This outcome underscores its considerable potential in cervical cancer treatment [140].

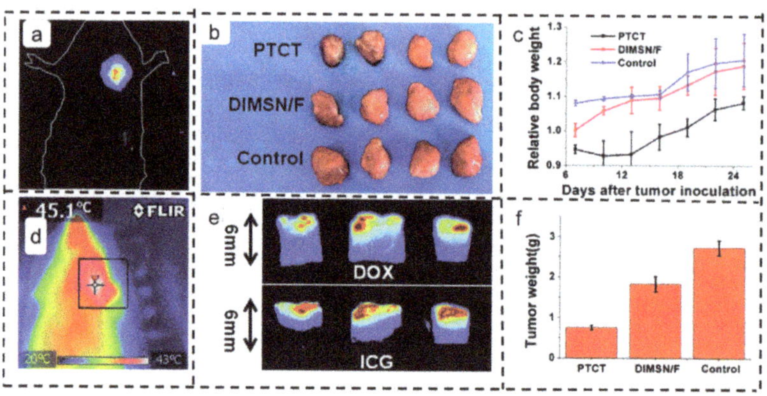

Figure 7. Application of chitosan/poly(vinyl alcohol) (CS/PVA) drug-loaded nanofiber scaffolds in the treatment of cervical cancer [140]. Copyright 2019, reproduced with permission from RSC. (**a**) Fluorescent (ICG) images of DIMSN/Fs subcutaneously implanted in the left axilla of mice. (**b**) Macroscopic tumor observation. (**c**) Relative tumor volume in each group. (**d**) Infrared thermal images of DIMSN/F mice after 808 nm laser irradiation. (**e**) Tumor weight. (**f**) Fluorescent images of tumor transects removed from the three mice receiving the first PTCT scan.

4.4. Colon Cancer

Colorectal cancer (CRC) stands among the most severe cancers, posing significant morbidity and mortality risks. Stent implantation, combined with radiotherapy and chemotherapy, has proven to cause effective relief in clinical practice. However, conventional metallic stents primarily address mechanical expansion needs, leading to temporary treatment. Postoperative complications, such as inward tumor growth and stent displacement due to stent stenosis, persist as unresolved issues. To address this, Xie et al. developed a fibroin-based nanofiber membrane for CRC treatment. The nanofiber membrane, loaded with a CUR/5-FU dual-drug system, exhibited enhanced antitumor effects. In practical application, a deflated balloon catheter and a compressed drug-carrying stent are inserted into a narrowed area. An inflatable balloon stent compresses the tumor, restoring the intestinal cavity and locally treating the tumor effectively (Figure 8) [141].

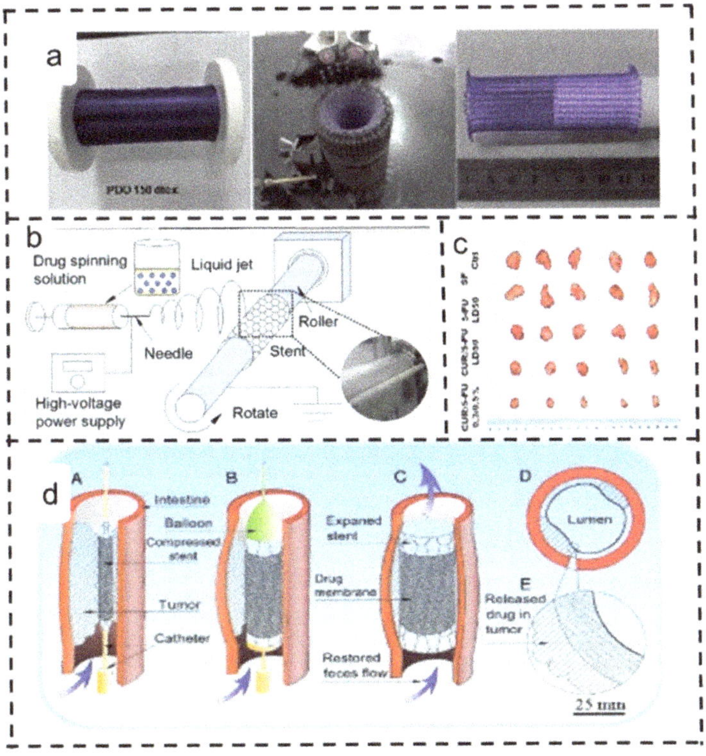

Figure 8. Application of expanded electrospun nanofiber scaffolds for colon cancer treatment [141]. Copyright 2018, reproduced with permission from WILEY. (**a**) Tubular stent diagram. (**b**) Schematic representation of a modified electrospinning machine. (**c**) Macroscopic images of exposed tumors. (**d**) Mechanism and process of drug-loaded stent placement and its role in colorectal cancer treatment. A, deflated balloon catheter and compressed drug-loaded stent are inserted into the stenosis site; B, inflated balloon expands the stent and compresses the tumor to restore the intestinal lumen; C, stent-widened intestine; D and E, the drugs are released from the coating membrane and the tumor is treated locally and effectively.

Another challenge in colon cancer treatment involves designing a fibrous membrane to deliver targeted drugs specifically into the intestine rather than into the stomach. This is typically achieved using a pH-sensitive substance to encapsulate and direct the drug to the precise location [65]. Rostami et al. prepared ultrathin, beadless, food-grade nanofibers using high concentration of CS, gel, and resveratrol. The presence of the gel significantly improved the electrospinning efficiency. The antioxidant activity of resveratrol-loaded nanofibers surpassed that of free resveratrol. In the gel-containing nanofibers, the release of resveratrol was hindered under acidic conditions, enabling an oral delivery mode to mitigate discomfort during implantation. Nearly half of the encapsulated resveratrol is delivered to the intestinal region [142].

4.5. Lung Cancer

Owing to the challenging location of lung cancer, achieving precise oral administration is unattainable, and implantable administration poses considerable pain. Hence, maximizing the drug release duration and minimizing the dosing frequency become pivotal. Mellatyar et al. investigated implantable nanofibers that enhanced the anticancer effect by inhibiting HSP90 expression and telomerase activity. PCL/PEG nanofibers loaded with 17-dimethylaminoethylamino-17-demethoxy geldanamycin (17-DMAG) exhibited superior efficacy in killing A549 lung cancer cells compared to free 17-DMAG. This was achieved by regulating heat shock protein 90 (HSP90) expression and inhibiting telomerase activity, presenting implantable 17-DMAG nanofiber scaffolds as promising tools to effectively eliminate residual lung cancer cells and prevent local tumor recurrence [143].

Samadzadeh et al. formulated nanofibers containing metformin hydro-chloride (MET) and elucidated its mechanism of action. MET primarily activates AMP-activated protein kinase (AMPK), leading to the inhibition of the mammalian target of the rapamycin (mTOR) pathway, thereby reducing protein synthesis and cancer proliferation. These fibers exhibited rapid drug release within the first day, continuing for over two weeks. Within 48 h, the MET-loaded nanofibers demonstrated significant cytotoxicity against A549 lung cancer cells. Moreover, these nanofibers effectively raised the intracellular ROS levels, inducing apoptosis in cancer cells. Compared to free MET, nanofibers with MET induced notable changes in the expression levels of Bax, Bcl-2, caspase-3, and -9 [144].

In a study by Li et al., the interaction between human lung epithelial A549 cells and electrospun nanofibers composed of PVA and type I collagen was examined by adjusting the diameter of the electrospun nanofibers. The A549 cells adhered to the nanofiber mats via a filopodia attachment and in an extended form, inducing the epithelial–mesenchymal transformation. This research into the tumor microenvironment and cancer helped elucidate the mechanism underlying lung cancer [145].

Yang et al. devised a lung chip using a PLGA electrospun nanofiber membrane as a substrate, creating a cell scaffold that emulates an alveolar microenvironment. The PLGA nanofiber membrane, approximately 3 μm thick, exhibited excellent porosity, molecular permeability, and biocompatibility. Using electrospinning technology, they investigated the potential causes of drug resistance in A549 cells co-cultured with HFL1 cells. They discovered that IGF-1 secreted by HFL1 cells sustained tumor cells by activating the PI3K/Akt signaling pathway post-drug inhibition of the EGFR-related signaling pathway. This led to the reduced sensitivity of tumor cells to chemotherapy drugs (Figure 9). This study, which simulated alveolar biochemical factors in the tumor microenvironment in vitro, particularly in the respiratory membrane, holds promise for significant contributions to the local treatment of lung tumors [146].

Numerous studies have highlighted the efficacy of electrospinning technology in simulating the tumor microenvironment, studying and inhibiting tumor pathogenesis, analyzing drug resistance mechanisms, and more [133,134]. Electrospinning-assisted cancer therapy represents a novel and promising therapeutic approach.

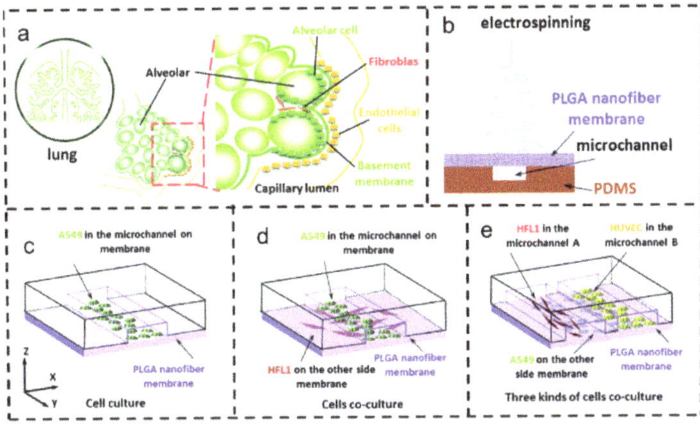

Figure 9. Application of electrospun fibers in exploring lung cancer mechanisms and simulating the tumor microenvironment [146]. Copyright 2018, reproduced with permission from RSC. (**a**) Schematic of alveolar structure. (**b**) Schematic of a microfluidic chip with PLGA nanofiber film as a substrate prepared by the electrospinning method. (**c**) Schematic representation of A549 cell culture on a chip. (**d**) A549 cells co-cultured with HFL1 cells. (**e**) A549 cells co-cultured with HFL1 cells and HUVECs.

4.6. Brain Cancer

The brain, the most intricate organ governing human behavior, poses significant challenges for treating tumors. Precision drug delivery using electrospun fibers emerges as a promising strategy for controlled drug loading. Glioblastoma (GBM), a prevalent malignant central nervous system tumor, notoriously resists treatment with a high recurrence rate. Temozolomide (TMZ), an alkylating agent capable of crossing the blood–brain barrier, holds promise for GBM treatment. M. Irani et al. prepared PCL-Diol-b-PU/Au composite nanofibers via electrospinning, observing the continuous delivery of TMZ in the nanofibers. An evaluation of the nanofibers' cytotoxicity on U-87 human glioblastoma cells revealed enhanced cytotoxic effects due to the gold coating on the nanofibers' surfaces [147].

Postoperative care for surgically resected glioblastomas typically necessitates intensive treatment, often involving implantable fibrous membranes. Molina-Peñaé et al. electrospun chitosan-mixed nanoparticles to create nanofiber scaffolds for entrapping GBM cells. These scaffolds encapsulated stromal cell-derived factor-1α (SDF-1α), sustaining the biological activity of cancer cells by inducing their migration. The scaffold facilitated the sustained release of SDF-1α for at least 5 weeks, with the nanofiber structure providing effective anchoring. This approach holds the potential for treating unresectable GBM tumors [148].

Combining photohyperthermia with chemotherapy presents numerous advantages in treating brain cancer, particularly in delicate operations. Yang et al. developed chitosan composite nanofibers incorporating self-made photosensitized CuSe nanoparticles via a green electrospinning method. After the craniotomy, these nanofibers were immediately detached upon resection, offering high accuracy (>90%) and rapid hemostasis (<8 s). Light-induced photothermal and photodynamic therapies demonstrated highly efficient bactericidal effects and effectively induced apoptosis in residual tumor cells, inhibiting tumor recurrence [149].

4.7. Oral Cancer

Oral cancers encompass various histological types, including adenocarcinoma, lymphoma, melanoma, and teratoma. Squamous cell cancers, accounting for around 90% of oral cancers, typically originate in the cheeks, bottom of the mouth, gums, lips, roof of the mouth, and tongue. Nam et al. employed PVA and Soluplus (SP) as raw materials to fabricate Angelica gigas Nakai (AGN)/PVA/SP nanomaterials via electrospinning, achieving an encapsulation rate of 80%. Both PVA and SP act as hydrophilic polymers and surfactants for nanofiber formulation through electrospinning processes. The developed electrospun nanofiber preparations loaded with AGN were administered multiple times into the mouth, demonstrating negligible toxicity to various organs and tissues in mice. The fiber membrane adopts an explosive release mode, ensuring drug release within the oral cavity [68].

Will et al. achieved the local release of diclofenac using electrostatically spun nanofibers made from poly(d,l-lactide-co-glycolide) polymer. Controlled experiments revealed that the implantation of these electrospun scaffolds reduced tumor recurrence and significantly improved survival over a 7-week study period post-tumor resection. This approach holds substantial potential for directly inhibiting tumor growth and curbing inflammation-mediated tumor progression [150].

Zhang et al. pioneered an astaxanthin-based polycaprolactone/gelatin (PCL/GT) nanofiber adhesive patch (PGA) engineered with a saliva-insoluble PCL backing to prevent drug loss post-oral administration (Figure 10). This patch was designed to achieve an optimal release rate of astaxanthin, ensuring a high local drug concentration and penetration into the oral mucosa. Notably, its high porosity facilitated excellent adhesion to wet tissue and allowed good air permeability. The bioactive adhesive tablets significantly fostered the recovery of oral premalignant lesions (OPLs) by suppressing the expressions of Ki67 and cyclooxygenase-2 (COX-2). Remarkably, these tablets exhibited efficacy comparable to clinical retinoic cream preparations but without the side effects observed in the latter, such as hair loss and mouth ulcers [151].

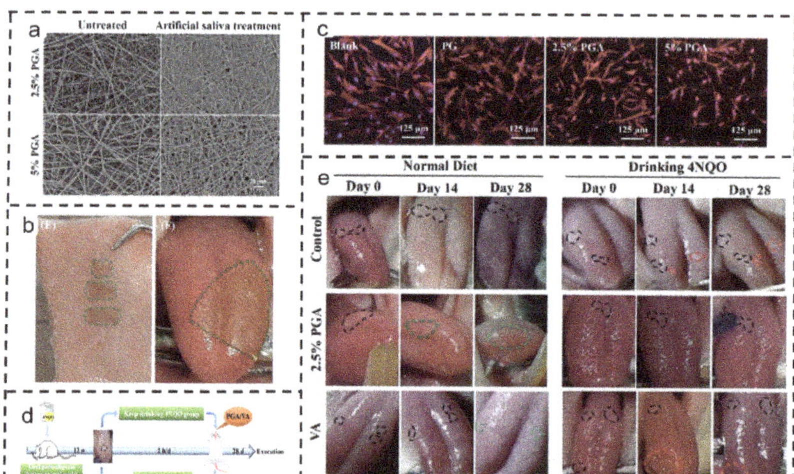

Figure 10. Application of polycaprolactone/gelatin (PCL/GT) nanofibers in oral cancer [151]. Copyright 2022, reproduced with permission from Elsevier. (**a**) Changes in fiber morphology of PGA electrospun membranes after soaking in artificial saliva. (**b**) Images of 2.5% PGA electrospun membrane adhesion to pig skin and rat tongue mucosa. (**c**) Fluorescence images of electrospun membranes with different ASX contents and hGFs co-cultured for 24 h. (**d**) Schematic representation of animal experiments. (**e**) Photographic representation of the PGA electrospun membrane and VA ointment treated with OPL on days 0, 14, and 28 (green dotted line indicates a good treatment effect; red dotted line indicates deterioration in treatment response; black dotted line indicates no change).

4.8. Other Cancers

The challenge in treating urothelial carcinoma lies in the restricted space of the ureter, making retention difficult. For low-risk upper tract urothelial carcinoma (UTUC), renal preservation surgery is the preferred strategy. Post-surgery, preventing ureteral cancer recurrence involves supporting the ureter with a stent. J. Wang et al. investigated the controlled delivery of epirubicin (EPI) using a gradient-degraded electrospun PCL/PLGA scaffold to assess its antitumor activity against UTUC. The PCL/PLGA electrospun fiber scaffold exhibited sustained and controlled degradation, enabling regulated drug release kinetics. Specifically, as the proportion of PCL increased, the degradation and release rates slowed, aiding in post-surgery stent retention to inhibit and eliminate participating cancer cells (Figure 11) [152].

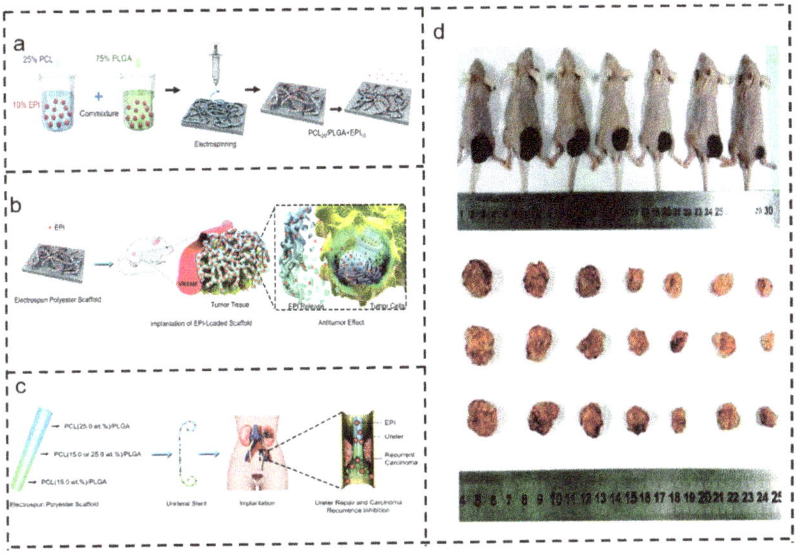

Figure 11. Application of fibrous membrane scaffolds in urothelial carcinoma [152]. Copyright 2019, reproduced with permission from RSC. (**a**) Stent preparation. (**b**) In vivo antitumor effect of electrospun polyester stent loaded with EPI implanted on the tumor surface. (**c**) Application diagram of electrospun polyester stent in preventing recurrence of upper urinary tract tumors and repairing the ureter in vivo. (**d**) Images of mice carrying T24 tumors after treatment and tumor resection. From left to right: control, PCL15/PLGA, PCL25/PLGA, PCL15/PLGA + EPI5, PCL15/PLGA + EPI10, PCL25/PLGA + EPI5, and PCL25/PLGA + EPI10 groups.

Electrospun fiber stents have been applied in urothelial carcinoma treatment. Musciacchio et al. reported that rifampicin (Rif)-supported membranes initially demonstrated rapid drug release within 24 h, followed by a slowed release while maintaining antibacterial efficacy by inhibiting bacterial proliferation and providing a bactericidal effect. The hydrophobic nature of the PCL/Rif membrane allowed for moistening by physiological liquids and cell media, ensuring cell adhesion, proliferation, and prolonged antibacterial activity against urothelial cells and smooth muscle cells, making it suitable for ureteral regeneration. This fibrous membrane stands as a potential ureteral stent for urothelial carcinoma [153].

Electrospun drug-loaded nanofibers have found diverse applications in treating various cancers. They have been employed in gastric retention administration for ascite cancer [154], continuous administration for liver cancer [155], simulating tumor microenvironments in osteosarcoma [156], detecting and locally delivering treatments in prostate cancer [157,158], and also in the prevention of pancreatic cancer [159] and gallbladder cancer [160]. Their versatility in drug delivery and targeting specific cancer types highlights their potential across a wide spectrum of oncological treatments. Unfortunately, the current application of drug-loaded electrospun nanofibers in cancer treatment is mostly limited to cell experiments and mouse experiments, with almost no relevant approved medical devices or clinical trials, as shown in Table S1.

5. Conclusions and Future Perspectives

Electrospinning stands as a versatile method for crafting nanoscale fibers through electrostatic forces. This technique allows for precise control over needle structures, facilitating the creation of specialized configurations like core–sheath, Janus, parallel, and eccentric structures. Moreover, adjusting the polymer composition, flow rates, and voltage settings enables the production of porous, hollow, and bead-on-string structures. These unique configurations significantly influence drug release dynamics, serving different roles at various stages of cancer treatment. However, challenges persist in scaling up electrospinning for mass production, representing a focal point for ongoing research and development efforts.

With the development of electrospinning technology, the range of raw materials that can be used to prepare nanofibers is gradually expanding. Organic compounds with good biocompatibility, various drug molecules, and even gene fragments can be used for electrospinning, and they can be combined to meet specific application requirements. Electrospun nanofibers, primarily composed of biodegradable polymers, offer versatility in cancer treatment without requiring removal, and they have been extensively studied for various applications.

Core–sheath structures, frequently utilized for gradual drug release, involve encapsulating drugs within the core layer, gradually releasing them through the hydrophobic polymer in the sheath. Fine-tuning the core–sheath thickness allows for precise control over the drug release rate, which is a prevalent strategy in cancer treatment. Recent trends include incorporating particles into electrospun fiber solutions. Magnetic electrospinning inclusion, for instance, allows for the integration of chemotherapy with magnetic, heat, or radiation therapies. This innovation significantly broadens cancer treatment options and helps combat drug resistance.

Additionally, the possibility of tailoring electrospinning to specific cancer types demonstrates its adaptability: it can be used with pH-sensitive polymers for precise delivery in colon cancer, hydrophilic polymers for rapid release in oral cancer, and floating drug-loaded fiber membranes for gastric cancer treatment. This method caters to diverse drug release and implantation needs in cancer therapy, positioning itself as a promising avenue for future applications.

Supplementary Materials: The following supporting information can be downloaded at: https://www.mdpi.com/article/10.3390/polym16040504/s1, Table S1. The research progress of cancer treatment applications listed in the article.

Author Contributions: Conceptualization, Y.Y. and R.Z.; funding acquisition, Y.Y.; project administration, Y.Y.; writing—original draft, R.Z., Z.L., J.G., B.C. and S.Z.; writing—review and editing, Y.Y. and D.Y. All authors have read and agreed to the published version of the manuscript.

Funding: This research was funded by the National Natural Science Foundation of China (Grant No. 51803121).

Institutional Review Board Statement: Not applicable.

Data Availability Statement: No new data were created or analyzed in this study. Data sharing is not applicable to this article.

Conflicts of Interest: The authors declare no conflicts of interest.

Abbreviations

Polyvinylpyrrolidone (PVP); polycaprolactone (PCL); cellulose acetate (CA); polyethylene glycol (PEG); quercetin (QUE); rutin (RUT); amlodipine besylate (AB); valsartan (VAL); poly(vinyl alcohol) (PVA); Soluplus (SP); angelica gigas Nakai (AGN); metformin hydrochloride (MET); N,N-Dimethylacetamide (DMAc); polylactic acid (PLA); ethyl cellulose (EC); ketoprofen (KET); acyclovir (ACY); polyacrylonitrile (PAN); ciprofloxacin (CIP); chloroform (CF); 2,2,2,-trifluoroethanol (TFE); key acid blue (KAB); key acid uranine (KAU); diclofenac sodium (DS); Eudragit S100 (ES100); polyethylene oxide (PEO); curcumin (CUR); Fixed-Dose Combinations (FDCs); chitosan (CS); poly(lactic-co-glycolic acid) copolymer (PLGA); resveratrol (RES); xanthohumol (XAN); γ-cyclodextrin (γ-

CD); Drug Delivery Systems (DDS); tumor microenvironment (TME); 3-(diethylamino)propylamine (DEAP); paclitaxel (PTX); lidocaine hydrochloride (Lid); silybin (SB); honokiol (HNK); fatty acid amide hydrolase (FAAH); epirubicin (EPI); doxorubicin (DOX); berberine (BBR); apatinib (AP); matrix metalloproteinases-2 (MMP-2); mesoporous silica nanoparticles (MSNs); 5-Fluorouracil (5-FU); Terminalia catappa (TC); amorphous calcium carbonate (ACC); gelatin/polycaprolactone (GP); spherical silver nanoparticles (AgNPs); indocyanine green (ICG); colorectal cancer (CRC); 17-dimethylaminoethylamino-17-demethoxy geldanamycin (17-DMAG); heat shock protein 90 (HSP90); AMP-activated protein kinase (AMPK); mammalian target of the rapamycin (mTOR); glioblastoma (GBM); temozolomide (TMZ); stromal cell-derived factor-1α (SDF-1α); gelatin (GT); oral premalignant lesions (OPL); cyclooxygenase-2 (COX-2); upper tract urothelial carcinoma (UTUC); rifampicin (Rif); poly(glycolide-ε-caprolactone) (PGCL); N-carboxymethyl chitosan (N-CMCS); prodigiosin (PG); sodium alginate (ALG); capsaicin powder (Cap); graphene oxide (GO); docetaxel(DTX); cetyltrimethylammonium bromide (CTAB); Tetraethoxysilane (TEOS); N-carboxymethyl chitosan (CMC); polycaprolactone diol (PCL-Diol); sodium alginate (SA); andrographolide (AG); 2,2-diphenyl-1-picrylhydrazyl (DPPH); thyme extract (TE); acid ferulic acid (FA); diclofenac sodium salt (DSS); polydioxanone (PDO); hexamethylene diisocyanate (HDI); stromal cell-derived factor-1α (SDF-1α); poly(d,l-lactide-co-glycolide) (PDLGA); poly(L-lactic acid) (PLLA); rectorite (REC); Cabazitaxel (CTX); gallic acid (GA).

References

1. Fawal, G.E.; Abu-Serie, M.M.; El-Gendi, H.; El-Fakharany, E.M. Fabrication, characterization and in vitro evaluation of disulfiram-loaded cellulose acetate/poly(ethylene oxide) nanofiber scaffold for breast and colon cancer cell lines treatment. *Int. J. Biol. Macromol.* **2022**, *204*, 555–564. [CrossRef]
2. Zhang, Z.; Yan, H.; Li, S.; Liu, Y.; Ran, P.; Chen, W.; Li, X. Janus rod-like micromotors to promote the tumor accumulation and cell internalization of therapeutic agents. *Chem. Eng. J.* **2021**, *404*, 127073. [CrossRef]
3. Yu, D.-G.; Huang, C. Electrospun Biomolecule-Based Drug Delivery Systems. *Biomolecules* **2023**, *13*, 1152. [CrossRef] [PubMed]
4. Yu, D.-G.; Xu, L. Impact Evaluations of Articles in Current Drug Delivery Based on Web of Science. *Curr. Drug Deliv.* **2023**, *20*, 360–367. [CrossRef]
5. Yu, D.G.; Zhou, J. How Can Electrospinning Further Service Well for Pharmaceutical Researches? *J. Pharm. Sci.* **2023**, *112*, 2719–2723. [CrossRef] [PubMed]
6. Kachwal, V.; Tan, J. Stimuli-responsive electrospun fluorescent fibers augmented with aggregation-induced emission (AIE) for smart applications. *Adv. Sci.* **2022**, *10*, 2204848. [CrossRef] [PubMed]
7. Rezaei, F.; Nikiforov, A.; Morent, R.; Geyter, N.D. Plasma modification of poly lactic acid solutions to generate high quality electrospun PLA nanofibers. *Sci. Rep.* **2018**, *8*, 2241. [CrossRef] [PubMed]
8. Duan, H.; Chen, H.; Qi, C.; Lv, F.; Wang, J.; Liu, Y.; Liu, Z.; Liu, Y. A novel electrospun nanofiber system with PEGylated paclitaxel nanocrystals enhancing the transmucus permeability and in situ retention for an efficient cervicovaginal cancer therapy. *Int. J. Pharm.* **2024**, *650*, 123660. [CrossRef] [PubMed]
9. Lv, H.; Guo, S.; Zhang, G.; He, W.; Wu, Y.; Yu, D. Electrospun structural hybrids of acyclovir-polyacrylonitrile at acyclovir for modifying drug release. *Polymers* **2021**, *13*, 4286. [CrossRef]
10. Khatami, F.; Baharian, A.; Akbari-Birgani, S.; Nikfarjam, N. Tubular scaffold made by gelatin/polylactic acid nanofibers for breast ductal carcinoma in situ tumor modeling. *J. Drug Deliv. Sci. Technol.* **2023**, *85*, 104606. [CrossRef]
11. Zhang, C.; Li, Y.; Wang, P.; Zhang, A.; Feng, F.; Zhang, H. Electrospinning of bilayer emulsions: The role of gum Arabic as a coating layer in the gelatin-stabilized emulsions. *Food Hydrocoll.* **2019**, *94*, 38–47. [CrossRef]
12. Yan, E.; Jiang, J.; Yang, X.; Fan, L.; Wang, Y.; An, Q.; Zhang, Z.; Lu, B.; Wang, D.; Zhang, D. pH-sensitive core-shell electrospun nanofibers based on polyvinyl alcohol/polycaprolactone as a potential drug delivery system for the chemotherapy against cervical cancer. *J. Drug Deliv. Sci. Technol.* **2020**, *55*, 101455. [CrossRef]
13. Qian, C.H.; Liu, Y.B.; Chen, S.; Zhang, C.Y.; Chen, X.H.; Liu, Y.H.; Liu, P. Electrospun core-sheath PCL nanofibers loaded with nHA and simvastatin and their potential bone regeneration applications. *Front. Bioeng. Biotechnol.* **2023**, *11*, 1205252. [CrossRef] [PubMed]
14. Yang, C.; Yu, D.; Pan, X.; Liu, X.; Wang, S.; Bligh, W.A.; Williams, G.R. Electrospun pH-sensitive core–shell polymer nanocomposites fabricated using a tri-axial process. *Acta Biomater.* **2016**, *35*, 77–86. [CrossRef] [PubMed]
15. Brako, F.; Luo, C.; Matharu, R.K.; Ciric, L.; Harker, A.; Edirisinghe, M.; Craig, D.Q.M. A Portable device for the generation of drug-loaded three-compartmental fibers containing metronidazole and iodine for topical application. *Pharmaceutics* **2020**, *12*, 373. [CrossRef] [PubMed]
16. Zhang, X.; Chi, C.; Chen, J.; Zhang, X.; Gong, M.; Wang, X.; Yan, J.; Shi, R.; Zhang, L.; Xue, J. Electrospun quad-axial nanofibers for controlled and sustained drug delivery. *Mater. Des.* **2021**, *206*, 109732. [CrossRef]
17. Zheng, X.; Kang, S.; Wang, K.; Yang, Y.; Yu, D.; Wan, F.; Williams, G.R.; Bligh, S.A. Combination of structure-performance and shape-performance relationships for better biphasic release in electrospun Janus fibers. *Int. J. Pharm.* **2021**, *596*, 120203. [CrossRef]

18. Chang, S.; Wang, M.; Zhang, F.; Liu, Y.; Liu, X.; Yu, D.; Shen, H. Sheath-separate-core nanocomposites fabricated using a tri-fluid electrospinning. *Mater. Des.* **2020**, *192*, 108782. [CrossRef]
19. Lv, J.; Gu, W.; Cui, X.; Dai, S.; Zhang, B.; Ji, G. Nanofiber network with adjustable nanostructure controlled by PVP content for an excellent microwave absorption. *Sci. Rep.* **2019**, *9*, 4271. [CrossRef]
20. Ignatova, M.; Anastasova, I.; Manolova, N.; Rashkov, I.; Markova, N.; Kukeva, R.; Stoyanova, R.; Georgieva, A.; Toshkova, R. Bio-based electrospun fibers from chitosan Schiff base and polylactide and their Cu^{2+} and Fe^{3+} complexes: Preparation and antibacterial and anticancer activities. *Polymers* **2022**, *14*, 5002. [CrossRef]
21. Contardi, M.; Kossyvaki, D.; Picone, P.; Summa, M.; Guo, X.; Heredia-Guerrero, J.A.; Giacomazza, D.; Carzino, R.; Goldoni, L.; Scoponi, G.; et al. Electrospun polyvinylpyrrolidone (PVP) hydrogels containing hydroxycinnamic acid derivatives as potential wound dressings. *Chem. Eng. J.* **2021**, *409*, 128144. [CrossRef]
22. Lian, H.; Meng, Z. Melt electrospinning vs. solution electrospinning: A comparative study of drug-loaded poly (ε-caprolactone) fibres. *Mater. Sci. Eng. C* **2017**, *74*, 117–123. [CrossRef]
23. Liu, Y.; Wang, Q.; Lu, Y.; Deng, H.; Zhou, X. Synergistic enhancement of cytotoxicity against cancer cells by incorporation of rectorite into the paclitaxel immobilized cellulose acetate nanofibers. *Int. J. Biol. Macromol.* **2020**, *152*, 672–680. [CrossRef]
24. Wu, S.; Li, J.; Mai, J.; Chang, M. Three-dimensional electrohydrodynamic printing and spinning of flexible composite structures for oral multidrug forms. *ACS Appl. Mater. Interfaces* **2018**, *10*, 24876–24885. [CrossRef]
25. Wang, Y.; Yu, D.-G.; Liu, Y.; Liu, Y.-N. Progress of Electrospun Nanofibrous Carriers for Modifications to Drug Release Profiles. *J. Funct. Biomater.* **2022**, *13*, 289. [CrossRef]
26. Yu, D.-G.; Zhao, P. The Key Elements for Biomolecules to Biomaterials and to Bioapplications. *Biomolecules* **2022**, *12*, 1234. [CrossRef]
27. Lian, H.; Meng, Z. Melt electrospinning of daunorubicin hydrochloride-loaded poly (ε-caprolactone) fibrous membrane for tumor therapy. *Bioact. Mater.* **2017**, *2*, 96–100. [CrossRef] [PubMed]
28. Zhou, K.; Wang, M.; Zhou, Y.; Sun, M.; Xie, Y.; Yu, D. Comparisons of antibacterial performances between electrospun polymer@drug nanohybrids with drug-polymer nanocomposites. *Adv. Compos. Hybrid Mater.* **2022**, *5*, 907–919. [CrossRef]
29. Prabu, G.T.V.; Dhurai, B. A novel profiled multi-pin electrospinning system for nanofiber production and encapsulation of nanoparticles into nanofibers. *Sci. Rep.* **2020**, *10*, 4302. [CrossRef] [PubMed]
30. Zhou, J.; Wang, L.; Gong, W.; Wang, B.; Yu, D.-G.; Zhu, Y. Integrating Chinese Herbs and Western Medicine for New Wound Dressings through Handheld Electrospinning. *Biomedicines* **2023**, *11*, 2146. [CrossRef] [PubMed]
31. Xi, H.; Zhao, H. Silk fibroin coaxial bead-on-string fiber materials and their drug release behaviors in different pH. *J. Mater. Sci.* **2019**, *54*, 4246–4258. [CrossRef]
32. Li, T.; Ding, X.; Tian, L.; Ramakrishna, S. Engineering BSA-dextran particles encapsulated bead-on-string nanofiber scaffold for tissue. engineering applications. *J. Mater. Sci.* **2017**, *52*, 10661–10672. [CrossRef]
33. Li, T.; Liu, L.; Wang, L.; Ding, X. Solid drug particles encapsulated bead-on-string nanofibers: The control of bead number and its corresponding release profile. *J. Biomater. Sci. Polym. Ed.* **2019**, *30*, 1454–1469. [CrossRef] [PubMed]
34. Ma, L.; Shi, X.; Zhang, X.; Li, L. Electrospinning of polycaprolacton/chitosan core-shell nanofibers by a stable emulsion system. *Colloid Surf. A* **2019**, *583*, 123956. [CrossRef]
35. Wang, L.; Ma, D.; Xu, C.; Gan, X.; Ge, P.; Zhu, L.; Wang, X.; Lv, Y. Preparation of flexible hollow TiO_2 fibrous membranes for thermal-insulation applications by coaxial electrospinning. *Ceram. Int.* **2023**, *49*, 22875–22881. [CrossRef]
36. Najafi, S.J.; Gharehaghaji, A.A.; Etrati, S.M. Fabrication and characterization of elastic hollow nanofibrous PU yarn. *Mater. Des.* **2016**, *99*, 328–334. [CrossRef]
37. Zhao, S.; Sun, J.; Qin, Z.; Li, Y.; Yu, H.; Wang, G.; Gu, X.; Pan, K. Janus-structural AIE nanofiber with white light emission and stimuli-response. *Small* **2022**, *18*, 2201117. [CrossRef]
38. Zhou, J.; Dai, Y.; Fu, J.; Yan, C.; Yu, D.-G.; Yi, T. Dual-Step Controlled Release of Berberine Hydrochloride from the Trans-Scale Hybrids of Nanofibers and Microparticles. *Biomolecules* **2023**, *13*, 1011. [CrossRef]
39. Chen, S.; Zhou, J.; Fang, B.; Ying, Y.; Yu, D.G.; He, H. Three EHDA Processes from A Detachable Spinneret for Fabricating Drug Fast Dissolution Composites. *Macromol. Mater. Eng.* **2023**, 2300361. [CrossRef]
40. Fang, D.; Qin, Z.; Zheng, L.; Yew, P.Y.M.; Jiang, X.; Kai, D.; Song, F.; Zhao, J. Ros-responsive nanocomposite scaffolds for sustained releasing puerarin toachieve chondroprotection in OA rats. *Mater. Des.* **2023**, *233*, 112214. [CrossRef]
41. Shen, S.; Zhu, L.; Liu, J.; Ali, A.; Zaman, A.; Ahmad, Z.; Chen, X.; Chang, M. Novel core-shell fiber delivery system for synergistic treatment of cervical cancer. *J. Drug Deliv. Sci. Technol.* **2020**, *59*, 101865. [CrossRef]
42. Zhao, L.; Orlu, M.; Williams, G.R. Electrospun fixed dose combination fibers for the treatment of cardiovascular disease. *Int. J. Pharm.* **2021**, *599*, 120426. [CrossRef]
43. Cao, X.; Chen, W.; Zhao, P.; Yang, Y.; Yu, D.-G. Electrospun Porous Nanofibers: Pore-Forming Mechanisms and Applications for Photocatalytic Degradation of Organic Pollutants in Wastewater. *Polymers* **2022**, *14*, 3990. [CrossRef]
44. Moghe, A.K.; Gupta, B.S. C-axial electrospinning for nanofiber structures: Preparation and applications. *Polym. Rev.* **2008**, *48*, 353–377. [CrossRef]
45. Yu, D.; Li, J.; Williams, G.R.; Zhao, M. Electrospun amorphous solid dispersions of poorly water-soluble drugs: A review. *J. Control. Release* **2018**, *292*, 91–110. [CrossRef] [PubMed]

46. Wang, H.; Lu, Y.; Yang, H.; Yu, D.-G.; Lu, X. The Influence of the Ultrasonic Treatment of Working Fluids on Electrospun Amorphous Solid Dispersions. *Front. Mol. Biosci.* **2023**, *10*, 1184767. [CrossRef] [PubMed]
47. Zheng, Q.; Xi, Y.; Weng, Y. Functional electrospun nanofibers: Fabrication, properties, and applications in wound-healing Process. *RSC Adv.* **2024**, *14*, 3359. [CrossRef] [PubMed]
48. Yu, D.G.; Zhou, J. Electrospun multi-chamber nanostructures for sustainable biobased chemical nanofibers. *Next Mater.* **2024**, *2*, 100119. [CrossRef]
49. Li, J.; Du, Q.; Wan, J.; Yu, D.G.; Tan, F.; Yang, X. Improved synergistic anticancer action of quercetin and tamoxifen citrate supported by an electrospun complex nanostructure. *Mater. Des.* **2024**, *238*, 112657. [CrossRef]
50. Alba-Perez, A.; Jayawarna, V.; Childs, P.G.; Dalby, M.J.; Salmeron-Sanchez, M. Plasma polymerised nanoscale coatings of controlled thickness for efficient solid-phase presentation of growth factors. *Mater. Sci. Eng. C* **2020**, *113*, 110966. [CrossRef]
51. Nagiah, N.; Murdock, C.J.; Bhattacharjee, M.; Nair, L.; Laurencin, C.T. Development of tripolymeric triaxial electrospun fibrous matrices for dual-drug delivery applications. *Sci. Rep.* **2020**, *10*, 609. [CrossRef]
52. Jouybari, M.H.; Hosseini, S.; Mahboobnia, K.; Boloursazd, L.A.; Moradie, M.; Irani, M. Simultaneous controlled release of 5-FU, DOX and PTX from chitosan/PLA/5-FU/g-C3N--DOX/g-C3N4-PTX triaxial nanofibers for breast cancer treatment in vitro. *Colloids Surf. B Biointerfaces* **2019**, *179*, 495–504. [CrossRef] [PubMed]
53. Chen, X.; Liu, Y.; Liu, P. Electrospun Core–Sheath Nanofibers with a Cellulose Acetate Coating for the Synergistic Release of Zinc Ion and Drugs. *Mol. Pharm.* **2024**, *21*, 173–182. [CrossRef] [PubMed]
54. Wang, Y.; Liu, L.; Zhu, Y.; Wang, L.; Yu, D.-G.; Liu, L.-y. Tri-Layer Core–Shell Fibers from Coaxial Electrospinning for a Modified Release of Metronidazole. *Pharmaceutics* **2023**, *15*, 2561. [CrossRef] [PubMed]
55. Liao, Q.; Kim, E.J.; Tang, Y.; Xu, H.; Yu, D.; Song, W.; Kim, B.J. Rational Design of Hyper-Crosslinked Polymers for Biomedical Applications. *J. Polym. Sci.* **2023**, *9*, 1–19. [CrossRef]
56. Han, D.; Steckl, A.J. Triaxial electrospun nanofiber membranes for controlled dual release of functional molecules. *ACS Appl. Mater. Interfaces* **2013**, *5*, 8241–8245. [CrossRef] [PubMed]
57. Sun, L.; Zhou, J.; Chen, Y.; Yu, D.G.; Liu, P. A Combined Electrohydrodynamic Atomization Method for Preparing Nanofiber/Microparticle Hybrid Medicines. *Front. Bioeng. Biotechnol.* **2023**, *11*, 1308004. [CrossRef] [PubMed]
58. Stoyanova, N.; Spasova, M.; Manolova, N.; Rashkov, I.; Georgieva, A.; Toshkova, R. Quercetin- and rutin-containing electrospun cellulose acetate and polyethylene glycol fibers with antioxidant and anticancer properties. *Polymers* **2022**, *14*, 5380. [CrossRef]
59. Xu, L.; He, H.; Du, Y.; Zhang, S.; Yu, D.-G.; Liu, P. Electrosprayed Core (Cellulose Acetate)–Shell (Polyvinylpyrrolidone) Nanoparticles for Smart Acetaminophen Delivery. *Pharmaceutics* **2023**, *15*, 2314. [CrossRef]
60. Ji, Y.; Zhao, H.; Liu, H.; Zhao, P.; Yu, D.-G. Electrosprayed Stearic-Acid-Coated Ethylcellulose Microparticles for an Improved Sustained Release of Anticancer Drug. *Gels* **2023**, *9*, 700. [CrossRef]
61. Yan, X.; Xu, B.; Xia, C.; Xu, M.; Zeng, B.; Zhang, R.; Zhu, L.; Zhang, C. Dual drug-loaded core-shell nanofibers membranes via emulsion electrospinning and their controllable sustained release property. *J. Drug. Deliv. Sci. Technol.* **2023**, *88*, 104909. [CrossRef]
62. Liu, M.; Hao, X.; Wang, Y.; Jiang, Z.; Zhang, H. A biodegradable core-sheath nanofibrous 3D hierarchy prepared by emulsion electrospinning for sustained drug release. *J. Mater. Sci.* **2020**, *55*, 16730–16743. [CrossRef]
63. Yang, G.; Wang, J.; Li, L.; Ding, S.; Zhou, S. Electrospun micelles/drug-loaded nanofibers for time-programmed multi-agent release. *Macromol. Biosci.* **2014**, *14*, 965–976. [CrossRef] [PubMed]
64. Yang, Y.; Chen, W.; Wang, M.; Shen, J.; Tang, Z.; Qin, Y.; Yu, D. Engineered shellac beads-on-the-string fibers using triaxial electrospinning for improved colon-targeted drug delivery. *Polymers* **2023**, *15*, 2237. [CrossRef]
65. Guo, H.; Tan, S.; Gao, L.J.; Wang, J. Sequential release of drugs form a dual-delivery system based on pH-responsive nanofibrous mats towards wound care. *Mater. Chem. B* **2020**, *8*, 1759–1770. [CrossRef] [PubMed]
66. Bukhary, H.; Williams, G.R.; Orlu, M. Electrospun fixed dose formulations of amlodipine besylate and valsartan. *Int. J. Pharm.* **2018**, *549*, 446–455. [CrossRef] [PubMed]
67. Nam, S.; Lee, J.; Lee, S.Y.; Jeong, J.Y.; Kang, W.; Cho, H. Angelica gigas Nakai extract-loaded fast-dissolving nanofiber based on poly(vinyl alcohol) and Soluplus for oral cancer therapy. *Int. J. Pharm.* **2017**, *526*, 225–234. [CrossRef] [PubMed]
68. Xu, H.; Xu, X.; Li, S.; Song, W.; Yu, D.; Bligh, S.W.A. The effect of drug heterogeneous distributions within core-sheath nanostructures on its sustained release profiles. *Biomolecules* **2021**, *11*, 1330. [CrossRef]
69. Um-i-Zahra, S.; Shen, X.X.; Li, H.; Zhu, L. Study of sustained release drug-loaded nanofibers of cellulose acetate and ethyl cellulose polymer blends prepared by electrospinning and their in-vitro drug release profiles. *J. Polym. Res.* **2014**, *21*, 602. [CrossRef]
70. Hai, T.; Wan, X.; Yu, D.; Wang, K.; Yang, Y.; Liu, Z. Electrospun lipid-coated medicated nanocomposites for an improved drug sustained-release profile. *Mater. Des.* **2019**, *162*, 70–79. [CrossRef]
71. Wang, M.; Hou, J.; Yu, D.; Li, S.; Zhu, J.; Chen, Z. Electrospun tri-layer nanodepots for sustained release acyclovir. *J. Alloys Compd.* **2020**, *846*, 156471. [CrossRef]
72. Liu, X.; Yang, Y.; Yu, D.; Zhu, M.; Zhao, M.; Williams, G.R. Tunable zero-order drug delivery systems created by modified triaxial electrospinning. *Chem. Eng. J.* **2019**, *356*, 886–894. [CrossRef]
73. Yu, D.; Li, X.; Wang, X.; Yang, J.; Bligh, S.W.A.; Williams, G.R. Nanofibers fabricated using triaxial electrospinning as zero order drug delivery systems. *ACS Appl. Mater. Interfaces* **2015**, *7*, 18891–18897. [CrossRef] [PubMed]

74. Lemraski, E.G.; Alibeigi, S.; Abbasi, Z. Ibuprofen@silver loaded on poly(vinyl alcohol)/chitosan co-polymer scaffold as a novel drug delivery system. *Mater. Today Commun.* **2022**, *33*, 104311. [CrossRef]
75. Tawfik, E.A.; Craig, D.Q.M.; Barker, S.A. Dual drug-loaded coaxial nanofibers for the treatment of corneal abrasion. *Int. J. Pharm.* **2020**, *581*, 119296. [CrossRef]
76. Zhang, X.; Han, L.; Sun, Q.; Xia, W.; Zhou, Q.; Zhang, Z.; Song, X. Controlled release of resveratrol and xanthohumol via coaxial electrospinning fibers. *J. Biomater. Sci. Polym. Ed.* **2020**, *31*, 456–471. [CrossRef]
77. Niiyama, E.; Uto, K.; Lee, C.M.; Sakura, K.; Ebara, M. Hyperthermia nanofiber platform synergized by sustained release of paclitaxel to improve antitumor efficiency. *Adv. Healthc. Mater.* **2019**, *8*, 1900102. [CrossRef]
78. Gao, S.; Zhou, A.; Cao, B.; Wang, J.; Li, F.; Tang, G.; Jiang, Z.; Yang, A.; Xiong, R.; Lei, J.; et al. A tunable temperature-responsive and tough platform for controlled drug delivery. *New J. Chem.* **2021**, *45*, 13056–13063. [CrossRef]
79. Khan, I.; Pandit, J.; Ahmed, S.; Zameer, S.; Nikita; Ahmad, S.; Bano, S.; Ansari, M.D.; Solanki, P.; Jahan, R.N.; et al. Development and evaluation of biodegradable polymeric lomustine nanofibres for the efficient tumor targeting: In vitro characterization, ex vivo permeation and degradation study. *J. Drug Deliv. Sci. Technol.* **2022**, *75*, 103685. [CrossRef]
80. Li, X.; He, Y.; Hou, J.; Yang, G.; Zhou, S. A time-programmed release of dual drugs from an implantable trilayer structured fiber device for synergistic treatment of breast cancer. *Small* **2020**, *16*, 1902262. [CrossRef] [PubMed]
81. García-García, P.; Reyes, R.; Segredo-Morales, E.; Pérez-Herrero, E.; Delgado, A.; Évora, C. PLGA-BMP-2 and PLA-17β-estradiol microspheres reinforcing a composite hydrogel for bone regeneration in osteoporosis. *Pharmaceutics* **2019**, *11*, 648. [CrossRef]
82. Bourdon, L.; Attik, N.; Belkessam, L.; Chevalier, C.; Bousige, C.; Brioude, A.; Salles, V. Direct-writing electrospun functionalized scaffolds for periodontal regeneration: In vitro studies. *J. Funct. Biomater.* **2023**, *14*, 263. [CrossRef]
83. Sun, M.; Liu, Y.; Jiao, K.; Jia, W.; Jiang, K.; Cheng, Z.; Liu, G.; Luo, Y. A periodontal tissue regeneration strategy via biphasic release of zeolitic imidazolate framework-8 and FK506 using a uniaxial electrospun Janus nanofiber. *J. Mater. Chem. B* **2022**, *10*, 765–778. [CrossRef]
84. Zhu, Y.; Song, F.; Ju, Y.; Huang, L.; Zhang, L.; Tang, C.; Yang, H.; Huang, C. NAC-loaded electrospun scaffolding system with dual compartments for the osteogenesis of rBMSCs in vitro. *Int. J. Nanomed.* **2019**, *14*, 787–798. [CrossRef]
85. Ullah, A.; Saito, Y.; Ullah, S.; Haider, M.K.; Nawaz, H.; Duy-Nam, P.; Kharaghani, D.; Kim, I.S. Bioactive Sambong oil-loaded electrospun cellulose acetate nanofibers: Preparation, characterization, and in-vitro biocompatibility. *Int. J. Biol. Macromol.* **2021**, *166*, 1009–1021. [CrossRef]
86. Ye, P.; Wei, S.; Luo, C.; Wang, Q.; Li, A.; Wei, F. Long-term effect against methicillin-resistant staphylococcus aureus of emodin released from coaxial electrospinning nanofiber membranes with a biphasic profile. *Biomolecules* **2020**, *10*, 362. [CrossRef] [PubMed]
87. Akbarzadeh, M.; Pezeshki-Modaress, M.; Zandi, M. Biphasic, tough composite core/shell PCL/PVA-GEL nanofibers for biomedical application. *J. Appl. Polym. Sci. A* **2020**, *137*, 48713. [CrossRef]
88. Su, Y.; McCarthy, A.; Wong, S.L.; Hollins, R.R.; Wang, G.; Xie, J. Simultaneous delivery of multiple antimicrobial agents by biphasic scaffolds for effective treatment of wound biofilms. *Adv. Healthc. Mater.* **2021**, *10*, 2100135. [CrossRef]
89. Alkaissy, R.; Richard, M.; Morris, H.; Snelling, S.; Pinchbeck, H.; Carr, A.; Mouthuy, P. Manufacture of soft-hard implants from electrospun filaments embedded in 3d printed structures. *Macromol. Biosci.* **2022**, *22*, 2200156. [CrossRef]
90. Sahu, D.K.; Pradhan, D.; Halder, J.; Biswasroy, P.; Kar, B.; Ghosh, G.; Rath, G. Design and optimization of gatifloxacin loaded polyvinyl alcohol nanofiber for the treatment of dry eye infection: In vitro and in vivo evaluation. *J. Drug Deliv. Sci. Technol.* **2022**, *76*, 103651. [CrossRef]
91. Khan, A.R.; Nadeem, M.; Bhutto, M.A.; Yu, F.; Xie, X.; El-Hamshary, H.; El-Faham, A.; Ibrahim, U.A.; Mo, X. Physico-chemical and biological evaluation of PLCL/SF nanofibers loaded with oregano essential oil. *Pharmaceuticsal* **2019**, *11*, 386. [CrossRef]
92. Bikiaris, N.D.; Koumentakou, I.; Michailidou, G.; Kostoglou, M.; Vlachou, M.; Barmpalexis, P.; Karavas, E.; Papageorgiou, G.Z. Investigation of molecular weight, polymer concentration and process parameters factors on the sustained release of the anti-multiple-sclerosis agent teriflunomide from poly(ε-caprolactone) electrospun nanofibrous matrices. *Pharmaceuticals* **2022**, *14*, 1693. [CrossRef] [PubMed]
93. Boncu, T.E.; Ozdemir, N. Electrospinning of ampicillin trihydrate loaded electrospun PLA nanofibers I: Effect of polymer concentration and PCL addition on its morphology, drug delivery and mechanical properties. *Int. J. Polym. Mater. Polym. Biomater.* **2022**, *71*, 669–676. [CrossRef]
94. Kabay, G.; Demirci, C.; Can, G.K.; Meydan, A.E.; Daşan, B.G.; Mutlu, M. A comparative study of single-needle and coaxial electrospun amyloid-like protein nanofibers to investigate hydrophilic drug release behavior. *Int. J. Biol. Macromol.* **2018**, *114*, 989–997. [CrossRef] [PubMed]
95. Schifino, G.; Gasparini, C.; Drudi, S.; Giannelli, M.; Sotgiu, G.; Posati, T.; Zamboni, R.; Treossi, E.; Maccaferri, E.; Giorgini, L.; et al. Keratin/Polylactic acid/graphene oxide composite nanofibers for drug delivery. *Int. J. Pharm.* **2022**, *623*, 121888. [CrossRef] [PubMed]
96. Zhou, J.; Yi, T.; Zhang, Z.; Yu, D.G.; Liu, P.; Wang, L.; Zhu, Y. Electrospun Janus Core (Ethyl Cellulose//Polyethylene Oxide) @ Shell (Hydroxypropyl Methyl Cellulose Acetate Succinate) Hybrids for An Enhanced Colon-Targeted Prolonged Drug Absorbance. *Adv. Compos. Hybrid Mater.* **2023**, *6*, 189. [CrossRef]
97. Altinbasak, I.; Kocak, S.; Colby, A.H.; Alp, Y.; Sanyal, R.; Grinstaff, M.W.; Sanyal, A. pH-Responsive nanofiber buttresses as local drug delivery devices. *Biomater. Sci.* **2023**, *11*, 813–821. [CrossRef] [PubMed]

98. Zhang, Y.; Lu, Y.; Li, Y.; Xu, Y.; Song, W. Poly(Glutamic Acid)-Engineered Nanoplatforms for Enhanced Cancer Phototherapy. *Curr. Drug Deliv.* **2024**, *21*, 326–338. [CrossRef]
99. Yu, H.S.; Lee, E.S. Honeycomb-like pH-responsive γ-cyclodextrin electrospun particles for highly efficient tumor therapy. *Carbohydr. Polym.* **2020**, *230*, 11556. [CrossRef]
100. Graham-Gurysh, E.G.; Murthy, A.B.; Moore, K.M.; Hingtgena, S.D.; Bacheldera, E.M.; Ainslie, K.M. Synergistic drug combinations for a precision medicine approach to interstitial glioblastoma therapy. *J. Control. Release* **2020**, *323*, 282–292. [CrossRef]
101. Anders, C.K.; Carey, L.A. Biology, metastatic patterns, and treatment of patients with triple-negative breast cancer. *Clin. Breast Cancer* **2009**, *9*, S73–S81. [CrossRef]
102. Al-Attar, T.; Madihally, S.V. Influence of controlled release of resveratrol from electrospun fibers in combination with siRNA on leukemia cells. *Eur. J. Pharm. Sci.* **2018**, *123*, 173–183. [CrossRef]
103. Cheng, H.; Yang, X.; Che, X.; Yang, M.; Zhai, G. Biomedical application and controlled drug release of electrospun fibrous materials. *Mater. Sci. Eng. C* **2018**, *90*, 750–763. [CrossRef] [PubMed]
104. Li, X.; Yu, N.; Li, J.; Bai, J.; Ding, D.; Tang, Q.; Xu, H. Novel "carrier-free" nanofiber codelivery systems with the synergistic antitumor effect of paclitaxel and tetrandrine through the enhancement of mitochondrial apoptosis. *ACS Appl. Mater. Interfaces* **2020**, *12*, 10096–10106. [CrossRef] [PubMed]
105. Lübtow, M.M.; Hahn, L.; Haider, M.S.; Luxenhofer, R. Title of the article. *J. Am. Chem. Soc.* **2017**, *139*, 10980–10983. [CrossRef]
106. Huo, M.; Wang, H.; Zhang, Y.; Cai, H.; Zhang, P.; Li, L.; Zhou, J.; Yin, T. Co-delivery of silybin and paclitaxel by dextran-based nanoparticles for effective anti-tumor treatment through chemotherapy sensitization and microenvironment modulation. *J. Control. Release* **2020**, *321*, 198–210. [CrossRef]
107. Wang, H.; Wang, S.; Wang, R.; Wang, X.; Jiang, K.; Xie, C.; Zhan, C.; Wang, H.; Lu, W. Co-delivery of paclitaxel and melittin by glycopeptide-modified lipodisks for synergistic anti-glioma therapy. *Nanoscale* **2019**, *11*, 13069. [CrossRef]
108. Wang, Z.; Li, X.; Wang, D.; Zou, Y.; Qu, X.; He, C.; Deng, Y.; Jin, Y.; Zhou, Y.; Zhou, Y.; et al. Concurrently suppressing multidrug resistance and metastasis of breast cancer by co-delivery of paclitaxel and honokiol with pH-sensitive polymeric micelles. *Acta Biomater.* **2017**, *62*, 144–156. [CrossRef]
109. Slivicki, R.A.; Xu, Z.; Mali, S.S.; Hohmann, A.G. Brain permeant and impermeant inhibitors of fatty-acid amide hydrolase suppress the development and maintenance of paclitaxel-induced neuropathic pain without producing tolerance or physical dependence in vivo and synergize with paclitaxel to reduce tumor cell line viability in vitro. *Pharmacol. Res.* **2019**, *142*, 267–282. [CrossRef] [PubMed]
110. Lin, F.; Chen, P.; Wei, K.; Huang, C.; Wang, C.; Yang, H. Rapid in situ MRI traceable gel-forming dual-drug delivery for synergistic therapy of brain tumor. *Theranostics* **2017**, *7*, 2524–2536. [CrossRef]
111. Mozaffari, S.; Seyedabadi, S.; Alemzadeh, E. Anticancer efficiency of doxorubicin and berberine-loaded PCL nanofibers in preventing local breast cancer recurrence. *J. Drug Deliv. Sci. Technol.* **2022**, *67*, 102984. [CrossRef]
112. Kumar, R.; Uppal, S.; Mansi, K.; Das, J.; Pandey, S.K.; Kaur, K.; Mehta, S.K. Ultrasonication induced synthesis of TPGS stabilized clove oil nanoemulsions and their synergistic effect against breast cancer cells and harmful bacteria. *J. Mol. Liq.* **2022**, *349*, 118130. [CrossRef]
113. Mehnath, S.; Chitra, K.; Karthikeyan, K.; Jeyaraj, M. Localized delivery of active targeting micelles from nanofibers patch for effective breast cancer therapy. *Int. J. Pharm.* **2020**, *584*, 119412. [CrossRef] [PubMed]
114. Fawal, G.E.; Abu-Serie, M.M.; Mo, X.; Wang, H. Diethyldithiocarbamate/silk fibroin/polyethylene oxide nanofibrous for cancer therapy: Fabrication, characterization and in vitro evaluation. *Int. J. Biol. Macromol.* **2021**, *193*, 293–299. [CrossRef] [PubMed]
115. Ignatova, M.; Manolova, N.; Rashkov, I.; Markova, N.; Kukeva, R.; Stoyanova, R.; Georgieva, A.; Oshkova, R.T. 8-Hydroxyquinoline-5-sulfonic acid-containing poly(vinyl alcohol)/chitosan electrospun materials and their Cu^{2+} and Fe^{3+} complexes: Preparation, Antibacterial, Antifungal and Antitumor Activities. *Polymers* **2021**, *13*, 2690. [CrossRef]
116. Rabionet, M.; Yeste, M.; Puig, T.; Ciurana, J. Electrospinning PCL scaffolds manufacture for three-dimensional breast cancer cell culture. *Polymers* **2017**, *9*, 328. [CrossRef]
117. Jaworska, J.; Smolarczyk, R.; Musiał-Kulik, M.; Cicho'n, T.; Karpeta-Jarząbek, P.; Włodarczyk, J.; Stojko, M.; Janeczek, H.; Kordyka, A.; Kaczmarczyk, B.; et al. Electrospun paclitaxel delivery system based on PGCL/PLGA in local therapy combined with brachytherapy. *Int. J. Pharm.* **2021**, *602*, 120596. [CrossRef]
118. Abasalta, M.; Asefnejad, A.; Khorasani, M.T.; Saadatabadi, A.R. Adsorption and sustained release of doxorubicin from N-carboxymethyl chitosan/polyvinyl alcohol/poly(ε-caprolactone) composite and core-shell nanofibers. *J. Drug Deliv. Sci. Technol.* **2022**, *67*, 102937. [CrossRef]
119. Li, X.; Xu, F.; He, Y.; Li, Y.; Hou, J.; Yang, G.; Zhou, S. A Hierarchical Structured Ultrafine Fiber Device for Preventing Postoperative Recurrence and Metastasis of Breast Cancer. *Adv. Funct. Mater.* **2020**, *30*, 2004851. [CrossRef]
120. Akpan, U.M.; Pellegrini, M.; Obayemi, J.D.; Ezenwafor, T.; Browl, D.; Ani, C.J.; Yipo, D.; Salifu, A.; Dozie-Nwachukwu, S.; Odusanya, S.; et al. Prodigiosin-loaded electrospun nanofibers scaffold for localized treatment of triple negative breast cancer. *Mater. Sci. Eng. C* **2020**, *114*, 110976. [CrossRef]
121. Ahmady, A.R.; Solouk, A.; Saber-Samandari, S.; Akbari, S.; Ghanbari, H.; Brycki, B.E. Capsaicin-loaded alginate nanoparticles embedded polycaprolactone-chitosan nanofibers as a controlled drug delivery nanoplatform for anticancer activity. *J. Colloid Interface Sci.* **2023**, *638*, 616–628. [CrossRef] [PubMed]

122. Yang, X.; Yang, N.; Zhang, L.; Zhao, D.; Lei, H.; Cheng, S.; Ge, J.; Ma, X.; Ni, C.; Liu, Z.; et al. Eddy current thermal effect based on magnesium microrods for combined tumor therapy. *Chem. Eng. J.* **2022**, *446*, 137038. [CrossRef]
123. Samadi, S.; Moradkhani, B.M.; Beheshti, H.; Irani, M.; Aliabadi, M. Fabrication of chitosan/poly(lactic acid)/graphene oxide/TiO_2 composite nanofibrous scaffolds for sustained delivery of doxorubicin and treatment of lung cancer. *Int. J. Biol. Macromol.* **2018**, *110*, 416–424. [CrossRef] [PubMed]
124. Sadeghi, M.; Falahi, F.; Akbari-Birgani, S.; Nikfarjam, N. Trilayer tubular scaffold to mimic ductal carcinoma breast cancer for the study of chemo-photothermal therapy. *ACS Appl. Polym. Mater.* **2023**, *5*, 2394–2407. [CrossRef]
125. Maleki, H.; Doostan, M.; Shojaei, S.; Doostan, M.; Stamatis, H.; Gkantzou, E.; Bonkdar, A.; Khoshnevisan, K. Nanofiber-based systems against skin cancers: Therapeutic and protective approaches. *J. Drug Deliv. Sci. Technol.* **2023**, *82*, 104367. [CrossRef]
126. Abasalta, M.; Asefnejad, A.; Khorasani, M.T.; Saadatabadi, A.R. Fabrication of carboxymethyl chitosan/poly(ε-caprolactone)/doxorubicin/nickel ferrite core-shell fibers for controlled release of doxorubicin against breast cancer. *Carbohydr. Polym.* **2021**, *257*, 117631. [CrossRef] [PubMed]
127. Mohebian, Z.; Babazadeh, M.; Zarghami, N.; Mousazadeh, H. Anticancer efficiency of curcumin-loaded mesoporous silica nanoparticles/nanofiber composites for potential postsurgical breast cancer treatment. *J. Drug Deliv. Sci. Technol.* **2021**, *61*, 102170. [CrossRef]
128. Farboudi, A.; Nouri, A.; Shirinzad, S.; Sojoudi, P.; Davaran, S.; Akrami, M.; Irani, M. Synthesis of magnetic gold coated poly (ε-caprolactonediol) based polyurethane/poly(N-isopropylacrylamide)-grafted-chitosan core-shell nanofibers for controlled release of paclitaxel and 5-FU. *Int. J. Biol. Macromol.* **2020**, *150*, 1130–1140. [CrossRef]
129. Munaweera, I.; Levesque-Bishop, D.; Shi, Y.; Di Pasqua, A.J.; Balkus, K.J.J. Radiotherapeutic bandage based on electrospun polyacrylonitrile containing holmium-166 iron garnet nanoparticles for the treatment of skin cancer. *ACS Appl. Mater. Interfaces* **2014**, *6*, 22250–22256. [CrossRef]
130. Stoyanova, N.; Spasova, M.; Manolova, N.; Rashkov, I.; Georgieva, A.; Toshkova, R. Antioxidant and antitumor activities of novel quercetin-loaded electrospun cellulose acetate/polyethylene glycol fibrous materials. *Antioxidants* **2020**, *9*, 232. [CrossRef]
131. Guo, M.; Zhou, G.; Liu, Z.; Liu, J.; Tang, J.; Xiao, Y.; Xu, W.; Liu, Y.; Chen, C. Direct site-specific treatment of skin cancer using doxorubicin-loaded nanofibrous membranes. *Sci. Bull.* **2018**, *63*, 92–100. [CrossRef]
132. Samadzadeh, S.; Babazadeh, M.; Zarghami, N.; Pilehvar-Soltanahmadi, Y.; Mousazadeh, H. An implantable smart hyperthermia nanofiber with switchable, controlled and sustained drug release: Possible application in prevention of cancer local recurrence. *Mater. Sci. Eng. C* **2021**, *118*, 111384. [CrossRef]
133. Muthulakshmi, L.; Prabakaran, S.; Ramalingam, V.; Rajulu, A.V.; Rajan, M.; Ramakrishna, S.; Luo, H. Sodium alginate nanofibers loaded Terminalia catappa scaffold regulates intrinsic apoptosis signaling in skin melanoma cancer. *Process Biochem.* **2022**, *118*, 92–102. [CrossRef]
134. Xue, C.; Li, M.; Sutrisno, L.; Yan, B.; Zhao, Y.; Hu, Y.; Cai, K.; Zhao, Y.; Yu, S.; Luo, Z. Bioresorbable scaffolds with biocatalytic chemotherapy and in situ microenvironment modulation for postoperative tissue repair. *Adv. Funct. Mater.* **2021**, *31*, 2008732. [CrossRef]
135. Çimen, C.G.; Dündar, M.A.; Kars, M.D.; Avcı, A. Enhancement of PCL/PLA electrospun nanocomposite fibers comprising silver nanoparticles encapsulated with thymus vulgaris l. molecules for antibacterial and anticancer activities. *ACS Biomater. Sci. Eng.* **2022**, *8*, 3717–3732. [CrossRef]
136. Aggarwal, U.; Goyal, A.K.; Rath, G. Development and characterization of the cisplatin loaded nanofibers for the treatment of cervical cancer. *Mater. Sci. Eng. C* **2017**, *75*, 125–132. [CrossRef]
137. Chen, Y.; Liu, Y.; LeeL, D.; Qiu, J.T.; Lee, T.; Liu, S. Biodegradable andrographolide-eluting nanofibrous membranes for the treatment of cervical cancer. *Int. J. Nanomed.* **2019**, *14*, 421–429. [CrossRef]
138. Yakub, G.; Ignatova, M.; Manolova, N.; Rashkov, I.; Toshkova, R.; Georgieva, A.; Markova, N. Chitosan/ferulic acid-coated poly(ε-caprolactone) electrospun materials with antioxidant, antibacterial and antitumor properties. *Int. J. Biol. Macromol.* **2018**, *107*, 689–702. [CrossRef] [PubMed]
139. Šišková, A.O.; Cková, M.B.; Kroneková, Z.; Kleinová, A.; Nagy, Š.; Rydz, J.; Opálek, A.; Sláviková, M.; Andicsová, A.E. The drug-loaded electrospun poly(ε-caprolactone) mats for therapeutic application. *Nanomaterials* **2021**, *11*, 922. [CrossRef] [PubMed]
140. Wang, X.; Wang, L.; Zong, S.; Qiu, R.; Liu, S. Use of multifunctional composite nanofibers for photothermal chemotherapy to treat cervical cancer in mice. *Biomater. Sci.* **2019**, *7*, 3846–3854. [CrossRef] [PubMed]
141. Xie, X.; Zheng, X.; Han, Z.; Chen, Y.; Zheng, Z.; He, X.; Wang, Y.; Kaplan, D.L.; Li, Y.; Li, G.; et al. A biodegradable stent with surface functionalization of combined-therapy drugs for colorectal cancer. *Adv. Healthc. Mater.* **2018**, *7*, 1801213. [CrossRef] [PubMed]
142. Rostami, M.; Ghorbani, M.; MohammadiM, M.A.; Delavar, M.; Tabibiazar, M.; Ramezani, S. Development of resveratrol loaded chitosan-gellan nanofiber as a novel gastrointestinal delivery system. *Int. J. Biol. Macromol.* **2019**, *135*, 698–705. [CrossRef] [PubMed]
143. Mellatyar, H.; Talaei, S.; Pilehvar-Soltanahmadi, Y.; Dadashpour, M.; Barzegar, A.; Akbarzadeh, A.; Zarghami, N. 17-DMAG-loaded nanofibrous scaffold for effective growth inhibition of lung cancer cells through targeting HSP90 gene expression. *Biomed. Pharmacother.* **2018**, *105*, 1026–1032. [CrossRef] [PubMed]

144. Samadzadeh, S.; Mousazadeh, H.; Ghareghomi, S.; Dadashpour, M.; Babazadeh, M.; Zarghami, N. In vitro anticancer efficacy of Metformin-loaded PLGA nanofibers towards the post-surgical therapy of lung cancer. *J. Drug Deliv. Sci. Technol.* **2021**, *61*, 102318. [CrossRef]
145. Li, X.; Yan, S.; Dai, J.; Lu, Y.; Wang, Y.; Sun, M.; Gong, J.; Yao, Y. Human lung epithelial cells A549 epithelial-mesenchymal transition induced by PVA/Collagen nanofiber. *Colloids Surf. B Biointerfaces* **2018**, *162*, 390–397. [CrossRef] [PubMed]
146. Yang, X.; Li, K.; Zhang, X.; Liu, C.; Guo, B.; Wen, W.; Gao, X. Nanofiber membrane supported lung-on-a-chip microdevice for anti-cancer drug testing. *Lab. Chip* **2018**, *18*, 486–495. [CrossRef] [PubMed]
147. Irani, M.; Sadeghi, G.M.M.; Haririan, I. The sustained delivery of temozolomide from electrospun PCL-Diol-b-PU/gold nanocomposite nanofibers to treat glioblastoma tumors. *Mater. Sci. Eng. C* **2017**, *75*, 165–174. [CrossRef]
148. Molina-Peña, R.; Mansor, M.H.; Najberg, M.; Thomassin, J.; Gueza, B.; Alvarez-Lorenzo, C.; Garcion, E.; Jérôme, C.; Boury, F. Nanoparticle-containing electrospun nanofibrous scaffolds for sustained release of SDF-1α. *Int. J. Pharm.* **2021**, *610*, 121205. [CrossRef]
149. Yang, J.; Xu, L.; Ding, Y.; Liu, C.; Wang, B.; Yu, Y.; Hui, C.; Ramakrishna, S.; Zhang, J.; Long, Y. NIR-II-Triggered Composite Nanofibers to Simultaneously Achieve Intracranial Hemostasis, Killing Superbug and Residual Cancer Cells in Brain Tumor Resection Surgery. *Adv. Fiber Mater.* **2023**, *5*, 209–222. [CrossRef]
150. Will, O.M.; Purcz, N.; Chalaris, A.; Heneweer, C.; Boretius, S.; Purcz, L.; Nikkola, L.; Ashammakhi, N.; Kalthoff, H.; Glüer, C.; et al. Increased survival rate by local release of diclofenac in a murine model of recurrent oral carcinoma. *Int. J. Nanomed.* **2016**, *11*, 5311–5321. [CrossRef]
151. Zhang, H.; Ji, Y.; Yuan, C.; Sun, P.; Xu, Q.; Lin, D.; Han, Z.; Xu, X.; Zhou, Q.; Deng, J. Fabrication of astaxanthin-loaded electrospun nanofiber-based mucoadhesive patches with water-insoluble backing for the treatment of oral premalignant lesions. *Mater. Design* **2022**, *223*, 111131. [CrossRef]
152. Wang, J.; Wang, G.; Shan, H.; Wang, X.; Wang, C.; Zhuang, X.; Ding, J.; Chen, X. Gradiently degraded electrospun polyester scaffolds with cytostatic for urothelial carcinoma therapy. *Biomater. Sci.* **2019**, *7*, 963–974. [CrossRef]
153. Musciacchio, L.; Mardirossian, M.; Guagnini, B.; Raffini, A.; Rizzo, M.; Trombetta, C.; Liguori, G.; Turco, G.; Porrelli, D. Rifampicin-loaded electrospun polycaprolactone membranes: Characterization of stability, antibacterial effects and urotheliocytes proliferation. *Mater. Design* **2022**, *224*, 111286. [CrossRef]
154. Anothra, P.; Pradhan, D.; Naik, P.K.; Ghosh, G.; Rath, G. Development and characterization of 5-fluorouracil nanofibrous film for the treatment of stomach cancer. *J. Drug Deliv. Sci. Technol.* **2021**, *61*, 102219. [CrossRef]
155. Tu, H.; Dai, F.; Cheng, G.; Yuan, M.; Zhou, X.; Wang, Y.; Zhang, R.; Zheng, Y.; Cheng, Y.; Deng, H. Incorporation of Layered Rectorite into Biocompatible Core–Sheath Nanofibrous Mats for Sustained Drug Delivery. *ACS Biomater. Sci. Eng.* **2021**, *7*, 4509–4520. [CrossRef] [PubMed]
156. Molina, E.R.; Chim, L.K.; Salazar, M.C.; Mehta, S.M.; Menegaz, B.A.; Lamhamedi-Cherradi, S.; Satish, T.; Mohiuddin, S.; McCall, D.; Zaske, A.M.; et al. Mechanically tunable coaxial electrospun models of YAP/TAZ mechanoresponse and IGF-1R activation in osteosarcoma. *Acta Biomater.* **2019**, *100*, 38–51. [CrossRef]
157. Rajendran, K.; Kokulnathan, T.; Chen, S.; Allen, J.A.; Viswanathan, C.; Therese, H.A. Nitrogen doped carbon nanofibers loaded with hierarchical vanadium tetrasulfide for the voltammetric detection of the non-steroidal anti-prostate cancer drug nilutamide. *Microchim. Acta* **2019**, *186*, 141. [CrossRef] [PubMed]
158. Jaworska, J.; Orchel, A.; Kaps, A.; Jaworska-Kik, M.; Hercog, A.; Stojko, M.; Włodarczyk, J.; Musiał-Kulik, M.; Pastusiak, M.; Bochenek, M.; et al. Bioresorbable nonwoven patches as taxane delivery systems for prostate cancer treatment. *Pharmaceutics* **2022**, *14*, 2835. [CrossRef] [PubMed]
159. Zhan, Q.; Shen, B.; Fang, Y.; Deng, X.; Chen, H.; Jin, J.; Peng, C.; Li, H. Drug-eluting scaffold inhibited in vivo pancreatic tumorigenesis by engaging murine CCR4+CD8+ T cells. *Colloid Surf. B* **2017**, *158*, 469–473. [CrossRef]
160. Acevedo, F.; Hermosilla, J.; Sanhueza, C.; Mora-Lagos, B.; Fuentes, I.; Rubilar, M.; Concheiro, A.; Alvarez-Lorenzo, C. Gallic acid loaded PEO-core/zein-shell nanofibers for chemopreventive action on gallbladder cancer cells. *Eur. J. Pharm. Sci.* **2018**, *119*, 49–61. [CrossRef] [PubMed]

MDPI AG
Grosspeteranlage 5
4052 Basel
Switzerland
Tel.: +41 61 683 77 34

Polymers Editorial Office
E-mail: polymers@mdpi.com
www.mdpi.com/journal/polymers

www.ingramcontent.com/pod-product-compliance
Lightning Source LLC
Chambersburg PA
CBHW040830250926
51850CB00007BA/1300
* 9 7 8 3 7 2 5 8 6 1 5 7 6 *